EPSA Philosophical Issues in the Sciences

Mauricio Suárez · Mauro Dorato · Miklós Rédei
Editors

EPSA Philosophical Issues in the Sciences

Launch of the European Philosophy of Science Association

Editors

Prof. Mauricio Suárez
Universidad Complutense
de Madrid
Fac. Filosofía
Depto. Lógica y Filosofía
de la Ciencia
28040 Madrid
Planta Sótano, Edificio B
Spain
msuarez@filos.ucm.es

Mauro Dorato
Via Ostiense 234
00144 Rome
Italy
dorato@uniroma3.it

Miklós Rédei
Houghton Street WC2 2AE
London
United Kingdom

ISBN 978-90-481-3251-5 e-ISBN 978-90-481-3252-2
DOI 10.1007/978-90-481-3252-2
Springer Dordrecht Heidelberg London New York

Library of Congress Control Number: 2010924525

© Springer Science+Business Media B.V. 2010
No part of this work may be reproduced, stored in a retrieval system, or transmitted in any form or by any means, electronic, mechanical, photocopying, microfilming, recording or otherwise, without written permission from the Publisher, with the exception of any material supplied specifically for the purpose of being entered and executed on a computer system, for exclusive use by the purchaser of the work.

Cover design: Boekhorst Design b.v.

Printed on acid-free paper

Springer is part of Springer Science+Business Media (www.springer.com)

Contents

Introduction ix
Mauricio Suárez, Mauro Dorato, and Miklós Rédei

1   Contingency and Inherency in Evolutionary
    Developmental Biology 1
    Werner Callebaut

2   Dualities and Intertheoretic Relations 9
    Elena Castellani

3   Are 'Identical Quantum Particles' Weakly Discernible
    Objects? 21
    Dennis Dieks

4   Wave–Particle Duality in Quantum Optics 31
    Brigitte Falkenburg

5   Remarks on a Structural Account of Scientific Explanation 43
    Laura Felline

6   Mathematical Knowledge and the Interplay of Practices 55
    José Ferreirós

7   Einstein, Kant, and the A Priori 65
    Michael Friedman

8   Causal Models and the Asymmetry of State Preparation 75
    Mathias Frisch

9   Bell-Type Inequalities from Separate Common Causes 87
    Gerd Graßhoff and Adrian Wüthrich

10 **Entanglement, Upper Probabilities and Decoherence in Quantum Mechanics** 93
    Stephan Hartmann and Patrick Suppes

11 **Gauge Symmetry and the Theta-Vacuum** 105
    Richard Healey

12 **The Chemical Bond: Structure, Energy and Explanation** 117
    Robin Findlay Hendry

13 **Randomness, Financial Markets and the Brownian Motion: A Reflection on the Role of Mathematics in Their Interaction with Financial Theory After 1973** 129
    Ghislaine Idabouk

14 **Causation Across Levels, Constitution, and Constraint** 141
    Max Kistler

15 **Epistemic Consequences of Two Different Strategies for Decomposing Biological Networks** 153
    Ulrich Krohs

16 **Matter(s) in Relativity Theory** 163
    Dennis Lehmkuhl

17 **Individual Particles, Properties and Quantum Statistics** 175
    Matteo Morganti

18 **Evolution and Directionality: Lessons from Fisher's Fundamental Theorem** 187
    Samir Okasha

19 **Substantive General Covariance: Another Decade of Dispute** 197
    Oliver Pooley

20 **Relativity, Locality and Tense** 211
    Steven Savitt

21 **A Weylian Approach Towards Theories of Matter: Dynamic Agents and Geometrisation** 219
    Norman Sieroka

22 **Mirroring and Understanding Action** 227
    Corrado Sinigaglia

23 **Absolute Objects and General Relativity: Dynamical
    Considerations** 239
    Adán Sus

24 **Empirical Foundation of Space and Time** 251
    László E. Szabó

25 **Making Contact with Observations** 267
    Ioannis Votsis

26 **The Formulation and Justification of Mathematical
    Definitions Illustrated By Deterministic Chaos** 279
    Charlotte Werndl

27 **Do We Need Some Large, Simple Randomized Trials
    in Medicine?** 289
    John Worrall

28 **Incontinence, Honouring Sunk Costs and Rationality** 303
    António Zilhão

29 **Causal Fundamentalism in Physics** 311
    Henrik Zinkernagel

Index 323

Introduction

Mauricio Suárez, Mauro Dorato, and Miklós Rédei

These two volumes contain a selection of the papers delivered at the first conference of the *European Philosophy of Science Association* (EPSA) which took place in Madrid, at Complutense University, from 14 to 17 November 2007. The first volume is entitled *Epistemology and Methodology*, and includes papers mainly concerned with general philosophy of science, rationality, and method. The second volume, devoted to *Philosophical Issues in the Sciences*, includes papers concerned with the philosophy of the sciences, particularly physics, economics, chemistry and biology. Overall the selection has been very severe and took place in two stages. The 30-strong conference programme committee chaired by Mauro Dorato and Miklós Rédei first selected 160 papers for presentation out of 410 abstracts submitted. After the conference the three of us went on to further select 60 papers among those delivered. The selection was made on the recommendation of the members of the programme committee and the chairs of the conference sessions, who were invited to nominate their favourite papers and provide reasons for their choices. Every paper included in these volumes has been independently nominated by at least two referees. There are thus good grounds to the claim that these essays constitute some of most significant and important research presently carried out in the philosophy of science throughout Europe.

The two volumes also represent the first tangible outcome of the newly born EPSA. Together with the conference they in effect constitute the launching of the Association. The resounding success of the conference and its call for papers bears testimony to the strong demand for an Association of this nature. EPSA was established in anticipation of such demand and it intends to be an institution that helps cultivate philosophy of science across Europe, where modern philosophy of science was born in the first half of the twentieth century. While based in Europe, EPSA is an association that welcomes members of any nationality – just like its more established, successful older sister, the Philosophy of Science Association. The varied

---

M. Suárez, M. Dorato, and M. Rédei
Madrid, Rome and London

range and outstanding quality of the papers in the present two volumes constitute a powerful signal of the healthy state and bright future of philosophy of science across Europe.

EPSA07, the Founding conference of EPSA, was organised by Mauricio Suárez and the members of his research group at Complutense. Mauricio Suárez would like to personally thank them for their help and work during the long and unnerving months before and after the conference. Thanks also to Julian Reiss for his support during the first few months of planning.

Financial help is acknowledged from the Spanish Ministry of Science and Education (grant HUM2007-29190-E), and the Vice-Rectorate of Research at Complutense University (programme OC36/07). Some of the funds allocated by the Government of Madrid's Autonomous Community (grant 930370-2007) were diverted to cover some unexpected last minute conference expenses too. Thanks are also due the Faculty of Philosophy at Complutense for its unconditional support. In particular the Dean of the Faculty of Philosophy, Juan Manuel Navarro Cordón, lent us the Faculty building and all its audiovisual facilities free of charge. This greatly reduced the cost of the conference, which a newly born association like EPSA would not have been able to cover.

In compiling these volumes we were fortunate to be able to rely on Iñaki San Pedro, who has been an efficient and responsible editorial assistant. In the last few months he has almost become a 'fourth editor' in the background – we are very grateful to him. *Springer* backed us from the beginning, and provided the indexes. Lucy Fleet was the always friendly and supportive first port of call at *Springer*. Ties Niejssen was a supportive editor in the last few stages. But our greatest debt is to Charles Erkelens, the lead Humanities editor at *Springer*. The level of support we have found in Charles is difficult to overemphasise. He contacted the conference organiser well ahead of the event, and made sure that there would be a strong *Springer* presence at EPSA07. He himself attended the conference in its entirety and continued to support us throughout the editorial work. Afterwards he responded with invariable efficiency to all kinds of requests, related not only to these volumes, but also the planned-for journal, the *European Journal of Philosophy of Science*, where these papers were originally intended to appear. It was the good fortune of EPSA to come across such a devoted Editor – at just the right time, and just the right place.

Madrid, Rome and London	Mauricio Suárez
25 May 2009	Mauro Dorato
	Miklós Rédei

Chapter 1
Contingency and Inherency in Evolutionary Developmental Biology

Werner Callebaut

1.1 Introduction

The 'modern' evolutionary synthesis (1936–1947) as developed by Theodosius Dobzhansky, R. A. Fisher, Julian Huxley, Ernst Mayr, George Gaylord Simpson, Sewall Wright, and others holds that genetic variation in populations arises by blind (i.e., not adaptively directed) mutation and recombination; that populations evolve by changes in gene frequency due to genetic drift, gene flow, and, most importantly, natural selection; that (most) phenotypic change is gradual; that speciation normally results from reproductive isolation among populations; and that these processes, if maintained long enough, may give rise to higher taxonomic levels (Futuyma 1986, p. 12). The modern synthesis, which made population genetics the core of evolutionary theorizing, continues to be the dominant 'paradigm'[1] in current biology, and, being a "moving target" (Burian 1988, p. 250), is likely to remain so in the foreseeable future (Callebaut 2010). Yet, it is becoming increasingly clear that the modern synthesis leaves open a number of important problems pertaining to macroevolution, neutral evolution, and morphogenesis (among others) and that its reigning modes of explanation, gene-selectionism and adaptationism, are biased and otherwise problematic (Callebaut et al. 2007; and references therein). Put differently, the modern synthesis is now widely acknowledged to have been incomplete. Specifically, it "black-boxed" development (Viktor Hamburger). This paper indirectly reflects on

---

W. Callebaut (✉)
Konrad Lorenz Institute for Evolution and Cognition Research (KLI), Adolf Lorenz Gasse 2, 3422 Altenberg, Austria
e-mail: werner.callebaut@kli.ac.at
and
Faculty of Sciences, Hasselt University, Agoralaan, Building, 3590 Diepenbeek, Belgium
e-mail: werner.callebaut@uhasselt.be

[1] The term 'paradigm' is used here loosely as in common scientific parlance, and should not be taken to refer specifically to Thomas Kuhn's theory (although many biologists use it in this way). In fact, rather than as a paradigm, the evolutionary synthesis may be more aptly characterized as a "treaty" that has allowed evolutionists and molecular biologists to work together under common presuppositions (Burian 1988; Callebaut 2010).

(i) the criticism that development (in the sense of ontogeny: the origin and differentiation of individual organisms from embryo to adult), understood as the unfolding of a 'genetic program', smacks of (long thought dead) *preformationism*, and (ii) the so-called existence problem: in order for natural selection to work, the necessary variation has to be in place already; however, the origins of variation reside not only in the selective milieu, but first and foremost in the variability of developmental processes (or at least so the proponents of evolutionary developmental biology or "EvoDevo" argue).[2]

1.2 The Historical Contingency of Evolution

George C. Williams, the austere defender of the view that natural selection operating at the level of individual genotypes is sufficient for the explanation of all evolutionary phenomena (a view that was widely popularized by Richard Dawkins as the 'gene's eye perspective'), insisted that biological explanations "should invoke no factors other than the laws of physics, natural selection, and the contingencies of history," thus admitting that "[t]he idea that an organism has a complex history through which natural selection has been in constant operation imposes a special constraint on evolutionary theorizing" (Williams 1985, pp. 1–2). The "special constraint" Williams refers to is usually explicated in terms of (historical) *contingency*.[3] Thus, philosopher John Beatty states his "evolutionary contingency thesis" as follows:

All generalizations about the living world:

(a) are just mathematical, physical, or chemical generalizations (or deductive consequences of mathematical, physical, or chemical generalizations plus initial conditions).
(b) are distinctively biological, in which case they describe contingent outcomes of evolution. (Beatty 1995, pp. 46–47)

The first claim acknowledges that "there are generalization about the living world whose truth values are not a matter of evolutionary history" (p. 47). An example of a mathematical generalization pertaining to the living world would be that all forms of life are subject to the laws of probability. Beatty's example of a physical generalization, Newton's laws of motion, although not biologically irrelevant (certain small animals, say, are attracted to larger ones by sheer gravity), may be constructed in such a way as not being a matter of evolutionary history. And his example of a chemical generalization, viz. that evolution will not result "in any carbon based

---

[2] According to the currently dominant adaptationist perspective, phenotypic variability, although a necessary condition of any evolutionary change, is not viewed as the *cause* of a particular adaptive shift; rather, it is treated as a (tacit) background assumption (Sterelny 2000, p. S373).

[3] Historical contingency is not to be confounded with contingency as understood in modal logic, where something is contingent if it is neither necessary nor impossible (which is *not* identical to "possible," since what is necessary must also be possible).

forms of life that are not subject to the principles of organic chemistry" (p. 47) is wisely chosen in that it leaves open the possibility of life forms not based on carbon, as dreamt of by some students of artificial life.

The second claim boils down to stating that biological generalizations "do not express any *natural necessity*; they may be true, but nothing in nature necessitates their truth" (p. 52; italics in original). Beatty thus joins the bandwagon of biologists (such as Ernst Mayr and Richard Lewontin) and philosophers of biology (such as John Dupré and Philip Kitcher) who deny the existence of laws in biology.[4] Lewontin's view seems to me especially illuminating here. Far from denying the main tenets of scientific realism ("Indeed, there is a real world out there"; Lewontin 1982, p. 162), he insists that "[i]t is a long way from the 'laws of nature' to the horse's hoof" (p. 163).[5] He insists that we do not further our understanding of evolution by general appeals to 'laws of nature' "to which all life must bend." Rather, we must ask how, within the general constraints of the laws of nature, organisms have constructed environments that are the conditions for their further evolution and reconstruction of nature into new environments. Organisms within their individual lifetimes and in the course of their evolution as a species do not *adapt* to environments; they *construct* them. They are not simply *objects* of the laws of nature, altering themselves to bend to the inevitable, but active *subjects* transforming nature according to its laws (Lewontin 1982, p. 163).[6]

Stephen Jay Gould's *Wonderful life* (1989) has become a watershed for students of the complexity of the living, and his image of rewinding the "tape of life" to the time of the Cambrian explosion and then replaying it has sparkled reflections in- and outside biology. Philosophically speaking, the notion of historical contingency is a minefield, as it has been associated with chance (Jacques Monod), irreversibility (Nicholas Georgescu-Roegen) or "lock-in" (Wallace Arthur), and nonrepeatability (Jon Elster), and equated with randomness and stochasticity (Eric Chaisson); see Callebaut et al. (2007) and references therin. Gould viewed historical explanations as taking "the form of narrative: E, the phenomenon to be explained, arose because D came before, preceded by C, B, and A. If any of these stages had not occurred, or had transpired in a different way, then E would not exist (or would be present

---

[4] My aim in this paper is not to articulate, let alone to defend, my personal stance on this issue (Elgin 2006 is a good discussion of the state of the art). Let me just say here that I sympathize with Van Fraassen's (1989) constructive empiricist position that there are no unproblematic criteria for laws of nature, and that we'd better dispense with the notion of laws at all in the philosophy of science. Giere (1999) offers additional arguments to the effect that we can understand the workings of science (his focus is on physics) without invoking laws of nature from a constructive realist perspective.

[5] This is in reference to the naïve correspondence view of fitness propounded by Konrad Lorenz, according to which "the hoof of the horse is already adapted to the ground of the steppe before the horse is born" (Lorenz [1962] 1982, pp. 124–125). More generally, Lewontin (1982, p. 162) rejects the "problem-solution model" according to which evolution is to be regarded as the "solution" by species of some predetermined "problems" on the grounds that "it is the life activities of the species themselves that determine both the problems and solutions simultaneously."

[6] Lewontin's (1982) article is justly regarded as pioneering the current view of niche construction (Odling-Smee et al. 2003).

in a substantially altered form, E', requiring a different explanation" (Gould 1989, p. 283). He thought that, given A–D, E "had to arise" and was in this sense nonrandom. Yet, "no law of nature enjoined E"; any variant E' arising from altered antecedents would have been equally explainable, "though massively different in form and effect." For Gould, then, contingency meant that the final result is "dependent ... upon everything that came before – the unerasable and determining [sic] signature of history."

In a first approximation, it may seem that Gould combined an epistemological notion of contingency – unpredictability – with an ontological one, which Oyama (2000, p. 116) has called "causal dependency" (cf. lock-in). But to me, the "unerasable and determining signature of history" suggests that the past leaves its traces in the present, which takes us back to epistemology (how much of the past do we have to know to understand the present?). Interpreting Gould's view of contingency in terms of stochasticity seems to me equally problematic, because one would then have to rule out stochastic processes that are memoryless (the Markov property: the present state of the system predicts future states as well as the whole history of past and present states) – assuming that models and theories that exhibit time lags play a more prominent role in biology than in the physical sciences, because the structural knowledge that would enable us to disregard the more ancient causes (A–C in Gould's schema) has not been (or cannot be) attained.

1.3 Inherency Versus Contingency?

Not unlike the evolutionary synthesis itself, the philosophy of biology that has become one of the most booming fields within philosophy of science in the last three decades has largely neglected developmental issues hitherto. Elliot Sober (2000, p. xvii), for one, writes: "For me, evolutionary biology is the center of gravity both for the science of biology and for the philosophy of that science." He also suggests that developmental biology is "full of unanswered questions" for which "[n]o adequate physicalistic explanation is available now" (p. 24; cf. also pp. 26–27 and 56). Similar views have been expressed in the writings of many other influential philosophers of biology. I have analyzed the phenomenon that philosophical, metalevel views tend to 'mirror' a certain scientific stance, bias, etc. – which is quite common in science – under the label "packages" elsewhere (Callebaut et al. 2007). Quite predictably, spokespersons for an emerging field such as EvoDevo and their philosophical allies are tempted to elaborate their 'own' philosophy. Müller and Newman's emphasis on the role of inherency in development and evolution, which they contradistinguish from contingency, is an example of this tendency. (Paleontologist Simon Conway Morris (1997) is another biologist who has made a great deal out of "inherency," which, however he understands quite differently than Müller and Newman's. Space limitations prevent me from comparing these respective approaches here; but see Sober (2003).)

Put most succinctly, *inherency* for Müller and Newman is the propensity of biological materials to assume preferred forms (e.g., Newman 1994, 2005; Müller and Newman 2003; Newman et al. 2006; Newman and Müller 2006). Most living tissues are 'soft matter' and as such subject (by virtue of inherent physical properties) to being molded, formed, and deformed by the external physical environment. All living tissues are also excitable media, viz., materials that employ stored chemical or mechanical energy to respond in active (and predictable) ways to the environment (Forgacs and Newman 2005). The concept of inherency is introduced along with *evolvability*, the intrinsic potential of certain lineages to change during the cause of evolution, and developmental and evolutionary *modularity*, which refers to the circumstance, ubiquitously observed in nature, that components of systems typically operate according to their own, intrinsically determined principles (see, e.g., Callebaut and Rasskin-Gutman 2007). In more philosophical parlance, something is inherent if it will always happen (e.g., entropy) or if the potentiality for it always exists, even if its actuality can be obstructed. In the scenario envisaged by the Organismal Systems Approach or OSA (Callebaut et al. 2007), morphological variation in response to the environment is a primitive, physically based property, carried over (to a limited extent) into modern organisms from the inherent plasticity and responsiveness to the external physical environment of the viscoelastic cell aggregates that constituted the first multicellular organisms.

In evolutionary terms, inherency suggests that the morphological motifs of contemporary organisms have their origins in the generic forms assumed by cell masses interacting with one another and their microenvironments, and were only later integrated into developmental repertoires by stabilizing and canalizing genetic evolution (cf. West-Eberhard 2003 on genes as "followers" in evolution; see also Jablonka 2006). In this view, the causal basis of phenotypic evolution is not reduced to gene regulatory evolution and population genetic events, but includes the *formative factors* inherent in the evolving organisms themselves, such as their physical material properties, self-organizing capacities, and reactive potential to external influence. From Monod to Williams and Gould, the modern evolutionary synthesis has emphasized the contingency of evolution (see also Brandon (2006) on drift as biology's "first law")[7]; OSA's watchword now becomes inherency.

In terms of the individual development of organisms, inherency defies the 'blueprint' or 'program' notions that currently abound. Cell collectives and tissue masses take on form not because they are instructed to do so but because of the inherent physical and self-organizational properties of interacting cells. The cell interactions of all organisms, ancient and modern, *do* depend on the

---

[7] Oyama (2000, Ch. 6, "The accidental chordate: Contingency in developmental systems") additionally argues for a notion of *development* "in which contingency is central and constitutive, not merely secondary alteration of more fundamental, 'preprogrammed' forms" (p. 116), thus departing from the common view that development is reoccurring regularly in nature and hence (to some extent) predictable, if not programmed, while evolution is taken to be neither (Sterelny 2000, pp. S369–S370). Oyama's postmodern celebration of developmental contingency differs from Müller and Newman's variety of EvoDevo in that the latter, and EvoDevo in general, extend the (nondeterministic!) programming of development by including epigenetic and environmental factors.

molecules that genes specify, but the resulting biological forms and specific cell arrangements are not encoded in any deterministic fashion in the genome. Inherency locates the causal basis of morphogenesis in the dynamics of interaction between genes, cells, and tissues, each endowed with their own, autonomous physical and functional properties.

1.4 Conclusion

Müller and Newman's deployment of inherency should be viewed properly as an illustration (one among others) of EvoDevo's moving away "from the external and contingent to the internal and inherent" (Müller 2007, p. 947). It is a plea to restore the balance after the externalist excesses of the modern synthesis (Godfrey-Smith 1996). Despite occasional suggestions to the contrary, this does not go counter at all to Beatty's evolutionary contingency thesis, for Müller and Newman gladly accept that natural selection has the last word, which is all Beatty requires.

A potentially more serious objection to 'internalist' explanations of living processes in general comes from the philosophy of physics, where the "containment metaphor" is being criticized in light of more or less recent scientific developments. A characteristic of efficient causation, on the containment metaphor, is that it is imagined to always 'flow' from 'inside out'. The ultimate constituents of the world that halt the regress of containment are then also taken to be the ultimate bearers of causal powers that "somehow support and determine the whole edifice of (often complex) causal relations that constitute the domain of observable dynamics" (Ladyman and Ross 2007, pp. 3, 4). The founding fathers of the current philosophy of biology threw out the physicists through the front door. Philosophers who listen to biophysicists may be tempted to open the back door for them, however slightly.

References

Beatty J (1995) The evolutionary contingency thesis. In: Wolters G, Lennox JG (eds) Concepts, theories, and rationality in the biological sciences, Universitätsverlag Konstanz, Konstanz; Pittsburgh University Press, Pittsburgh, pp 45–81

Brandon RN (2006) The principle of drift: Biology's first law. J Philos 102:319–335

Burian RM (1988) Challenges to the evolutionary synthesis. In: Hecht MK, Wallace B (eds) Evolutionary biology, vol 23. Plenum, New York, pp 247–269

Callebaut W (2010) The dialectics of dis/unity in the Evolutionary Synthesis and its extensions. In: Pigliucci M, Müller GB (eds) Toward an extended synthesis. MIT Press, Cambridge, MA, pp 443–481

Callebaut W, Rasskin-Gutman D (eds) (2005) Modularity: Understanding the development and evolution of natural complex systems. MIT Press, Cambridge, MA

Callebaut W, Müller GB, Newman SA (2007) The organismic systems approach: Streamlining the naturalistic agenda. In: Sansom R, Brandon RN (eds) Integrating evolution and development: From theory to practice. MIT Press, Cambridge, MA, pp 25–92

Conway Morris S (1997) Molecular clocks: Defusing the Cambrian 'explosion'? Curr Biol 7: R71–R74
Elgin M (2006) There may be strict empirical laws in biology, after all. Biol Philos 21:119–134
Forgacs G, Newman SA (2005) Biological physics of the developing embryo. Cambridge University Press, Cambridge
Futuyma DJ (1986) Evolutionary biology, 2nd edn. Sinauer, Sunderland, MA
Giere RN (1999). Science without laws. University of Chicago Press, Chicago, IL
Godfrey-Smith P (1996) Complexity and the function of mind in nature. Cambridge University Press, Cambridge
Gould SJ (1989) Wonderful life: the Burgess shale and the nature of history. Norton, New York
Jablonka E (2006) Genes as followers in evolution – a post-synthesis synthesis. (Review of West-Eberhard 2003.) Biol Philos 21:143–154
Ladyman J, Ross D (2007) Everything must go: metaphysics naturalized. Oxford University Press, Oxford
Lewontin RC (1982) Organism and environment. In: Plotkin HC (ed) Learning, development, and culture. Wiley, Chichester, pp 151–170
Lorenz K (1982) Kant's doctrine of the a priori in the light of contemporary biology. In: Plotkin HC (ed) Learning, development, and culture. Wiley, Chichester, pp 121–143. Originally published in General Systems 7:23–35 (1962)
Müller GB (2007) Evo-devo: Extending the evolutionary synthesis. Nat Rev Genet 8:943–949
Müller GB, Newman SA (2003) Origination of organismal form: The forgotten cause in evolutionary theory. In: Müller GB, Newman SA (eds) Origination of organismal form: beyond the gene in developmental and evolutionary biology. MIT Press, Cambridge, MA, pp 3–10
Newman SA (1994) Generic physical mechanisms of tissue morphogenesis: a common basis for development and evolution. J Evol Biol 7:467–488
Newman, SA (2005) The pre-Mendelian, pre-Darwinian world: shifting relations between genetic and epigenetic mechanisms in early multicellular evolution. J Biosci 30:75–85
Newman SA, Müller GB (2006) Genes and form: Inherency in the evolution of developmental mechanisms. In: Neumann-Held EM, Rehmann-Sutter C (eds) Genes in development: rereading the molecular paradigm. Duke University Press, Durham, NC, pp 38–73
Newman SA, Forgacs G, Müller GB (2006) Before programs: the physical origination of multicellular forms. Int J Dev Biol 50:289–299
Odling-Smee J, Laland KL, Feldman MW (2003) Niche construction: the neglected process in evolution. Princeton University Press, Princeton, NJ
Oyama S (2000) Evolution's eye: a systems view of the biology-culture divide. Duke University Press, Durham, NC
Sober E (2000) Philosophy of biology, 2nd edn. Westview Press, Boulder, CO
Sober E (2003) Contingency or inevitability? What would happen if the evolutionary tape were replayed? New York Times, November 30
Sterelny K (2000) Development, evolution, and adaptation. Philos Sci 67:S369–S387
Van Fraassen BC (1989) Laws and symmetry. Clarendon, Oxford
West-Eberhard MJ (2003) Developmental plasticity and evolution. Oxford University Press, Oxford
Williams GC (1985) A defense of reductionism in evolutionary biology. Oxford Surveys Evol Biol 2:1–27

Chapter 2
Dualities and Intertheoretic Relations

Elena Castellani

2.1 Introduction

The idea of duality is at the core of the most relevant developments in recent fundamental physics. During the last 40 years theoretical physics has used the notion of duality in different ways and frameworks: in the so-called dual resonance model of the late sixties, which gave birth to early string theory; in the context of quantum field theory, where a groundbreaking generalization of electromagnetic duality was conjectured by Claus Montonen and David Olive in 1977; in supersymmetric string theory, where various sorts of dualities are playing a key role in the theoretical elaboration.

This paper is concerned with the significance of physical dualities from the viewpoint of philosophy of science. The idea is that, for its peculiarity, this 'new' ingredient in theory construction can open unexpected perspectives for the current philosophical reflection on contemporary physics.[1] In particular, dualities represent an unusual type of intertheory relation, the meaning of which deserves to be investigated. It is the aim of the paper to show how discussing this point brings into play, at the same time, discussing what is intended by a 'theory' and in which sense dualities are to be considered 'symmetries' (if they are).

Considering the role and meaning of physical dualities in general poses immediately a problem. The dualities applied in recent fundamental physics are of different forms and status. While some of them seem to have a sound basis, others are just theoretical conjectures and a good part of the last developments grounded on dualities are still at a work-in-progress stage. Nonetheless, in most of the cases where dualities are applied in a quantum framework it is possible to individuate some common relevant characteristic features. A duality type that results particularly representative from this point of view is the so-called *electromagnetic duality* (EM duality). EM

E. Castellani (✉)
Department of Philosophy, University of Florence, via Bolognese 52, 50139 Firenze, Italy
e-mail: elena.castellani@unifi.it

[1] The philosophical literature on physical dualities is still very meagre. The philosophers of physics are just starting to turn their attention to string theory and its forms of dualities. One of the few contributions in this direction is Dawid (2007).

duality also represents the first form of duality explicitly applied in twentieth century physics: namely, in P. A.M. Dirac's famous two papers (published, respectively, in 1931 and 1948) on his 'theory of magnetic poles'. It therefore offers an appropriate, however specific, case study to begin with. Starting to investigate the significance of physical dualities by focussing on this case study is the object of the paper.

2.2 The Case of Electromagnetic Duality

Electromagnetic duality as formulated by Dirac is, in a sense, the prototype of today's physical dualities. In this section we present a brief survey of the development of this duality idea from the classical to the quantum context.

2.2.1 EM Duality (1): Classical Electrodynamics

EM duality is grounded on the idea that there is a substantial symmetry between electricity and magnetism. This is an old idea, going back to Michael Faraday, and first made more precise with the formulation by James Clerk Maxwell of his famous equations regulating the behaviour of electric and magnetic fields.

In current notation (using a unit system for which $c = 1$), *Maxwell's equations* read:

$$\vec{\nabla} \cdot \vec{E} = \rho_e,$$
$$\vec{\nabla} \cdot \vec{B} = 0,$$
$$\vec{\nabla} \wedge \vec{E} = -\frac{\partial \vec{B}}{\partial t},$$
$$\vec{\nabla} \wedge \vec{B} = \vec{j}_e + \frac{\partial \vec{E}}{\partial t},$$

where $\vec{E}$ is the electric field, $\vec{B}$ the magnetic field, ρ_e the density of electric charge and $\vec{J}_e$ the density of electric current.

There is an evident similarity in the role of electric and magnetic fields in these equations, apart from the presence of the electric source terms. In the absence of such terms – that is, in the case of *free Maxwell's equations* – the similarity becomes complete.

2.2.1.1 EM Duality in the Absence of Sources

In fact, when there are no charges and current ($\rho_e = \vec{J}_e = 0$), the equations read:

$$\vec{\nabla} \cdot \vec{E} = 0,$$
$$\vec{\nabla} \cdot \vec{B} = 0,$$

$$\vec{\nabla} \wedge \vec{E} = -\frac{\partial \vec{B}}{\partial t},$$
$$\vec{\nabla} \wedge \vec{B} = \frac{\partial \vec{E}}{\partial t}.$$

As is immediately apparent, the free Maxwell's equations are invariant under the following *duality transformation*:

$$D: \quad \vec{E} \to \vec{B}, \quad \vec{B} \to -\vec{E}.$$

In the sourceless case EM duality, expressed by the invariance of the equations under the duality transformation D, is thus an *exact symmetry*.

Notice that the duality transformation D can be generalized to *duality rotations* parameterized by an arbitrary angle θ as follows:

$$\vec{E} \to \cos\theta \vec{E} + \sin\theta \vec{B},$$
$$\vec{B} \to -\sin\theta \vec{E} + \cos\theta \vec{B}.$$

EM duality can then be expressed as the invariance of the sourceless Maxwell's equations under 'rotations' of the electric and magnetic fields. This can be better visualized by introducing the *complex vector field* $\vec{E} + i\vec{B}$, in terms of which Maxwell's equations can be written in the following concise form:

$$\vec{\nabla} \cdot (\vec{E} + i\vec{B}) = 0,$$
$$\vec{\nabla} \wedge (\vec{E} + i\vec{B}) = i\frac{\partial}{\partial t}(\vec{E} + i\vec{B}).$$

These equations remain invariant under the *duality rotation*

$$\vec{E} + i\vec{B} \to e^{i\theta}(\vec{E} + i\vec{B}).$$

In these terms, it is easy to see that the energy and momentum densities of the electromagnetic field, represented respectively by the following two expressions,

$$\mathcal{E} = \frac{1}{2}|\vec{E} + i\vec{B}|^2 = \frac{1}{2}(E^2 + B^2),$$
$$\mathcal{P} = \frac{1}{2i}(\vec{E} + i\vec{B})^* \wedge (\vec{E} + i\vec{B}) = \vec{E} \wedge \vec{B},$$

are invariant with respect to the EM duality transformations.

To sum up:

- When no source terms are present, the duality D exchanges the roles of the electric and magnetic fields while leaving the 'physics' – that is, the Maxwell's equations and the physical relevant quantities such as the energy and momentum densities of the electromagnetic field – invariant.

- When electric source terms are present, the Maxwell equations are no longer invariant under the duality D and EM symmetry is broken.

2.2.1.2 Restoring EM Duality in the Presence of Sources

There is a way to restore the symmetry between the electric and magnetic fields in the presence of sources: that is, by including *magnetic source terms*. Assuming the existence of a magnetic density of charge ρ_g and magnetic current $\vec{j}_g$, in addition to the usual electric charge density ρ_e and electric current $\vec{j}_e$, the Maxwell's equations take the form

$$\vec{\nabla} \cdot \vec{E} = \rho_e,$$
$$\vec{\nabla} \cdot \vec{B} = \rho_g,$$
$$-\vec{\nabla} \wedge \vec{E} = \vec{j}_g + \frac{\partial \vec{B}}{\partial t},$$
$$\vec{\nabla} \wedge \vec{B} = \vec{j}_e + \frac{\partial \vec{E}}{\partial t}.$$

These equations are invariant under the following duality transformation, interchanging the roles of the electric and magnetic fields and – at the same time – the roles of the electric and magnetic charges and currents:

$$\vec{E} \to \vec{B}, \qquad \vec{B} \to -\vec{E},$$
$$\rho_e, \vec{j}_e \to \rho_g, \vec{j}_g, \qquad \rho_g, \vec{j}_g \to -\rho_e, -\vec{j}_e.$$

In terms of the complex vector field $\vec{E} + i\vec{B}$, the above equations can be written concisely as:

$$\vec{\nabla} \cdot (\vec{E} + i\vec{B}) = \rho_e + i\rho_g,$$
$$\vec{\nabla} \wedge (\vec{E} + i\vec{B}) = i\left[\left(\vec{j}_e + i\vec{j}_g\right) + \frac{\partial}{\partial t}\left(\vec{E} + i\vec{B}\right)\right].$$

These equations are invariant under the *duality rotations*:

$$\vec{E} + i\vec{B} \to e^{i\theta}\left(\vec{E} + i\vec{B}\right),$$
$$\rho_e + i\rho_g \to e^{i\theta}\left(\rho_e + i\rho_g\right),$$
$$\vec{j}_e + i\vec{j}_g \to e^{i\theta}\left(\vec{j}_e + i\vec{j}_g\right).$$

Maxwell's equations can thus be modified to accommodate the inclusion of magnetic charges and currents. The problem is that isolated magnetic charges, the so-called *magnetic monopoles* (or, in Dirac's terminology, *magnetic poles*), have never been observed. If we break a magnet bar in two parts, we always obtain two smaller magnets and never an isolated North pole and an isolated South pole. Quoting Dirac (1948, p. 817): "The field equations of electrodynamics are symmetrical between electric and magnetic forces. The symmetry between electricity and magnetism is, however, disturbed by the fact that a single electric charge may occur on a particle, while a single magnetic pole has not been observed to occur on a particle."

If, in order to save the EM symmetry, we nevertheless assume the existence of isolated magnetic poles, the question is: why are isolated magnetic poles not observed? As we shall see in the next Section, Dirac investigated the problem in the context of quantum electrodynamics, arriving at the following answer: because an enormous energy is needed to produce a particle with a single magnetic pole.

2.2.2 EM Duality (2): Quantum Electrodynamics

Dirac's solution to the problem posed by EM symmetry is contained in his theory of magnetic poles. The theory was first proposed in his seminal 1931 paper *Quantised Singularities in the Electromagnetic Field* (Dirac 1931). In his second paper on the subject, appeared in 1948 with the title *The Theory of Magnetic Poles* (Dirac 1948), Dirac completed the theory by providing "all the equations of motion for magnetic poles and charged particles interacting with each other through the medium of the electromagnetic field in accordance to quantum mechanics" (Dirac 1948, p. 817–818).

In his 1931 paper Dirac put forward the idea of magnetic pole as "quantised singularities of the EM field", working out the consequences of this idea in the formalism of quantum mechanics. Declared object of his paper was "to show that quantum mechanics does not really preclude the existence of isolated magnetic poles" (Dirac 1931, p. 71). Why did quantum mechanics present a specific problem for the existence of isolated magnetic poles? The issue at stake was the following: turning from the classical to the quantum formulation of electromagnetic theory with magnetic sources posed a *consistency problem*. On the one hand, the electromagnetic vector potential $\vec{A}$ plays a central role in coupling electromagnetism to quantum mechanics.[2] On the other hand, the vector potential $\vec{A}$ is introduced in standard electromagnetism by taking advantage of the absence of magnetic source terms:

$$\vec{\nabla} \cdot \vec{B} = 0 \rightarrow \vec{B} = \vec{\nabla} \wedge \vec{A} \quad (\text{for all } \vec{A}, \quad \vec{\nabla} \cdot (\vec{\nabla} \wedge \vec{A}) = 0).$$

---

[2] In the canonical quantization procedure followed by Dirac, the electromagnetic potentials are required for putting the equations of motions into the form of an action principle. In general, the standard way of describing the electromagnetic couplings of the matter wave functions is in terms of the so-called minimal coupling prescription (requiring to replace the momentum operator $\vec{p} = -i\vec{\nabla}$ by its 'covariant' generalization $-i(\vec{\nabla} - ie \cdot \vec{A})$, where e is the electric charge). In other words, the vector potentials $\vec{A}$ explicitly enter the covariant derivative of the wave function of the electrically charged particle and therefore are needed to determine its evolution.

This seems to imply that quantum mechanics is inconsistent with the presence of magnetic charge. Dirac had thus to address the following consistency issue: whether it was possible to include particles carrying a magnetic charge without disturbing the consistency of the coupling of electromagnetism to quantum mechanics.

The argument he proposed in his 1931 paper for solving this apparent inconsistency is remarkable under many aspects. In particular, it represents one of the first example of an explicit use of *topological considerations* in the early twentieth century physics. In developing his argument, centered on the relation between the phase change of the wave functions round closed curves and the flux of the magnetic field $\vec{B}$ through closed surfaces, Dirac in fact applied ideas involving the structure of the space in the large (what is now known as global topology).[3] The result he obtained was the following: the introduction of magnetic charge can be consistent with the quantum theory provided its values are 'quantized'. In his own words (Dirac 1931, p. 68): "Our theory thus allows isolated magnetic poles, but the strength of such poles must be quantised, the quantum μ_0 being connected with the electronic charge e by $\hbar c/e\mu_0 = 2$."

In current notation (denoting magnetic charge by g and using the unit system $\hbar = c = 1$), Dirac's result is that a magnetic charge g can occur in the presence of an electric charge e if the following condition, known as *Dirac quantization condition*, is satisfied:

$$eg = 2\pi n \qquad n = 0, \pm 1, \pm 2, \ldots.$$

This condition has an immediate striking consequence: the mere existence of a magnetic charge g somewhere in the universe implies the *quantization of electric charge*, since any electric charge must then occur in integer multiples of the unit $2\pi/g$. In Dirac's words (*ibid.*), "The theory also requires a quantisation of electric charge, since any charged particle moving in the field of a pole of strength μ_0 must have for its charge some integral multiple (positive or negative) of e, in order that wave functions describing the motion may exist."

The quantization of electric charge was a fact of observation, but theoretically unexplained. For Dirac, it was indeed the possibility of obtaining an explanation of this fact to constitute one of the main reason of interest in his theory of magnetic poles. As he wrote in his 1948 paper (Dirac 1948, p. 817), "The interest of the theory of magnetic poles is that it forms a natural generalization of the usual electrodynamics and *it leads to the quantization of electricity*. [...] The quantization of electricity is one of the most fundamental and striking features of atomic physics, and there seems to be no explanation for it apart from the theory of poles. This provides some grounds for believing in the existence of these poles."

In substance, according to Dirac, even if magnetic charges are not observed the theory provides a good reason for believing in their existence. In fact, the theory also provides an explanation of why isolated magnetic poles are not observed. The explanation is based on the great difference between the numerical values for the quantum of electric charge e_0 and the quantum of magnetic pole g_0. In the notation used by

---

[3] On Dirac's anticipation of topological ideas in physics see, for example, Olive (2003).

Dirac in his 1948 paper, if we take the experimental value for the *fine structure constant*, i.e., $\alpha = e_0^2/(\hbar c) = 1/137$, and we use the quantization condition (in its original form: $e_0 g_0 = (1/2)\hbar c$), we can infer that the value of the quantum of magnetic pole is $g_0^2 = (137/4)\,\hbar c$, that is much greater than the numerical value for the quantum of electric charge, $e_0^2 = (1/137)\,\hbar c$.[4]

Thus, Dirac notes, "although there is symmetry between charges and poles from the point of view of general theory, there is a difference in practice" (Dirac 1948, p. 830). For example, two one-quantum poles of opposite sign attract one another with a force $(137/2)^2$ times as great as that between two one-quantum charges at the same distance. "It must therefore be very difficult to separate poles of opposite sign", Dirac continues, and his conclusion is that "this explains why electric charges are easily produced and not magnetic poles" (*ibid.*).

2.3 The Meaning of EM Duality

In *classical electrodynamics* (with magnetic source terms included), we have seen that the EM duality transformation

$$\vec{E} \to \vec{B}, \qquad \vec{B} \to -\vec{E},$$
$$\rho_e, \vec{j}_e \to \rho_g, \vec{j}_g, \qquad \rho_g, \vec{j}_g \to -\rho_e, -\vec{j}_e,$$

exchanges, at the same time, the roles of the electric and magnetic fields and the roles of the electric and magnetic charges and currents, while leaving the physics invariant. 'The physics' means the Maxwell's equations and the relevant physical quantities (such as the energy and momentum densities of the electromagnetic field). EM duality is thus a symmetry of the theory, expressing the equivalence of the following *dual ways* of describing the same physics:

(1) *Description*$_1$. The physics is described in terms of:

- The electric field $\vec{E}_1$ and the magnetic field $\vec{B}_1$;
- The electric charge and current densities ρ_{e_1} and $\vec{j}_{e_1}$, and the magnetic charge and current densities ρ_{g_1} and $\vec{j}_{g_1}$.

(2) *Description*$_2$. The physics is described in terms of:

- The electric field $\vec{E}_2 = \vec{B}_1$ and the magnetic field $\vec{B}_2 = -\vec{E}_1$;
- The electric charge and current densities $\rho_{e_2} = \rho_{g_1}$ and $j_{e_2} = j_{g_1}$, and the magnetic charge and current densities $\rho_{g_2} = -\rho_{e_1}$ and $\vec{j}_{g_2} = -\vec{j}_{e_1}$.

This means, in concrete, that a calculation of a physical quantity in the framework of *description*$_1$ can be obtained by means of another calculation in the dual framework

---

[4] With respect to the quantization condition (13), the quantization condition in the form originally given by Dirac uses definitions of the electric charge and the magnetic charge differing by a factor 4π.

of *description*$_2$. For example, calculating the force of the electric field $\vec{E}_1$ on a particle with electric charge e_1 in the framework of *description*$_1$ is the same as calculating the force of the magnetic field $\vec{B}_2$ on a particle with magnetic charge $g_2 = -e_1$ in the framework of *description*$_2$.

For the duality issue of concern here, this does not say much. The idea of a symmetry between electricity and magnetism is, of course, much more profound then what the above consideration can show. In particular, it has played a very important heuristic role in the history of pre-quantum electrodynamics – think about its influence on Faraday's discovery of electromagnetic induction or Einstein's 1905 work on special relativity. But it is only in the quantum context that the full theoretical significance of physical dualities does actually emerge.

In order to have a complete grasp on the real meaning of EM duality in quantum physics, we should follow the development of this idea in quantum field theory and string theory. In this paper we pursue a more modest scope. We remain in the conceptual range of the preceding Section, and consider what can be extracted from Dirac's seminal work for the issue at stake. In fact Dirac anticipated so much that, on the basis of his results, it is possible to get an idea of some general features of today's physical dualities. Here we focus on the most striking of these features: that is, the fact that dualities typically interrelate weak and strong coupling. This is known, in the physics literature, as *weak-strong duality*.

In the framework of Dirac's theory of magnetic monopoles, it is easy to see how the weak-strong interchange naturally follows from assuming EM duality and the quantization condition. As we have seen, EM duality implies interchanging electric and magnetic charges:

- EM duality: $\quad e \to g, \quad g \to -e,$

while Dirac's quantization condition implies that the electric and magnetic charges (that is, the electric and magnetic coupling constants) are so related:

- Quantization condition: : $\quad eg = 2\pi n.$

Putting the two together, we obtain:

$$e \to g = \frac{2\pi n}{e}, \quad g \to -e = -\frac{2\pi n}{g}.$$

This means that if the charge e is small, the charge g into which it is transformed is strong and *vice versa*. That is: in quantum physics, EM duality relates weak and strong coupling.

In general, turning to the more appropriate context of quantum field theory and string theory, what happens is that dualities typically relate a theoretical description concerning a strong-coupling regime to another description concerning a weak-coupling regime (while leaving the 'physics' invariant). That is, dualities exchange physical regimes that are very different, with the remarkable consequence that calculations involving strong forces in one theoretical description can be obtained from calculations involving weak forces in the dual theoretical description.[5] This is not all: at the same time, dualities also typically exchange elementary quanta ('electric charges') with collective excitations ('solitons' or 'magnetic charges'), with the consequence that what was viewed as *fundamental* in one theoretical description becomes *composite* in the dual description.[6]

2.4 Concluding Remarks: Dualities and Physical Theories

From a philosophical point of view, the above illustrated features are rather unusual, especially if dualities are to be considered as intertheoretic relations. Physical theories are generally intended to describe a given range of phenomena: they have specific domains of application, defined in correspondence to some range or level of the adopted physical scale (for example, the energy scale). In the cases of intertheoretic relations usually discussed in the philosophy of science – in connection, for example, with such issues as reductionism and continuity across theory change – the theories considered are either on the same level or on successive levels. In this latter case, the two theories are typically so related that one can be seen as 'emerging' from the other. But dualities show that another type of situation is possible: the two interrelated theoretical descriptions can be on very different scale levels. Moreover, by means of dualities the 'same physics' is described by two theoretical formulations presenting apparently different ontologies: the fundamental objects in one formulation become composite objects in the dual formulation, and viceversa.

A first question is then: what do dualities indeed relate? Two different theories or just two different formulations of the same theory? The answer surely depends on the sort of duality we are considering. But also on what we intend by a 'theory', and this is also closely connected to the question of what sort of symmetry is represented by dualities, if these are indeed symmetries (as is commonly assumed).

It is usually said that dual theories, or dual theoretical descriptions, are connected with one another by transformations 'leaving the physics invariant': dualities are in this sense 'symmetries'. This can be made more precise by specifying the meaning of the expression 'leaving the physics invariant'. If by this we intend that the dynamical equations of the theory remain invariant, as in the EM duality case discussed in Section 2.2 (where the Maxwell's equations are invariant under the duality transformation D), then the duality is a symmetry of the theory in the precise sense normally used in contemporary physics. That is, the sense according to which G is a symmetry group of a theory if the dynamical equations (or the 'action') of the theory

---

[5] This is what makes dualities particularly interesting and useful in the context of quantum field theory and string theory, as we usually know only the perturbative part of a theory, that is its 'weak coupling' regime. Dualities thus relate what is still unknown to what can be calculated.

[6] To be honest, this cannot be seen in the context of Dirac's theory of magnetic poles. It is important to underline that this feature could emerge only with the extension of dualities in the framework of quantum field theory.

are invariant under the transformations (that are the elements) of the group G. The symmetries postulated through the so-called invariance principles of physics, such as the space-time symmetries and the gauge symmetries of the Standard Model of particle physics, are properties of physical theories in this sense.

But, in general, the dualities used in today physics relate two different theoretical descriptions that concern different scale levels and present apparently different ontological scenarios. These descriptions can even involve different actions (or Hamiltonians) and different fields. In which sense, then, they are just two different formulations of the same underlying theory, as is commonly maintained? This clearly depends on the meaning attributed to the notion of *theory*. The clue is given by the extended sense in which duality is considered a symmetry. That is: the 'theory' is identified on the basis of what remains invariant under the duality transformations, the 'same physics' that is differently described by means of the dual formulations. And this 'same physics', according to the physicists working on the subject, is given by the spectra and the transition amplitudes.[7]

We thus arrive at an apparently 'phenomenological' understanding of the notion of a theory that may seem paradoxical in such a highly mathematized and far away from common (and, for now, possible) experience as is string theory. Note that such a notion is not new in the history of quantum physics: think about the ideology behind Heisenberg's matrix mechanics in the 1920s or the S-matrix approach dominating in the 1960s (which was, it is worth noting, at the basis of the so-called 'dual resonance model' from which early string theory was born in the late 1960s).[8]

Summing up, physical dualities pose a dilemma to the philosophers of science: either the physicists's 'received view' that dualities relate different formulations of the same theory is accepted, but this implies a notion of what is a 'physical theory' which is quite different from the common idea that a theory is identified on the basis of its fundamental dynamical equations and ontology; or, on the contrary, dualities are understood as relations between different physical theories, but then it is difficult to understand the real meaning of such inter-theory relations and to see in which sense they can be considered 'symmetries'

Acknowledgments Many thanks to Andrea Cappelli and Jos Uffink for precious feedback. Parts or earlier versions of the paper were presented on various occasions (2005: Boston; 2006: Florence; 2007: Irvine, Banff, Madrid). I am grateful to the audiences for useful comments and questions.

References

Cappelli A, Castellani E, Colomo F, Di Vecchia P (eds) (2010) The Birth of string theory. Cambridge University Press, Cambridge

Dawid R (2007) Scientific realism in the age of string theory. Phys Philos 11:1–35

---

[7] See, for example, Polchinski (1999), Section 2.4

[8] See, on this point, the introduction and part I of Cappelli et al. (2010).

Dirac PAM (1931) Quantised singularities in the electromagnetic field. Proc Roy Soc Lond A 133:60–72
Dirac PAM (1948) The theory of magnetic poles. Phys Rev 74:817–830
Olive D (2003) Paul Dirac and the pervasiveness of his thinking. arXiv: hep-th/03041133
Polchinski J (1999) Quantum gravity at the Planck length. Int. J. Mod. Phys. A 14:2633–2658

Chapter 3
Are 'Identical Quantum Particles' Weakly Discernible Objects?

Dennis Dieks

3.1 Introduction

According to classical physics the world consists of *individuals*, i.e., distinct objects that can bear their own characteristic names. In physics this individuality is standardly seen not as something primitive, but as based on qualitative physical differences – in accordance with Leibniz's principle of the identity of indiscernibles (PII).

In quantum theory the status of individual objects is notoriously more controversial than in classical physics. The symmetrization postulates that apply to the states of so-called identical particles appear to show that these "particles" do not obey Leibniz's principle; and this in turn raises doubts about whether they are objects at all. Conventional wisdom among physicists is that a field-theoretical picture is actually more appropriate than a particle one (according to this point of view there are no particles, but only *field quanta*); by contrast, philosophers of physics seem to tend more to the view that quantum mechanics *is* about particles but that their individuality defies PII (see Muller and Saunders 2008, for relevant quotations).

There is a growing recent literature, however, in which it is claimed that the difference between quantum theory and classical physics concerning this individuality issue is not at all as drastic as it was assumed to be (Saunders 2003, 2006; Muller and Saunders 2008; Muller and Seevinck 2010). Indeed, the argument goes, even in classical physics there are situations in which Leibniz's principle seems to fail, namely situations in which there are qualitatively similar objects in symmetrical situations. A famous example, introduced by Black (1952), concerns two spheres of exactly the same form and material constitution, alone in the universe, and at two miles from each other. There are *two* spheres; however, there are no physical features that distinguish between them. A closer look reveals, however, that PII can be salvaged in classical situations of this kind. The key idea is to introduce the notion of *weak discernibility* (see below for an explanation of this concept). The claim then

---

D. Dieks (✉)
Institute for History and Foundations of Science Utrecht University, P.O. Box 80.010, 3508 TA Utrecht, The Netherlands
e-mail: d.dieks@uu.nl

is that 'identical quantum particles' obey PII in the same way as classical objects in symmetrical configurations; they are 'weakly discernible' objects.

The aim of this paper is to critically evaluate this conclusion (see for more detail on some of the points to be discussed (Dieks and Versteegh 2008)). As I shall argue, an important difference between quantum mechanics and classical physics is that it is not clear that "systems of many identical particles" in quantum mechanics consist of more than one components at all; the very *applicability* of the notion of weak discernibility remains therefore moot. Admittedly, this is an interpretation dependent issue; for the sake of the argument of this article I shall assume standard interpretational ideas. There are certainly alternative interpretational possibilities, and it is true that some of these clearly make identical quantum particles into individual objects (think of Bohm's theory, for example). But these alternative interpretations do not need the concept of weak discernibility for achieving this, and are not the focus of those who defend the claim that quantum particles are weakly discernible. Accordingly, my purpose here is not to defend standard interpretational ideas but rather to show that the introduction of the concept of weak discernibility does not make the difference to the discussion that it has been claimed to make. Indeed, it will turn out that within the framework of standard interpretational ideas, combined with PII, there is no reason to withdraw from the traditional view that there are no individual particles at all but only field quanta.

3.2 Weak Discernibility

As already pointed out, there are cases in classical physics in which PII seems to fail. One famous example, proposed by Black (1952), concerns two spheres of identical chemical composition and two miles apart (in an otherwise empty relational space à la Leibniz, not in Newtonian absolute space where one could resort to absolute positions in order to label the spheres). Another example is provided by a universe consisting of two hands that are each other's mirror images (Kant's enantiomorphic hands). There are also many examples from mathematics (Keränen 2001): think, for example, of the points in the Euclidean plane. The essential feature in these cases is that the objects in question have all their properties in common, and still there are more than one of them. It seems to follow that we are employing concepts of object and individuality that are independent of the presence of distinguishing qualitative differences – in violation of PII.

As Hawley points out (Hawley 2006), defenders of PII can respond to such examples in a variety of ways. First, they may query whether the described situations are possible at all – but the above situations all are possible as far as the relevant science is concerned. Second, they can dispute that these situations are best described in terms of distinct but indiscernible individuals. This can take two forms: they may either argue that if a correct analysis of discernibility is employed the objects are discernible after all, or they may claim that there were no distinct objects to start with, that there is only one undivided whole.

In the symmetrical situations from classical physics the most plausible response is to say that we are dealing with distinct objects that obey PII, but that these objects are only *weakly discernible*. In this we follow Saunders (2003, 2006), who takes his clue from Quine (1981), in noting that in such cases *irreflexive* relations are instantiated: relations entities cannot bear to themselves. This irreflexivity is the key to proving that (a generalized version of) PII is satisfied after all: if an entity stands in a relation that it cannot have to itself, there must be at least two entities.

To see in logical detail how this works, let us formalize the argument. PII can be formulated as follows, with = denoting identity:

$$s = t \Leftrightarrow \forall P(P(s) \leftrightarrow P(t)). \tag{3.1}$$

The universal quantifier here ranges over all physical predicates P. The right-hand side of the equation stipulates that s and t can replace each other, *salva veritate*, in *any P*.

There can now be various kinds of discernibility. Two objects are *absolutely discernible* if there is a (physical) one-place predicate that applies to only one of them. They are *weakly discernible* if an *irreflexive* two-place predicate relates them. The latter possibility is relevant to our examples. If there is an irreflexive but symmetric two-place predicate $P(.,.)$ that is satisfied by s and t, PII in the form (3.1) requires that if s and t are to be identical, we must have:

$$\forall x (P(s, x) \leftrightarrow P(t, x)). \tag{3.2}$$

But this is false: in any valuation in which $P(s,t)$ is true, $P(t,t)$ cannot be satisfied by virtue of the fact that P is irreflexive. It follows therefore that PII is satisfied by any two individual objects that stand in an irreflexive qualitative relation.

3.3 Remaining Worries

There is a remaining worry, though. Aren't we begging the question by already assuming that there are one or more individuals between which the irreflexive relations hold? To avoid circularity we seem to need some independent assurance that the whole we are thinking of consists of elements of which there could be more than one, in order to even start testing the validity of PII. But this possibility of "splitting up the domain" is not at all evident in the context of discussions of the applicability of PII. As we have already seen, one possible stance in such discussions is to argue that there is no multiplicity at all: that there is only *one* undivided physical system. If this is assumed, no questions about the individuation of elements of the domain by means of PII have to be answered, and a simple parsimonious ontological picture is achieved. If there is no good reason to think of the domain as consisting of several things in the first place, this option surely recommends itself.

There is one form of this circularity worry that does not need to detain us here, namely the fear that it does not make sense to speak about relations *at all* if there has not been a prior identification of the *relata*. This concern can be defused by appealing to structuralist analyses, according to which relata need not be ontologically prior to the relations they have to each other: the relata can be conceived of as *determined by the relations*, as a kind of nodes in a relational network (see French and Krause 2006; Dieks and Versteegh 2008; Esfekd and Lam 2008, for details). If we follow this path, we need not assume that there is a division of labor between "objecthood providers" that come first and relations that come into play only subsequently. It is possible without contradiction that the relational structure that is used in PII is at the same time our only access to Leibnizean objecthood, i.e., objecthood with a qualitative grounding. The relations in our two physical examples – being at a spatial distance from each other, being each other's mirror image – indeed give us information about the presence of objects in this way. These are obviously physically meaningful relations that pertain to relata that can be displaced with respect to each other, or that can be reflected and whose orientations can be compared. It clearly makes physical sense to speak about such relata as actual things, objects (that differ from each other because of their mutual distance or their mirror-image relation, respectively).

It is important that the relations themselves are physically meaningful; the mere possibility of speaking about a domain in terms of irreflexive relations does not by itself suffice to ensure that the domain is split up into different objects. There are situations in which it is possible and even usual to employ properties or relations talk, in spite of the fact that it is clear from the outset that there are no different objects at all – in such cases considerations about the irreflexitivity of the relations and about weak discernibility obviously do not help making PII apply to the "objects".

The standard example to illustrate possibilities of this kind is that of money in a bank account. Imagine a situation in which by virtue of some financial regulation the Euros in a particular account can only be transferred to different one-Euro accounts. So, in a complete money transfer an account with five Euros, say, could be emptied and five different one-Euro accounts would result. In this case the Euros in the original account stand in an irreflexive relation to each other, namely "only transferable to different accounts". But this does not make them into different objects. We could of course make an attempt to exploit the irreflexive relations for the purpose of distinguishing individual Euros, by labelling the Euros via the accounts they end up in. However, this means looking at the situation *after* the money transfer, which does not achieve anything for the purpose of distinguishing between the Euros in the account they are actually in, before the transfer. The essential point is that the relations here do not relate occurrent physical characteristics of the situation; they do not connect actual relata. The case of more than one money units in one bank account is the standard example to illustrate absence of individuality; it is a case in which only the account itself, with the total amount of money in it, can be treated as possessing individuality (Schrödinger 1998; Teller 1998). Although we are accustomed to using relations and things *talk* here, there is nothing in the actual physical situation that corresponds to this.

Thus, we have found an indispensable clause in the argument for PII-based individuality in the presence of irreflexive relations. Such relations can only be trusted to be significant for the individuality issue if they are "physically meaningful", i.e., of the sort to be able to connect actual physical relata.

3.4 "Physically Meaningful"

Not all predicates and relations that occur within a physical theory are physically meaningful in the sense intended here (cf. Muller and Saunders 2008, [4.3]). In particular, the mere occurrence of *labels*, as in the quantum mechanics of "many particles", does not suffice to decide the issue of whether there indeed *are* individual particles. Some correlation between these labels and one or another physical quantity of the kind to be used in PII should be established before the labels become physically respectable. To get a clue about possible criteria here, let us first have a look at classical physics.

In situations in classical physics without particular symmetries, a feature of such qualitative physical relations is that they can be used to distinguish and name different relata. For example, in an arbitrary configuration of classical particles the distances with respect to other particles will unambiguously characterize each individual particle. Changing the configuration so that it becomes *more* symmetrical (but not yet completely symmetrical) will obviously change the values of the distances, but not the number of individual objects. The distance relations thus clearly are the kind of relations that connect actual physical objects. The possibility of discerning and naming actual objects in asymmetrical situations thus provides us with a test for the physical meaningfulness of the distance relations. The completely symmetrical situation is a degenerate situation, in which it turns out that the distance relations are still sufficient to establish *weak* discernibility. The breaking of the symmetry, in thought, is like introducing a coordinates origin in describing a completely symmetrical figure. If a physically meaningful mapping between points and natural numbers exists in the presence of such an origin, which proves Leibniz-like individuality, this individuality will still be present in "weakened" form when the reference point has been removed. What changes in the transition from absolute to weak discernibility is the *constructibility* of the mapping: the possibility of actually naming and distinguishing by means of the involved relations disappears when we end up in the fully symmetrical situation, but this does not collapse the different objects into one. They still are physical entities that have distances to other entities in the structure, even though the symmetry makes it impossible to use this for assigning names.

Indeed, why are we so sure that there are two Blackean spheres and two Kantian hands? Our mind's eye sees Black's spheres at different distances, and Kant's hands with different orientations, before us; when we break the symmetry of the configurations in these cases by imagining ourselves or some other standard as points of reference, the relations with respect to this reference point make it possible to distinguish the entities. Thus we can name Black's spheres via their unequal distances to a fixed point and in Kant's universe we may imagine a reference hand conventionally called "left". Another example, relevant for our subsequent discussion of quantum objects, is furnished by two oppositely directed arrows in an otherwise empty Leibnizean world. If we fix a standard of being "up", we break the symmetry and the individual arrows become absolutely discernible as up or down.

These cases are to be contrasted with the case of the Euros in a bank account. Even though the Euros have different destinations this cannot be used to distinguish actual Euros in the account. There is a total amount of money in the account, but this amount does not consist of individual Euros.

3.5 The Quantum Case

In "many-particles quantum mechanics" labels occur, which seem to refer to different individual particles. But as we have seen, something more is needed to turn this into a basis for assigning individuality: the indices should be physically meaningful, i.e., some meaningful physical property or relation should be associated with them. It should be made clear that the indices do not only possess their uncontroversial mathematical significance, in that they number different Hilbert spaces in a tensor product space, but that they also correlate to something physical. Of course, it is simple enough to define (irreflexive) relations between indices, for instance the relation of *inequality*, or the relation of *referring to different Hilbert spaces*. It is also easy to couch such relations in seemingly physical language: e.g., any two hermitean operators belonging to different Hilbert spaces commute, whereas this is not the case for an arbitrary pair of such operators in one Hilbert space (a proposal very much like this occurs in Muller and Seevinck [2010]). But in spite of the fact that it is uncontroversial that hermitean operators can function as representatives of physical quantities, in this case the manoeuvre only pays off if it can be shown that *operators labelled by particle indices* have physical meaning to begin with.

Of course, the intuitive appeal of thinking of the indices in the formalism as particle labels is very understandable. Indeed, the formalism of one-particle quantum mechanics (e.g., with a one-electron Hilbert space) can without difficulty be interpreted as giving information about the behavior of one single physical particle – it is consequently only natural to think of the tensor product formalism with its indices as referring to labelled copies of this one-particle case. That there are nevertheless grave difficulties in this interpretation is due to the (anti-)symmetrization postulates that apply to the states of a "many-particles" system composed of so-called identical particles. These symmetrization rules have the effect that all indices occur symmetrically, so that none of them is physically distinct from the others. To see how this complicates matters, think of a one-particle position measurement carried out on a many-particles system described by such a symmetrized state. The result found in such a measurement (for example, the click of a Geiger counter or a black spot on a photographic plate) is not linked to *one* of the "particle labels"; it is, in symmetrical fashion, linked to *all* of them. This already demonstrates how the classical limit of quantum mechanics does not simply connect the classical particle concept to individual indices in the quantum formalism.

Nevertheless, it is true that the quantum situation is to some extent reminiscent of the symmetric configurations of classical objects described in the previous section, and one might hope to escape the just-mentioned problems by making use of this analogy. Indeed, as we have seen, symmetry is not decisive for proving the absence of Leibniz-style individuality: we may be dealing with weakly discernible individuals. Could it therefore not be that in the quantum case there are irreflexive physical relations between particles that guarantee their individuality in the same way as they did for Black's spheres, Kant's hands and Euclid's points? This is the position adopted by Saunders et al. (Saunders 2006; Muller and Saunders 2008; Muller and Seevinck 2010). For example, the anti-symmetry of the state of many-fermions systems seems to imply the existence of irreflexive relations between components of the total system (labelled by indices): intuitively speaking, the "fermions" in any pair stand in the relation of "occupying different one-particle states", even though the particles do not receive individually different quantum mechanical state descriptions. We already noted that this approach can only work if the relations in question possess physical significance – but let us now look at the proposal in a more detailed way.

The technical details of the argument can be illustrated by the example of two fermions in the singlet state. If $|\uparrow\rangle$ and $|\downarrow\rangle$ stand for states with spins directed upwards and downwards in a particular direction, respectively, the anti-symmetrization principle requires that a typical two-fermion state looks like

$$\frac{1}{\sqrt{2}}\{|\uparrow\rangle_1|\downarrow\rangle_2 - |\downarrow\rangle_1|\uparrow\rangle_2\}, \tag{3.3}$$

in which the subscripts 1 and 2 refer to the one-particle state-spaces of which the total state-space (a Hilbert space) is the tensor product. The anti-symmetry of the total state implies that the state restricted to state-space 1 is the same as the restricted state defined in state-space 2. (The "partial traces" are $\frac{1}{2}\{|\uparrow\rangle\langle\uparrow| + |\downarrow\rangle\langle\downarrow|\}$ in both cases.) The *total spin* has the definite value 0 in state (3.3); that is, state (3.3) is an eigenstate of the operator $S_1 \otimes I + I \otimes S_2$. Therefore, it seems natural to say that we are dealing with two spins that are oppositely directed. On the other hand, we cannot associate definite spin directions with the indices because the up and down states occur symmetrically in each of the Hilbert spaces 1 and 2, respectively.

This situation may make the impression of being essentially the same as the one of the two arrows mentioned earlier: the *oppositeness* of spins seems to guarantee, via weak discernibility, that there are *two individual spins*.

3.6 Quantum Individuals?

On closer examination the similarity starts to fade away, however. One should already become wary by the observation that the irreflexive relations in the quantum case have a theoretical representation that is quite different from that of their classical counterparts. There the relations could be formalized by ordinary predicates that can be expressed as *functions* of occurrent properties of the individual objects (like "up" and "down" with respect to a conventionally chosen standard, or $+1$ and -1), with the correlation expressed by the fact that the sum of these two quantities has a fixed value. By contrast, in quantum theory the correlation is expressed in a more complicated way: the state of the total system is an *eigenstate* of a linear *operator* in the total system's Hilbert space. The standard interpretation (remember that we are studying the status of weak discernibility within a standard interpretational framework) says that quantum states should be interpreted in terms of possible *measurement results* and their probabilities, rather than in terms of occurrent physical properties. In the case at hand, the singlet state (3.3), the prediction of quantum mechanics is that individual spin *measurements* will with certainty yield opposite results, summing up to 0; but on the pain of running into well-known paradoxes and no-go theorems it cannot be maintained that these results reveal oppositely directed spins that were already there before the measurements. This is an illustration of the notorious "holism" of quantum mechanics: definite properties of a composite system do not always supervene on definite properties of its parts.

This suggests that the correct analogue to the quantum case is not provided by two oppositely directed classical arrows but, if anything, rather by the example of a two-Euro account that can be *transformed* (upon "measurement", i.e., the *intervention* brought about by a money transfer) into two distinct one-Euro-accounts.

We already observed that the indices in the many-particle formalism do not have a direct particle interpretation via the classical limit. Weak discernibility focuses on *relations*, though, so let us investigate further whether or not the relations between indices have physical significance. In order to do so we may copy the strategy followed in the classical case, namely breaking the symmetry and seeing whether in the resulting situation these relations can serve as name-givers. This cannot work, however, as long as we stay within a "many identical particles" system: quantum mechanics strictly forbids such systems that are not in a (anti-)symmetric state. This is a significant difference from the symmetrical classical cases, in which the symmetry was contingent and the theory allowed evolutions from symmetrical to asymmetrical configurations. In quantum mechanics the mutual relations between "identical particles" cannot serve to distinguish individual component systems *as a matter of principle*. The theory does not allow any asymmetrical situations with which to approach the symmetrical situation, and our earlier tests fail.

It is true that this is not conclusive: compare the situation in a hypothetical classical world in which *laws* stipulate that spheres can only occur in symmetric configurations. In such a world we would still have good reasons to think in terms of individual spheres, because our theories allow for an external object whose relations to the spheres makes them discernible. Analogously we can try to break the symmetry in the quantum case by the introduction of a standard that is external to the quantum system itself. Quantum mechanics does not require symmetry of such a total state ("identical particles" plus something else), and with such an external standard in hand we may hope to be able to distinguish individual identical particles.

To see the inevitability of a negative outcome of any such test, consider an arbitrary system of identical quantum particles to which a gauge system has been added without any disturbance of the original system (i.e., the total state is the product of the original symmetrical or anti-symmetrical identical particles state and the state of the gauge system). Let the new total state be denoted by $|\Psi\rangle$. Any quantum relation in this state between the gauge system g and one of the identical particles, described in subspace j, say, has the form $\langle\Psi|A(g,j)|\Psi\rangle$. Here $A(g,j)$ is a hermitian operator working in the state-spaces of the gauge system g and identical particle j. We can now use the (anti)-symmetry of the original identical particles state to show that the gauge system stands in exactly the same relations to *all* "identical particles". The (anti)-symmetry entails that $P_{ij}|\Psi\rangle = \pm|\Psi\rangle$, where P_{ij} stands for the operator that permutes identical particle indices i and j. Now,

$$\langle\Psi|A(g,j)|\Psi\rangle = \langle P_{ij}\Psi|A(g,j)|P_{ij}\Psi\rangle =$$
$$\langle\Psi|P_{ij}^{-1}A(g,j)P_{ij}|\Psi\rangle = \langle\Psi|A(g,i)|\Psi\rangle.$$

In other words, any quantum relation the gauge system has to j, it also has to i, for arbitrary values of i and j. That means that these quantum relations have no discriminating value.

All evidence points into the same direction: "identical quantum particles" behave like money units in a bank account rather than like Blackean spheres. It does not matter what external standards we introduce, they will always possess the same relations to all (hypothetically present) entities. The irreflexive relations used by Saunders and others to argue that identical quantum particles are weakly discernible individuals lack the physical significance required to make them suitable for the job.

3.7 Conclusion

The analogy between quantum mechanical systems of "identical particles" and classical collections of weakly discernible objects is only superficial. There is no sign within standard quantum mechanics that "identical particles" are things at all: there is no ground for the supposition that relations between the indices in the formalism possess physical significance in the sense that they connect actual objects. Consequently, the irreflexivity of these relations is not important either. Conventional wisdom appears to have it right after all.

References

Black M (1952) The identity of indiscernibles. Mind 61:153–164
Dieks D, Versteegh MAM (2008) Identical particles and weak discernibility. Found Phys 38: 923–934

Esfeld M, Lam V (2008) Moderate structural realism about space-time. Synthese 160:27–46
French S, Krause D (2006) Identity in physics: A historical, philosophical, and formal analysis. Oxford University Press, Oxford
Hawley K (2006) Weak discernibility. Analysis 66:300–303
Hawley K (2007) Identity, indiscernibility and number. Unpublished manuscript
Keränen J (2001) The Identity problem for realist structuralism. Philos Math 9:308–330
Muller FA, Saunders S (2008) Discerning fermions. Br J Philos Sci 59:499–548
Muller FA, Seevinck M (2010) Discerning elementary particles. Philos Sci, to be published
Quine WV (1976) Grades of discriminability. J Philos 73:113–116, Reprinted in Quine WV (1981) Theories and things. Harvard University Press, Cambridge
Saunders S (2003) Physics and Leibniz's principles. In: Brading K, Castellani E (eds) Symmetries in physics: Philosophical reflections. Cambridge University Press, Cambridge
Saunders S (2006) Are quantum particles objects? Analysis 66:52–63
Schrödinger E (1952) Science and humanism. Cambridge University Press. Partly reprinted as: What is an elementary particle? In: Castellani E (ed) Interpreting bodies: Classical and quantum objects in modern physics. Princeton University Press, Princeton (1998), pp 197–210
Teller P (1998) Quantum mechanics and haecceities. In: Castellani E (ed) Interpreting bodies: Classical and quantum objects in modern physics, pp 114–141. Princeton University Press, Princeton

Chapter 4
Wave–Particle Duality in Quantum Optics

Brigitte Falkenburg

Philosophers of science are inclined to think that *wave–particle duality* is an obsolete concept, because according to quantum mechanics there are neither waves nor particles in a classical sense. But in physical practice, wave–particle duality is alive. The concept is crucial in order to understand the recent *which-way* experiments of quantum optics. First, several aspects of the concept will be sketched. Then I explain why the experimenters say that they *prepare waves* but *detect particles*. Indeed, their pragmatic attitude helps to understand a prominent thought experiment of Scully, Englert and Walther and the *which-way* experiments that realised it. Finally, I discuss a simple polarizer experiment. The experiment shows that no realistic interpretation of particles can cope with wave–particle duality, whereas the causal relevance of the quantum waves can not be denied.[1]

4.1 What Is Wave–Particle Duality?

The concept of wave–particle duality dates back to the beginnings of quantum theory. It was based on Einstein's light quantum hypothesis (1905), its integration into relativistic kinematics (1916), and de Broglie's (1923) hypothesis of matter waves. The crucial laws were the Planck–Einstein relation $E = h\nu$ and the Einstein–de Broglie relation $p = \hbar k$. They relate particle properties to wave properties, i.e., the energy E to the frequency ν and the momentum p to the wave number k or wavelength $\lambda = 2\pi/k$.

Both relations have a clear operational content. In the Compton effect (1923), light causes a momentum kick, i.e., a momentum transfer from a photon to an electron; whereas the well-known diffraction phenomena indicate light waves. In the experiment of Davisson and Germer (1927), an electron beam sent through a crystal causes a diffraction pattern; whereas the curved particle tracks observed in a cloud chamber with magnetic field indicate massive charged particles.

B. Falkenburg (✉)
Technische Universität Dortmund, Fakultät 14, Institut für Philosophie und Politikwissenschaft, D-44221 Dortmund

[1] In Falkenburg (2007), the argument is presented in more detail.

So far, wave–particle duality means that *operational* particle properties are attributed to light waves and *operational* wave properties to matter particles. The above relations $E = h\nu$ and $p = \hbar k$ and the phenomena supporting them are an uncontroversial formal and operational basis of wave–particle duality up to the present day.

Later, these operational relations were spelled out in quantum mechanical terms. Heisenberg (1925) developed matrix mechanics. He skipped the particle trajectories of the electrons and recommended metaphysical abstinence about what happens inside an atom. Schrödinger (1926) developed his famous wave equation and suggested a wave interpretation of the electron which, however, failed. According to Born's (1926a, b) probabilistic interpretation, Schrödinger's wave function Ψ has no direct physical meaning. It is a probability amplitude, i.e., its square $|\Psi|^2$ is a probability density predicting the distribution of the individual measurement outcomes. For position measurements, it predicts the spatial distribution of particle detections.

Born himself was puzzled about the relation between the wave-like and particle-like features of quantum phenomena. He had classical particles and waves (or fields) in mind when he expressed wave–particle duality as follows:

> The guiding field which is represented by a scalar function Ψ [...] spreads according to Schrödinger's differential equation. Energy and momentum, however, are transferred as if corpuscles were really flying around.[2]

Here, Born characterizes Schrödinger's wave function Ψ in terms of a ghost-like particle-guiding field or pilot wave. These terms were due to Einstein.[3] Later, they gave rise to Bohm's hidden variable approach (1952). In contradistinction to the probabilistic meaning of $|\Psi|^2$, the ideas of guiding field and corpuscle propagation are fictitious, i.e., they do not have any operational content. Pilot waves or particle trajectories cannot be *measured*. Born related this lack of operational content to the probabilistic structure of quantum mechanics:

> From the standpoint of our quantum mechanics there is no quantity which in any *individual case* causally fixes the consequence of a collision; but also experimentally we have so far no reason to believe that there are some inner properties of the atoms which condition a definite outcome for the collision.[4]

Born's probabilistic interpretation was generalized by von Neumann (1932). The probabilistic interpretation of quantum field theory is completely analogous. Quantum field theory matches the experimental results in the S-matrix. The S-matrix elements are probability amplitudes too. Their square gives the transition probabilities of quantum processes such as the particle reactions of high energy physics.

---

[2] Born (1926b, 803). My translation.
[3] See Jammer (1966, 41).
[4] Born (1926a, 51); translation from Wheeler and Zurek (1983, 54).

On the grounds of the probabilistic interpretation, wave–particle duality is the duality of *probability waves* and *particle detections*. From an operational point of view, however, the particle-like and wave-like properties of electrons or photons are very distinct. In *individual* subatomic position measurements, particle tracks, or scattering events, only the *particle-like* properties of electrons, photons, and other subatomic particles are measured. The *wave-like* properties only show up in a *probabilistic ensemble* of many particle detections. For a large number of position measurements, quantum waves and their squared amplitudes predict the relative frequencies of particle detections. Therefore, the wave-like properties of electrons or photons are measured at the ensemble level, but the particle-like properties at the level of the individual measurement results. Indeed, the Compton effect (in which light shows particle-like behaviour) is observed as a momentum kick in an *individual* photon–electron scattering event, whereas electron diffraction (in which matter shows wave-like behaviour) is observed with an electron *beam*. In order to observe the diffraction pattern behind a double slit or a thin crystal, one needs *many* particle detections.

Finally, the *uncertainty principle* and *complementarity* added to the concept of wave–particle duality. Heisenberg's uncertainty principle $\Delta p \Delta q \geq \hbar/2$ (1927) tells that it is impossible to determine the wave-like and particle-like properties of quantum objects simultaneously with sharp results. Bohr's (1927) Como lecture claimed that these properties are complementary, i.e., that the wave-like and particle-like properties cannot be attributed at the same time to something like a quantum object. In his 1930 book, Heisenberg explained Bohr's views in terms of analogies between quantum phenomena and the corresponding classical pictures of wave or particle. According to the Copenhagen interpretation (which should not be confused with the 'orthodox', probabilistic Born–von Neumann interpretation), there are neither subatomic particles nor subatomic waves. Subatomic realism is refuted, as far as particles or waves are conceived as substances on their own. There are only complementary *pictures* of particles or waves, i.e., models which capture the quantum phenomena in mutually exclusive terms of *either* particle properties *or* wave properties. Particle tracks or the Compton effect are described in particle terms. But electron diffraction or light interference are described in wave terms.

Later interpretations of quantum mechanics *beyond* the probabilistic standard interpretation made the confusion about wave–particle duality complete. Einstein added his instrumentalist ensemble view. According to the famous EPR paper (1935), quantum mechanics does not give a complete description of subatomic physical reality. In 1949, Einstein emphasized that the wave function Ψ does not describe individual quantum processes but only the statistical ensemble. Finally, realism showed up again, following Einstein's and Born's remarks about a guiding field. Bohm (1952) attempted to re-establish subatomic particles. In order to do so, however, he had to attribute non-local properties to them.

4.2 The Pragmatic Attitude: Prepare Waves and Detect Particles

In physical practice, Born's probabilistic interpretation and the operational Planck–Einstein–de Broglie relations are predominant. But in the philosophy of physics, only the possibilities of interpreting quantum mechanics *beyond* the probabilistic interpretation have been discussed and physical practice has been neglected. However, the practice of quantum physics has an important lesson to teach. It concerns a certain *asymmetry* between the preparation and the detection of quantum states. In most experiments of quantum physics, the preparation and the detection of a quantum state are completely distinct empirical procedures. Many physicists would support the following words of the Nobel prize winner Ketterle (2003):

> It is very hard to understand quantum mechanics but after several years of physical practice one gets used to *preparing* waves and *detecting* particles.

It is easy to *prepare* a *momentum state* of subatomic particles and to keep it stable enough to perform an experiment. To prepare a sharply localized particle state needs much more efforts. In the scattering experiments of particle physics, a particle beam of well-defined momentum $p = \hbar k$ is prepared in order to scatter it at a fixed target or at another beam of well-defined momentum. The experiments of quantum optics, too, prepare particle beams: approximately monochromatic light with energy $E = h\nu$ or beams of atoms with well-defined velocity. The beams may be damped down to such a *low intensity* that in average only one particle or field quantum remains in the apparatus, but they remain to be *beams of well-defined momentum and energy*. – In order to obtain *sharply localized particle states* from the beam, *position measurements* have to be performed. According to quantum mechanics, the localized state of a massive particle has dispersion, it spreads into something described as a wave packet and after a while the particle is no longer sharply localized. According to quantum field theory, there are no strictly localized particle states. And for a low intensity beam of well-defined momentum, the occupation number of the corresponding field mode is fluctuating. In order to obtain a well-defined single particle state from a low intensity light beam (a quantum field state of occupation number 1) it is necessary to *prepare an entangled photon pair* and to *detect one* of them. Only the measurement of the first photon prepares the second in a well-defined 1-photon state.

The *measurement* of a quantum state is usually performed by a *particle detector*. In particle physics, particle tracks are measured by means of a cloud chamber, the bubble chamber, drift chambers, etc. Particle tracks are sequences of position measurements. In the experiments of quantum optics, such as the famous double slit experiment with single photons, it is crucial to measure the *counts of a photo detector*. In order to measure entangled photon pairs, two photo counters make coincidence measurements. Even a wave-like quantum phenomenon (e.g., the interference pattern of the electron diffraction in the Davisson–Germer experiment mentioned above) is measured by the detection of particles, by means of a photo plate or a screen made of scintillating material which detects the absorption of single photons or electrons.

Why are the preparation and the detection of a quantum state asymmetric? Not for theoretical, but for several pragmatic reasons. In a certain sense, quantum mechanics and quantum field theory *prefer* momentum states. According to the Schrödinger equation, localized states do have dispersion and momentum states do not. Hence, the momentum states are stable and the position states are not. According to quantum field theory, there are sharp field modes but no sharply localized photon, and the occupation number of a field mode is in general not sharp. Without experimental tricks that include some kind of particle detection, no sharp 1-photon states can be prepared. Hence, for many experiments it is better to prepare momentum states. Experiments aim at preparing *stable, reproducible* states. For many experiments it also better to measure the states via of particle detection. The precision of a measurement depends on its spatio-temporal resolution. The experiments of particle physics aim at detecting particles of well-defined mass, charge, and spin within a *very* small space-time region. The experiments of quantum optics aim at counting rates of very high temporal resolution, above all when coincidence counters are involved in the experiment.

The claim that waves are prepared and particles are detected is based on the probabilistic standard interpretation. According to quantum mechanics and quantum field theory, the waves are nothing but probability amplitudes and the particles are nothing but the events detected by single measurements. Hence, the relation between the waves and the particles of quantum physics is as follows: The waves determine the quantum probabilities which approximately correspond to the relative frequencies of the particle detections.

4.3 The *Which-Way* Experiments of Quantum Optics

The *which-way* experiments of quantum optics do not re-establish hidden particle trajectories. Quite on the contrary, they are in perfect agreement with the above operational interpretation of quantum mechanics and quantum field theory. Their name has historical roots. They stand in the tradition of Einstein's recoiling slit thought experiment from the Bohr–Einstein debate.[5] Einstein's idea was to let single photons pass through a double slit and measure their path from the slit recoil. Bohr argued that the measurement will wash out the interference fringes. Feynman (1965) suggested another version of the thought experiment, namely to measure the path of single electrons behind a double slit by means of light scattering.[6]

In recent quantum optics, it became possible to realize experiments with single photons and electrons. Scully et al. (1991) proposed a refined version of a *which-way* experiment in Einstein's and Feynman's spirit. They proposed to proceed in three steps: (1) to prepare interference, (2) to mark the path, and (3) to erase the path information. In order to realise these steps, they suggested a double slit experiment

---

[5] Bohr (1949).
[6] Feynman et al. (1965, 1–4 to 1–9).

with single atoms. The atoms are excited by a laser beam. Behind the double slit, there are two cavities. The experimental arrangement is such that the exited state of the atom decays to the ground state by photon emission inside the cavity region, with a probability close to 1. Therefore, the atom will deposit a photon in one of both cavities. In this way, *the path of the atom through one of both slits is marked*, and the *path information is stored* in one of the cavities. The stored path information may be *erased* by opening a shutter in the wall that separates the cavities.

The general quantum mechanical scheme behind the thought experiment is as follows:

(1) *Preparing Interference* The quantum system is prepared in a sharp momentum state or plane wave. It propagates through a double slit, a Mach–Zehnder interferometer, or a similar interference device. A superposition of two quantum states $|\Psi_1>$ and $|\Psi_2>$) with different paths and with phase difference is prepared, giving rise to a quantum state $|\Psi^V>$ of maximally visible interference fringes ($V = visibility$):

$$|\Psi^V> = \sqrt{1/2}(|\Psi_1> + |\Psi_2>)$$

(2) *Path Marking* Then, some internal degree of freedom of the propagating quantum system is entangled with *orthogonal detector states* $|1>, |2>$. The resulting quantum state is still a superposition. Orthogonal detector states do *not* interfere. So, the interference terms cancel. Now, the preparation gives rise to a quantum state $|\Psi^D>$ of maximally distinguishable paths ($D = distinguishability$):

$$|\Psi^D> = \sqrt{1/2}(|\Psi_1>|1> + |\Psi_2>|2>)$$

(3) *Erasing the Path Information* Finally, an additional device is added in order to prepare a superposition of $|1>$ and $|2>$. In the thought experiment of Scully et al., this is achieved by opening the shutter between the cavities. In general, a preparation procedure is needed that changes the wave function $|\Psi^D>$ into some superposition of the detector states $|1>$ and $|2>$, giving again rise to visible interference fringes:

$$|\Psi^{V'}> = \sqrt{1/2}\{(|\Psi_1>(a|1> + b|2>) + |\Psi_2>(c|1> + d|2>)\}$$

The quantum mechanical scheme is in perfect accordance with the pragmatic attitude towards wave–particle duality described above. Each of the steps (1)–(3) prepares another wave-like quantum state or kind of superposition. The so-called *path marking* is nothing but the preparation of a superposition of two *non*-entangled states. All these states propagate wave-like through the double slit or interferometer. During the propagation, there are no particles, only waves. Only the final measurement by means of particle counters gives rise to local particle-like detections. Hence, the name of a *which-way* experiment is highly misleading. Even though a *path*-like quantum state is prepared in step (2), the path itself is not measured.[7] As usual, only the relative frequency of particle detections is measured.

The interpretation of Scully et al. agrees with the quantum mechanical scheme. They predict that the storage of the path information will destroy the interference fringes, even if the information is not read out; and that the interference fringes will reappear if the stored path information is erased, even if the photon from the cavities is detected long after the passage of the atom. In addition, they claim that the storage of path information can not be interpreted in (Heisenberg's) terms of disturbing a momentum measurement by a position measurement. They emphasized that the path marking is not due to any disturbance of the atom's momentum by "scattering or otherwise introducing large uncontrolled phase factors into the interfering beams", but due to "correlations between the measuring apparatus and the systems being observed".[8] Here, "correlations" mean "quantum entanglement". This claim is also in perfect accordance with the quantum mechanical scheme.

However, it gave rise to a confusing debate about the question of what is more fundamental: Heisenberg's uncertainty relations or Bohr's complementarity?[9] The debate only came to an end when Dürr et al. (1998a, b) published the results of a *which way* experiment with an atomic beam. Dürr and Rempe (2000a, b) interpreted the results in terms of "duality relations" which hold for the visibility V of interference fringes and the distinguishability D of the particle path. Indeed, these "duality relations" were *generalised uncertainty relations* in Heisenberg's sense, which also expressed a *generalised complementarity* between the wave property V and the path property D of the measurement outcomes.

From a quantum mechanical point of view, the debate was not substantial. From a historical point of view, it may be understood as an example of missing communication between certain sub-communities of current physics. But from a philosophical point of view, the debate shows that the picture of classical particle kicks which destroy the interfering momentum states remained influential within the scientific community of physics up to the present day, even though it is at odds with the correct quantum mechanical scheme of *which-way* experiments.

Dürr et al. used an atom interferometer. A beam of ^{85}Rb atoms was generated by means of a magneto-optical trap, split, and made interfere with itself. The atomic beam was not sent through a double slit but through two pulsed standing light waves which splitted the beam twice into a transmitted and a Bragg-reflected beam. The path was marked by 'sandwiching' the first standing light wave with microwave pulses, resulting in a phase shift of the reflected beam and in different internal atom states of the transmitted and the reflected beam. Finally, quantum erasure was realised by measuring an observable with eigenvectors that belong to orthogonal superpositions of these internal atom states. Finally, the atoms were detected in the far field. As predicted by the above quantum mechanical scheme, the counting

---

[7] In addition, the quantum state is not *particle*-like because the photon number is not well-defined. See below Section 4.4.
[8] Scully et al. (1991, 111).
[9] See Falkenburg (2007, pp. 296–305).

rates showed interference patterns without path marking, no interference patterns with path marking and again interference patterns with quantum erasure of the path marking.

In 2003, Walburn et al. (2003) realized a similar experimental scheme with photons and a double slit. In their experiment, the polarization state of the photons was used as the internal degree of freedom needed to store the path information. A double slit was equipped with quarter-wave plates. They had the effect of generating a phase shift of $\lambda/4$ between the components of the electric field strength and preparing circularly polarized light. The experimental setup was such that a light wave got right-handed polarization by passing through the first slit, but left-handed polarization by passing through the second slit. Marking the path in this way made the interference pattern disappear, whereas inserting a horizontal polarizer behind the quarter-wave plates erased the path information. In order to study a *quantum eraser with delayed choice*, Walborn et al. designed their experiment as follows. They generated an entangled photon pair by means of a non-linear crystal that splits one photon in two entangled photons of lower energy. Now the path marking and its quantum erasure could be prepared by means of a polarizer put in the *other* branch of the experiment. In this way, quantum erasure with delayed choice was combined with an EPR-like arrangement. But here, too, all "causal" paradoxes disappear once the experimental results are interpreted in terms of *quantum waves* that propagate through the apparatus and *quantum particles* that only show up in the clicks of the coincidence detectors.

4.4 On What There Is

So far, the moral of the *which-way* experiments is: The name is misleading. *There is no particle path*. The experimental arrangement prepares a wave-like low-intensity beam that propagates through the apparatus. Only in the end, single particle clicks are measured.

One may nevertheless argue that the quantum wave propagation and particle detection might *finally* be explained in terms of underlying *real* particles. As far as I see, this possibility is ruled out by the following simple but striking polarizer experiment.

Let a low intensity light beam from a short pulsed laser pass three subsequent polarizers $P_|$, $P_/$, and P_- crossed with respect to one another at angles of $45°$ and $90°$. A photo counter behind the third polarizer P_- measures the light that passes through. Behind the first polarizer $P_|$, the light is polarized in vertical direction, behind the second $P_/$ it is polarized at $45°$ relative to $P_|$, and behind P_- it is polarized horizontally. If $P_/$ is removed, the remaining polarizers are perpendicular to each other, no photons pass P_-, and nothing is detected by the photon counter. If $P_/$ is re-inserted, behind P_- again some photons are detected. (Even in the analogous classical experiment with ordinary white light it is amazing to see the light on a screen appear and disappear when the second polarizer is put in and out.)

4 Wave–Particle Duality in Quantum Optics

The beam intensity should be so low that on average at most one photon at a time is in the radiation field. This does *not* necessarily mean that the quantized Maxwell field is in a well-defined number state with occupation number 1. This depends on the preparation. Even if by appropriate devices a field mode with occupation number 1 is prepared, any polarizer destroys this preparation. Any polarizer prepares a superposition rather than a well-defined number state of the photon field. A state of well-defined polarization is a state of well-defined phase difference. But the occupation number and phase of a quantum field are not simultaneously sharp or well-defined. In a sharp polarization state, the occupation number of the field mode is no longer well-defined.

Quantum field theory describes the polarizer effects in terms of field operators for the annihilation and creation of polarized field quanta. However, any field mode of vertical or horizontal polarization may be described as a superposition of two orthogonal diagonally polarized field states, and vice versa. This is decisive.

Let the photon state $\Psi_{k,1}$ represent a field mode of wave number k and occupation number 1. Behind the first polarizer $P_|$, the photon is in a quantum state $\Psi_|$ of vertical polarisation. $P_|$ reduces the wave function Ψ of the short pulsed laser beam to a state of well-defined, vertical polarization, which no longer corresponds to a well-defined occupation number. In terms of the second polarizer $P_/$, this state is a superposition of a photon wave or field mode $\Psi_/$ which can pass $P_/$ (polarization of 45°, relative to $P_|$) and a photon wave or field mode Ψ which *cannot* pass it (polarization perpendicular to $P_/$). Behind the second polarizer the photon field is in the state $\Psi_/$, due to reduction of the wave function. In the basis of eigenstates of the polarizers $P_|$ and $P_{\rule{1em}{0.4pt}}$, $\Psi_/$ is a superposition of $\Psi_|$ and $\Psi_{\rule{1em}{0.4pt}}$, with the state $\Psi_{\rule{1em}{0.4pt}}$ of horizontal polarization being orthogonal to $\Psi_|$. The quantum state $\Psi_/$ corresponds to a superposition of photons that *could* pass the first polarizer $P_|$ and photons that could *not* pass it but could only pass the perpendicular polarizer $P_{\rule{1em}{0.4pt}}$.

Finally, let the photon wave pass through the third polarizer $P_{\rule{1em}{0.4pt}}$ which is perpendicular to the first one. Behind $P_{\rule{1em}{0.4pt}}$ single photons are detected if and only if $P_/$ is between $P_|$ and $P_{\rule{1em}{0.4pt}}$. (The observation corresponds exactly to the classical analogue.) With $P_|$ and $P_{\rule{1em}{0.4pt}}$ alone, without the second polarizer $P_/$, no light passes. Not a single photon is detected. In this case, the effect of $P_{\rule{1em}{0.4pt}}$ on $\Psi_|$ is the vacuum state Ψ_{k0}. But if $P_/$ is put in between them, some photons are again detected at the screen. In this case, inserting the polarizer results in a superposition of a photon state of occupation number 1 and the vacuum state:

$$P_{\rule{1em}{0.4pt}}\Psi_| = 0\Psi_{\rule{1em}{0.4pt}} = \Psi_{k,0}$$

Obviously, each polarizer reduces the wave functions to a well-defined polarization state, as in a measurement. This gives rise to the absorption of some photons. Each polarizer damps the amplitude of the wave function or field mode by a factor $1/\sqrt{2}$. This has the result of damping the photon intensity (i.e., the average photon number given by the expectation value of the occupation number operator) by a factor $1/2$. The resulting photon intensity or average photon number is $1/4$ of the beam intensity behind the first polarizer $P_|$. In terms of the transition probabilities or counting rates, this means that *on average*, 3 out of 4 photons get lost or do not arrive at the photon counter. However, the counter detects the photons with a completely *irregular* counting rate, due to the ubiquitous vacuum state of the quantum field which predicts ubiquitous probabilistic quantum field fluctuations.

With the classical particle concept in mind, one may be inclined to ask: But what happens to the single photons, given that each of them must *either be absorbed* at one of the polarizers *or detected* behind P—? Quantum electrodynamics tells us that this is the wrong question. Due to the effect of the polarizers, the occupation number of the photon field is not sharp. One has to be agnostic about photon absorption at any polarizer as long as it is not measured there.

In a particle picture of the process, the polarizers have the physical effect of selecting photons of a given polarization. They absorb all photons with perpendicular polarization. This seems to be a measurement. But, how can *three* absorbers let the single photons pass given that they cannot pass *two* of them? The answer has already been given above. The polarizers prepare wave-like modes in which the occupation number of the quantum field is *not* well-defined. They prepare completely *different* field modes with or without the second polarizer $P_/$. But the photon counter only measures the *average* occupation number by detecting single photons. The preparation of the photon state by the polarizers differs substantially from the measurement by the photo counter behind P—, i.e., the detection of single photons. Otherwise, it was not possible to undo the preparation by inserting the second polarizer.

The transition probabilities of quantum field theory tell us that *on average* 3 out of 4 photons are absorbed at one of the polarizers. If the probability of photon detection is $1/2$ behind $P_/$, it is $1/4$ behind P—. However, this statement only makes sense at the probabilistic level. It does not apply to the individual photon detections. Obviously, no polarizer can damp the intensity of a single photon by a factor $1/2$. And obviously, no *third absorber* $P_/$ is able to *generate* photons in a field left *without any photon* due to the *two absorbers* $P_/$ and P—. These considerations preclude any realistic interpretation of the photon propagation in terms of individual particles, localized photon wave packets, or whatever individual causes of the single photon counts.

However, the effects are real. According to the principle of causality, their causes should be too. Indeed, the way in which the photon polarization is prepared has causal relevance for the relative frequency of photon counts. According to Hacking's reality criterion *If you can spray them, they exist*[10] we have to assume the *quantum waves prepared in order to achieve certain experimental results* must be *real*. But, *real in which sense?* The causal relevance of the preparation is only expressed in terms of conditional probabilities of the final photon counting rates. Operationally, there are only particle clicks. But according to the theoretical description in terms of field modes, there are only fields or waves with the usual probabilistic meaning.

According to Hacking's reality criterion, quantum states exist, even though neither for the quantum waves nor for the quantum particles realistic interpretations in a classical sense are tenable. There are *neither real waves* (i.e., oscillating field strengths of sharp phase and amplitude) *nor are there real particles* (i.e.,

---

[10] See Hacking (1983, 22–25).

microscopic projectiles with classical trajectory). Only the operational interpretation remains. For the quantum wave, it is restricted to the unconditional and conditional probabilities predicted by the evolution of the quantum state through the apparatus. For the quantum particles, it is restricted to the click of the detector.

What are the particles localized in such a click? Quantum theory tells that they are collections of dynamic magnitudes (mass–energy, spin, charge). Their connection to a quantum dynamics is as follows. The propagation of a quantum state obeys conversation laws (such as mass–energy conservation, charge conservation, etc.). These conservation laws belong to symmetry groups. The symmetries, in turn, characterise the quantum dynamics. Quantum probabilities and particle clicks are connected by Wigner's (1939) famous group theoretical definition of a particle or field, according to which the elementary particles correspond to the irreducible representations of symmetry groups. Beyond the usual probabilistic interpretation, the only *well-understood theoretical tie* between the propagation of a quantum wave and the detection of a quantum particle is the relation between the conservation laws and the symmetry groups of particle physics.

References

Bohm D (1952) A suggested interpretation of the quantum theory in terms of "hidden variables", I & II. Phys Rev 85:166–179, 180–193
Bohr N (1927) Como lecture. The quantum postulate and the recent development of atomic theory. Nature 121(1928):580–590. In Bohr's collected works 6, pp 109–158
Bohr N (1949) Discussion with Einstein on epistemological problems of atomic physics. In: Schilpp PA (ed) Albert Einstein: Philosopher–scientist. Evanston, IL, pp 115–150
Born M (1926a) Zur Quantenmechanik der Stoßvorgänge. Z Physik 37:863–867
Born M (1926b) Quantenmechanik der Stoßvorgänge. Z Physik 38:803–827
Compton AH (1923) A quantum theory of the scattering of X-rays by light elements. Phys Rev 21:483–502
Davisson C, Germer LH (1927) Diffraction of electrons by a crystal of nickel. Phys Rev 30:705–740
de Broglie L (1923) Ondes et Quanta. Comptes rendus 177:507–510
Dürr S et al (1998a) Origin of quantum-mechanical complementarity probed by 'which-way' experiment in an atom interferometer. Nature 395:33–37
Dürr S et al (1998b) Fringe visibility and which-way information in an atom interferometer. Phys Rev Lett 81(26):5705–5709
Dürr S, Rempe G (2000a) Wave–particle duality in an atom interferometer. Adv Atom Mol Optic Phys 42:29–71
Dürr S, Rempe G (2000b) Can wave–particle duality be based on the uncertainty relation? Am J Phys 68(11):1021–1024
Einstein A (1905) Über einen die Erzeugung und Verwandlung des Lichts betreffenden heuristischen Gesichtspunkt. Annalen der Physik 17:132–148
Einstein A (1916) Zur Quantentheorie der Strahlung. First published in 1916. Repr.: Physikal. Zeitschrift 18:121–128
Einstein A et al (1935) (=EPR paper) Can quantum mechanical description of reality be considered complete? Phys Rev 47:777–780
Falkenburg B (2007) Particle metaphysics. A critical account of subatomic reality. Springer, Berlin
Feynman RP, et al (1965) The Feynman lectures on physics, vol III. Addison-Wesley, Reading, MA

Hacking I (1983) Representing and intervening. Cambridge University Press, Cambridge
Heisenberg W (1925) Über quantentheoretische Umdeutung kinematischer und mechanischer Beziehungen. Z. Physik 33:879–893
Heisenberg W (1927) Über den anschaulichen Inhalt der quantentheoretischen Kinematik und Mechanik. Z. Physik 43:172–198
Jammer M (1966) The conceptual development of quantum mechanics. McGraw-Hill, New York
Ketterle W (2003) Public talk on the Bose–Einstein condensate. Annual meeting of the DPG (German Physical Society), Hannover
Schrödinger E (1926) Quantisierung als Eigenwertproblem I-IV. Annalen der Physik 79:361–376, 489–527. Annalen der Physik 80:734–756. Annalen der Physik 81:109–139
Scully MO et al (1991) Quantum optical tests of complementarity. Nature 351:111–116
von Neumann J (1932) Mathematische Grundlagen der Quantenmechanik. Springer, Berlin
Walborn St. P et al (2003) Quantum erasure. Am Sci 91:336–343
Wheeler JA, Zurek WH (1983) Quantum theory and measurement. Princeton University Press, Princeton, NJ
Wigner EP (1939) On Unitary Representations of the Inhomogeneous Lorentz Group. Annals of Mathematics 40:149–204

Chapter 5
Remarks on a Structural Account of Scientific Explanation

Laura Felline

5.1 Introduction

The pervasive role of mathematics in modern science has cross-fertilized the philosophy of science in many ways. Among them, a topic of growing interest is the epistemological status of mathematical explanations of natural phenomena. An extensive literature can be found on this subject, for instance in cognitive science – concerning the so-called computational explanations (McCulloch and Pitts 1943; Piccinini 2006), where the mental capacities of the brain are explained by its computations – and in more recent times a significant number of papers have investigated the role of mathematical explanations also in biology (Berger 1998).

Since the role that mathematics plays in the explanation of natural phenomena can hardly be overrated, it seems remarkably odd that such a topic has been hitherto neglected in the philosophy of physics, the mathematised science *par excellence*.

The current state of scientific knowledge and within it of the relationship between mathematics and explanation is well illustrated by Ruth Berger:

> "Today's science is often concerned with the behavior of extremely complicated physical systems and with huge data sets that can be organized in many different ways. To deal with this, scientists increasingly rely on mathematical models to process, organize, and generate explanatory information. Since much of the understanding produced by contemporary science is gathered during the process of mathematical modelling, it is incumbent upon philosophical accounts of explanation to accommodate modelling explanations. This is recognized by the semantic view of theories, which identifies mathematical modelling as one of the mains explanatory engines of science." (Berger 1998, p. 308).

But the acknowledgement of the central role of models in science did not correspond to the recognition of a similar role in the more restricted field of scientific explanation:

> Although many philosophers accept the basic features of the semantic view of theories, there have been surprisingly few attempts to reconcile it with our best philosophical accounts of scientific explanation. [...] [C]ausal accounts cannot illuminate precisely those

L. Felline (✉)
Department of Pedagogical and Philosophical Sciences, University of Cagliari
e-mail: felline@uniroma3.it

explanatory features of science which the semantic view deems most important. Specifically, causal accounts of explanation cannot accommodate, and often obscure, the crucial role which mathematical modelling plays in the production of explanatory information. Moreover, evidence from modelling explanations indicates that causal relevance is neither a necessary nor a sufficient condition for explanatory relevance. (ibid. pp. 308, 309)

Berger's specific target in the article just cited is modelling explanations in biological sciences. However, the need for a deeper investigation on this subject becomes especially urgent if we take a look at quantum mechanics, which currently represents the *bête noire* of the theory of scientific explanation.

The problems that exist in relating quantum mechanical phenomena to classical concepts like properties, causes, or entities like particles or waves are well-known and still open, so that there is not yet an agreement on what kind of metaphysics lies at the foundations of quantum mechanics.

It is for this reason that many philosophers say that they are not ready to take lessons from quantum theory until its interpretation is sorted out, and, in particular, that before we can draw any conclusion towards explanation in quantum theory we have to wait for the interpretational problem to be solved (Salmon 1984, pp. 254, 255).

Contrary to this last attitude, the program of Structural Explanation (SE, henceforth) tries to account for the fact that, in spite of the lack of a clear *categorial framework*[1] of reference for the theory, physicists constantly use the formal resources of quantum mechanics in order to explain quantum phenomena.

SE hinges on the following main points (possibly conflated in Berger's quote, but which we will keep separate):

(i) Scientific models are central in scientific explanation.
(ii) In some cases the relevant information for the explanation/understanding of a phenomenon P consists in the sole *structural* properties of the (models displayed by the) theory.
(iii) In these cases, the interpretation of the formalism in terms of a categorial framework is unessential for the explanation of P and a *mathematical* model[2] can be at the base of an objective and effective scientific explanation.

Robert Clifton provides the following definition of SE:

> We explain some feature B of the physical world by displaying a mathematical model of part of the world and demonstrating that there is a feature A of the model that corresponds to B, and is not explicit in the definition of the model.

---

[1] "[A] categorial framework is a set of fundamental metaphysical assumptions about what sorts of entities and what sorts of processes lie within the theory's domain" (Hughes 1989b, p. 175).

[2] For what the definition of the models utilized in SE is concerned, a 'mathematical model' is a set-theoretic structure (Suppes 1967). A set-theoretic structure $S = <U, O, R>$ is a triple consisting of (i) a non-empty set U of individuals called the domain (or universe) of the structure S, (ii) an indexed set O (i.e., an ordered list) of operations on U (which may be empty), and (iii) a non-empty indexed set R of relations on U. The fact that M is the *model of a theory*, in the sense that the theory of reference must be true of it, guarantees that the explanation is anchored to the actual scientific knowledge, and is not an arbitrary invention.

It is natural to call explanations based on this maxim *structural* to emphasize that they need not be underpinned by causal stories and may make essential reference to purely mathematical structures that display the similarities and connections between phenomena. (Clifton 1998)

The structural account of explanation was first formulated by R.I.G. Hughes (Hughes 1989a, b) and then taken up in (Clifton 1998). A more recent attempt to develop and support this program is proposed in (Dorato and Felline, forthcoming).

The present paper will carry a reflection about some issues arising from these works in the attempt to outline some details of SE.

5.2 Structural Explanation of the Uncertainty Relations

One of the weak points of R.I.G. Hughes' deserving work on SE is probably that it never provides a convincing example of SE in quantum mechanics.[3] In the present section we will present a case study which, we will argue, represents a significant case of SE in quantum mechanics: the explanation of Heisenberg's Uncertainty Relations. This case study is already proposed in (Dorato and Felline, forthcoming), however in the following we will pursue that analysis in an attempt to highlight some important points left unexplored.

The importance of this example relies on its representing a well known case of physical phenomenon for which there is no universally accepted account of the processes leading to its occurrence (or of the kind of entities underlying such processes), but which, at the same time, is nowadays conceived as a perfectly intelligible aspect of the world and of which quantum mechanics provides a clear insight. In (Dorato and Felline, forthcoming) the special case of the Uncertainty relation between position (p) and momentum (q) is examined, and it is argued that the modern understanding provided by quantum theory of such a relation is gained through a SE. More exactly, the existence of a minimum for the product of the uncertainties of these two measurements, or the non-simultaneous sharpness possessed by the two observables, represented formally by the equation:

$$\Delta x \cdot \Delta p \geq \frac{\hbar}{2} \quad (5.1)$$

is explained structurally by showing that, in the Hilbert space of square summable functions (the mathematical model M), the formal representative $\Psi(p_x, p_y, p_z)$ of the observable momentum is the Fourier transform of the function $\Psi(x, y, z)$, formal representative of the position. Consequently, the SE of the uncertainty relation

---

[3] In (Hughes 1989a) Hughes proposes a presumed SE of the EPR correlations which, however, necessarily hinges on an interpretation *à la* Bub of quantum mechanics, where measurements are ultimately treated as black boxes. Hughes' example is therefore bound to this interpretation of quantum mechanics.

exploits the well-known *mathematical* property of the Fourier transform on the basis of which the narrower the interval in which one of the two functions differs significantly from zero, the larger is the interval in which its Fourier transform differs from zero, in such a way that Eq. 5.1 must be satisfied.

We still maintain that the above is a good example of SE, however we will now propose a further analysis with the aim of showing that the case of the Uncertainty Relations provides an even more vivid illustration of SE in physics. It could be argued that we do in effect have an account of the mechanisms underlying the holding of Eq. 5.1 – account provided by Heisenberg himself in its illustration of the thought experiment for the measurement of the position of the electron (Heisenberg 1927). We have argued elsewhere (Dorato and Felline, forthcoming) against the reliability of such an account, however in the following we will take another direction in order to answer this objection.

An essential step towards today's understanding of Heisenberg's relations was taken some years after Heisenberg's first derivation, in particular with the more general derivation provided in 1929 by Robertson (Robertson 1929). This step was essentially the achievement of a deeper understanding of Heisenberg's relations *via* a different SE. The *new* formal representative of the *explanandum* phenomenon is here

$$\Delta\alpha \cdot \Delta\beta \geq \frac{1}{2} |\langle\Psi, [A, B]\Psi\rangle| \qquad (5.2)$$

where A and B are any two non-commuting operators (also spin in different directions, for example), $[A, B]$ is their commutator, so that for every state Ψ, and every pair of non-commuting observables α and β and corresponding operators A and B, the product of the uncertainties is greater than the expression on the right hand side. Accordingly, the new SE of the Uncertainty relations shows that the Uncertainty relations hold for any pair of non-commuting observable, i.e., it shows how the more general relation (5.2) is part and parcel of the models displayed by quantum mechanics.

Notice that the generality and insight typical of the current understanding of the Uncertainty relations are independent of any analogical model of the phenomenon. For instance, there is no mechanical or visualizable model which can render more intelligible to us the fact that there is no spin state in which one could predict with certainty the result of both a z-spin and a x-spin measurement.

To be sure, the SE based on the properties of the Fourier transform for the case of the momentum/position uncertainty relation continues to be a valid explanation and very effective, due to its intuitiveness and the less abstract model exploited for the explanation of this singular relation. However, Robertson's general derivation has not only shown how the Uncertainty Relations are built into the fundamental structure of quantum theory, but has also provided a modern understanding of them. Such a new understanding, common to all the pairs of non-commutable observables, is clearly not the result of the reflection on the mechanisms responsible for the holding of the relations. In which sense, in fact, could the mechanism underlying the loss of a determinate position in a particle with definite momentum be said to be the same as the one leading to the loss of spin-x in a particle with determinate spin-z? Such a new understanding must instead be the result of a reflection on the *common formal* properties of non-commutable observables.

Finally, the fact that this modern understanding of the Uncertainty Relations is also common to all the different interpretations of quantum mechanics, also shows that the former is independent on the question of what kind of ontology underlies quantum theory.

5.3 Models and Explanation

Clifton's minimalist definition quoted before was actually borrowed from R.I.G. Hughes' definition of *theoretical explanation*, according to which:

> "We explain some feature X of the world by displaying a model M of part of the world and demonstrating that there is a feature Y of the model that corresponds to X, and is not explicit in the definition of M." (Hughes 1993, p. 133).

An important virtue of the above definition is that it acknowledges the essential role that scientific models also play within scientific explanation. As asserted above ((i) Section 5.1), the emergent view of scientific explanation well suits the central place occupied by models in the current philosophical picture of scientific theories and scientific practice.

Secondly, this definition also allows for a plurality of explanations, depending on how broad our definition of scientific model is. Contrary to Berger's case, the notion of "scientific model" used here includes but is not equivalent to that of set-theoretic structure. Less abstract, analogical models are also admitted, whose elements are interpreted in terms of physical entities, relations, properties or processes. In this sense, for instance, causal explanation can be seen as a special case of theoretical explanation within which, roughly, M is an analogical model and Y is shown to be part of M, as causally following from other known elements of M.

Obviously, Hughes' definition allows also *mathematical* models to be explicative with respect to physical *explananda* – it is due to this reason that we will treat SE as a particular case of Hughes' theoretical explanation. Doing this also allows us to highlight both the points of connection and of divergence between causal and structural explanations.[4]

Given the crucial role played by the concept of model within SE, this point must be dealt with briefly.

---

[4] One could counter at this point that in this way SE looses its peculiarity and it is not clear whether there is the need of a theory of SE in contrast to a theory of causal explanation. However, the non trivial question we want to face here is, again, if mathematical models, not supported by an underlying categorial framework, can be explicative towards physical phenomena. A specific account of SE (as one of causal explanation) has then the aim of clarifying when and why mathematical models can be explicatively effective in physics and what kind of understanding of the physical world mathematical modelling provides us with.

In the growing literature about scientific models, Hughes' theoretical explanation probably better fits with the view that sees the unifying features of scientific models in their allowing us to acquire knowledge about the world. This requirement corresponds to what we shall call, following Chris Swoyer, *surrogative reasoning* (Swoyer 1991). Accounts of scientific modelling hinging on the notion of surrogative reasoning were proposed by Hughes himself (Hughes 1997) and more recently in Suarez (2004) and Contessa (2007).

Surrogative reasoning is obviously not an end unto itself: it is instead aimed at the achievement of a given epistemic aim: explanation, prediction, description, etc. Once this has been considered, it naturally follows that surrogative reasoning can only serve its function provided that the model used by the agent satisfies some specific requirement.

Depending on the epistemic aim to be reached, one can obviously favour different strategies of inquiry and these can obviously be better supported by different kinds of models and *styles* of representation (Frigg 2006). Just to cite an example provided by Hughes (Hughes 1997), take the two-slit interference experiment. We can model it either with the mathematics of wave functions, or with a real ripple tank. Both models have an internal dynamics, represented respectively by geometry and algebra, and by the physical processes which are involved in the propagation of water waves. Both dynamics allow us to conclude that the "distance between interference fringes varies inversely with the separation of the sources, and also with the frequency of the waves" (ibid. p. 332). Obviously, if surrogative reasoning is aimed at obtaining the most precise predictions, then the mathematical model would be the most appropriate one to achieving this aim. However, sometimes analogical models are the most apt to support a scientific activity – as in the case, say, of the billiard balls model of ideal gases, which is particularly effective for the explanation of Boyle's law.

In this sense it is possible to explain why empirical adequacy is typically an important aspect of scientific representation (for no scientific activity can be performed independently from empirical data) though sometimes dispensable: in many cases, for instance when the theory is too complicated, accuracy ends up being an obstacle for the manipulation of relevant information.

In other words, since the core of scientific modelling and representation is surrogative reasoning, and since (depending on the epistemic aim to which it tends) the latter can privilege different styles of representation, it follows that, in the view we propose, the requirement for a good scientific representation depends contextually on the epistemic aim to be reached by the cognitive agent.

The same argument applies, obviously, to scientific explanation and can serve to better explain the above example of the SE of the Uncertainty Relations. An adequate model for the explanation of a phenomenon must necessarily present all the explanatorily relevant information, but can omit all the details that render the description of P more accurate, but which are irrelevant in its explanation. In our example, the fact that the same understanding of the Uncertainty Relations is basically common to all the various interpretations of quantum mechanics shows that *the question of what kind of entities or processes underlie the hold of the relations is irrelevant and therefore dispensable for the achievement of a genuine understanding of the latter*, and that a mathematical model is instead sufficient. The crucial question of *when* a categorial framework is relevant, and therefore not dispensable, for the explanation of a phenomenon P, is treated more in detail in (Dorato and Felline, forthcoming) and will be further alluded to in Section 5.6 of this paper.

5.4 Understanding and Explanation

What kind of understanding of the world can SE provide? In the last part of Clifton's definition of SE it is specified that A must not be "explicit in the definition of the model". Clifton argues that such an informal requirement is meant to avoid cases of spurious unifications by mere cataloguing the phenomena to be explained – in doing so he follows Kitcher's unificationist account according to which explaining means reducing the number of laws covering the phenomena to be explained (Kitcher 1989). However Clifton does not clarify why SE should be related to the unificationist view. Moreover, since one of the problematic features of Kitcher's unification is exactly spurious unification, the reference to Kitcher's theory in this case could transfer the same problem on SE. The importance of the discussed requirement comes rather from the fact that it is the act of *making explicit* (Brandom 1998) the place of A within the model that provides an understanding of B. In other words, SE works by exploiting, and therefore highlighting, the relations linking A to the other elements of M, or by showing the place of A within the web of relations that constitute the structure of the theory.

If conceived in this way, the kind of understanding involved by SE displays some points of convergence with Schurz and Lambert's unificationist theory of understanding (Schurz and Lambert 1994), where the process of understanding P involves the capacity to fit (the sentence expressing) P into the cognitive corpus C (containing all statements known or believed by the inquirer).

In Schurz and Lambert's theory, a failure in the understanding of P can happen also when all the necessary descriptive information is possessed. There are cases, in fact, in which the lack of understanding is due to the ignorance of, or the inability to master, some new inference. In these cases the additional information required in order to make P understandable contains no new fact or law, but consists in inferring P from some premises X already known in C. Notice how this account suits the analysed case of the SE of the Uncertainty Relations based on Robertson's derivation. Here, all the elements necessary for 'putting P into M' are already present in M and what is needed is an inference showing how P is connected to these elements.

This, however, does not compel the structural account of explanation, as so far illustrated, to a unificationist theory of scientific explanation and understanding. Unification is an important element in the scientific enterprise as a whole and obviously also in scientific explanation, however this does not imply that unification constitutes *the essence* of scientific explanation in general or of SE in particular.[5] What the illustrated example shows, in our view, is that the process of understanding a physical phenomenon structurally involves reflecting on the defining properties of its formal counterpart and that, therefore, it also displays many similarities to the way we typically grasp mathematical notions. As we understand mathematical objects (by means, say, of the implicit definitions typically provided by manuals of mathematics) as relational objects (Shapiro 2000, p. 283), in the same way we understand a physical *explanandum* as a relational object, by means of the cluster of relations that its representative holds with the other elements of the mathematical models.

On the other hand, one important difference to be noticed is the fact that within Shapiro, of all the different structures that can characterize a system, only one is actually the real structure of the system. According to (Dorato and Felline, forthcoming) this does not apply to SE, since different SE, based on different structures, can exist of the same phenomenon.

5.5 Structural Explanation, Structural Realism

For someone approaching the theory of SE it could seem natural to see a connection with the program of structural realism, or even that the efficacy of the former can represent an argument in favour of the latter. It would therefore be useful also to clarify how effectively the two theories relate with each other.

First of all, SE is legitimately connected with structural realism, as it represents a clear example of how current scientific knowledge hinges on structures, however we think it unlikely that it could help in demonstrating any realist stance. An argument in defence of structural realism grounded on the effectiveness of SE should be based on some inference to the best explanation, and the legitimacy of such kind of arguments is subject to well known controversies in the realism/antirealism debate (see Psillos 1999, Chapter 4).

However even once a realist stance is taken, to what extent could SE, as characterized so far, be legitimately said to be part and parcel of a structuralist view of scientific theories? The accord is obvious in the case of Worral's Epistemic Structural Realism, within which SE could find a natural place – under the assumption that the same mathematical structures that remain stable in theory changing also have explanatory power, while Worral's 'hidden natures' are irrelevant to the understanding of (at least some) phenomena.

What about the *ontic* version of structural realism (OSR)? First of all, even if both SE and OSR essentially hinge on the central role of structures, they refer to two different kinds of structure, respectively mathematical and physical. Secondly, it has to be considered that, although the assumption that the world is ultimately

---

[5] If some affinity is to be found between structural (and theoretical) explanation and another account of explanation, this is surely with Nancy Cartwright's *simulacrum* account of explanation.

entirely structural is surely the most straightforward explanation of the effectiveness of SE, the latter does not presuppose a structural ontology and can also be perfectly coherent with a 'traditional' object-based realism.

Keeping in mind these two points, it is interesting to notice that for an advocate of OSR the issue of the range of application of SE has a straightforward solution. We are not thinking about the universality of SE in physics, since OSR could be compatible with the admittance of explanations relying on an object-based account of phenomena. This is due to the fact that OSR does not claim the reconceptualization of all the macroscopic processes in terms of structure, but relies on the assumption that macroscopic processes and objects can be reduced to quantum processes and that the latter can in turn be reconceptualized in terms of structures (French 2006). It follows then that an advocate of OSR would most likely claim that the explanations provided by fundamental physical theories, reconceptualized in structuralist terms, are only SE and arguably also that SE only occur at the level of fundamental theories.

Interestingly enough, this is also Hughes' position (see Hughes 1989b, p. 257), which, however, has always been left unwarranted. While so far this position seems to us far from obvious, it could represent an interesting point of convergence between SE and OSR, and is surely a promising idea to be further pursued.

5.6 Structural Explanation and Causality

The question of the relationship between SE and structural realism then introduces us to the issue of the relationship between causal and structural explanations.

The dichotomy causal/structural explanation is first of all deeply connected to their metaphysical/antimetaphysical character: contrary to SE, causal explanation needs a categorial framework of reference in order to individuate where the 'active principle' responsible for change (Chakravartty 2003) is located. Such an active principle can lie within an ontology of objects as well as of structures (French 2006), though it essentially needs *some* categorial framework within which to articulate the causal discourse.

It could be questioned at this point whether the dichotomy between SE and causal explanation assumed thus far is not more apparent than real, i.e., whether the effectiveness of SE comes from its hinging, even if not explicitly, on causal relations. In other words, one could wonder if SE is in effect a causal explanation, hidden behind a mathematical language. To prevent this kind of objection, one should first of all consider another fundamental feature of causal explanation. A causal history has a natural direction, from cause to effect, responsible for the asymmetry of causal explanation. This, on the other hand, is not the case within SE. In the latter, for a given mathematical model M with A and B its elements, there is no objective arrow of explanation connecting A and B and individuating one as the *explanandum* and the other as the *explanans*. The direction of the explanation is instead individuated contextually by the state of knowledge and the aims of the cognitive agent. But the possibility of formulating two equally acceptable explanations 'from A to B' and 'from B to A' is incompatible with these explanations hinging on causal relations, since the latter necessarily individuate only one objective direction.

As a second, related, proof that SE does not reduce to causal explanation, consider that, under the assumption that the definition of a fact as 'brute' is relative to the kind of explanation one requires (Fahrbach 2005), structural and causal explanation individuate different brute facts. Brute facts, in relation to a causal explanation (i.e., fact conceived as 'natural', or uncaused) can be structurally explainable. In this sense, for instance, we can explain a (causally) brute fact such as the constancy of the speed of light, with the Theory of Special Relativity – by exploiting the mathematical models of space-time displayed by the theory. It is, again, in this sense, that with quantum mechanics we can explain what by many is called the Uncertainty Principle (a brute fact relative to causal explanation), by exploiting the mathematical models displayed by the theory.

Notice that the limits of the applicability of SE do not necessarily coincide with a commitment to the objective existence of causal relations – i.e., these limits do not necessarily imply realism about causation. Also with respect to this point, one can as well place the picture just proposed of the applicability of SE within a general view, neutral with respect to the realist-antirealist debate. One could, for instance, rely on a view of explanation and understanding where causation is a tool for achieving understanding, and whose utility depends contextually on the beliefs and skills of the scientists demanding the explanation (de Regt and Dieks 2005). From this perspective, not only the request for a structural rather than a causal explanation seems to be a matter that depends crucially on contextual factors, but two different, causal and structural, explanations could legitimately cohabit in science of the same phenomenon.

Acknowledgments My research has been funded by a grant from the Master and Back program of Regione Sardegna. I am extremely grateful to Mauro Dorato, who has provided me with guidance throughout this project. I am also extremely grateful to Angelo Cei, Marco Giunti and Matteo Morganti for their comments on a previous version of this paper.

References

Berger R (1998) Understanding science: why causes are not enough. Philos Sci 65:306–332
Brandom R (1998) Making it explicit: reasoning, representing and discursive commitments. Harvard University Press, Harvard
Chakravartty A (2003) The structuralist conception of objects. Philos Sci 70:867–878
Clifton R (1998) Structural explanation in quantum theory. Philos Sci archive. http://philsci-archive.pitt.edu/archive/00000091/00/explanation-in-QT.pdf
Contessa G (2007) Scientific representation, interpretation, and surrogative reasoning. Philos Sci 74:48–68
De Regt H, Dennis D (2005) A contextual approach to scientific understanding. Synthése 144(1):137–170
Dorato M, Felline L, Scientific explanation and scientific structuralism. Forthcoming in the Boston Studies in the Philosophy of Science
Fahrbach L (2005) Understanding brute facts. Synthese 145(3):449–466

French S (2006) Structure as a weapon of the realist. Proc Aristotel Soc 106(2):167–185(19)
Frigg R (2006) Scientific representation and the semantic view of theories. Theoria 55:37–53
Heisenberg W (1927) Ueber den anschaulichen Inhalt der quantentheoretischen Kinematik and Mechanik. Zeitschrift für Physik. 43:172–198. English translation in 1983. Quantum theory and measurement. In Wheeler and Zurek (eds). Princeton University Press, Princeton NJ, pp 62–84
Hughes RIG (1989a) Bell's theorem, ideology, and structural explanation. In: Cushing J, McMullin J (eds) Philosophical consequences of quantum theory. University of Notre Dame Press, Notre Dame, IN
Hughes RIG (1989b) The structure and interpretation of quantum mechanics. Harvard University Press, Cambridge, MA
Hughes RIG (1993) Theoretical explanation. Midwest studies in philosophy XVIII. University of Notre Dame Press, Notre Dame, IN, pp 132–153
Hughes RIG (1997) Models and representation. Philos Sci 64:325–336
Kitcher P (1989) Explanatory unification and the causal structure of the world. In: Kitcher P, Salmon W (eds) Scientific explanation. University of Minnesota Press, Minneapolis, MN, pp 410–505
McCulloch WS, Warner P (1943) A logical calculus of the ideas immanent in nervous activity. Bull Math Biophys 7:115–133
Piccinini G (2006) Computational explanation in neuroscience. Synthése 153(3):343–353
Psillos S (1999) Scientific realism: how science tracks truth. Routledge
Robertson HP (1929) The uncertainty principle. Phys Rev 34:163–164
Salmon W (1984) Scientific explanation and the causal structure of the world. Princeton University Press, Princeton, NJ
Shapiro S (2000) Thinking about mathematics. Oxford University Press, Oxford
Schurz G, Lambert K (1994) Outline of a theory of scientific understanding. Synthese 101:65–120
Suppes P (1967) What is a scientific theory? In: Morgenbesser S (ed) Philosophy of science today. Basic Books, New York, pp 55–67
Suarez M (2004) An inferential conception of scientific representation. Philos Sci 71:767–779
Swoyer C (1991) Structural representation and surrogative reasoning. Synthese 87:449–508

Chapter 6
Mathematical Knowledge and the Interplay of Practices

José Ferreirós

The aim of this paper is to offer a brief presentation of the approach to the analysis of mathematical knowledge that I am developing in a forthcoming book entitled "*Mathematical Knowledge and the Interplay of Practices*".

My approach can be said to be (i) *cognitive*, due to the emphasis on math as knowledge produced by human agents, on the basis of their biological and cognitive abilities; (ii) *pragmatic*, because of my emphasis on the practical roots of math, i.e., roots in everyday practices, technical practices, and scientific practices); and (iii) *historical* since I emphasize the need to analyze math's historical development, and to accept the presence of what may be called contingent elements in modern mathematics.

6.1 On the Notion of Mathematical Practice

There is no question about the importance of the notion of mathematical practice today, in the field of philosophy of mathematics (see Mancosu 2008). At the same time, while the idea of practice is frequently employed, there exist a diversity of views on its scope and meaning. Hence we need to start with some clarifications, and I have found it useful to focus on Kitcher's pioneering views for a contrast (Kitcher 1984).

My own model could – on a first, rough approximation – be presented as a modification of Kitcher's quintuple <L, S, R, Q, M>. Kitcher analyzed the historical development of mathematical knowledge as a sequence of "rational transitions" from one practice to the next. There was some resemblance between this model of mathematical change, and Kuhn's ideas (1962) about the evolution of scientific disciplines in terms of "normal science" and "revolutions". We come back to this Kuhnian ingredient in Section 6.2 below.

By a mathematical practice Kitcher understood a quintuple <L, S, R, Q, M>, where L is a certain language, formal or informal, S is a set of accepted statements,

---

J. Ferreirós
Instituto de Filosofía, CCHS - CSIC. Albasanz, 26-28. Madrid 28037
e-mail: josef@us.es

R is a collection of established forms of reasoning, Q is the set of open problems, and by M one understands a collection of "metamathematical" views. Much of Kitcher's attention was drawn towards an attempt to classify kinds of historical transitions, and to assess their rationality. But here we shall focus on the basic idea of a practice.

To begin with, Kitcher's notion of a practice is too abstract and disembodied. Responding critically to Kitcher, I claim that there cannot be a practice without practitioners: the actor or agent must be centrally placed in any perspective on practice. One can thus propose a couple:

Framework – Agent

where the Framework can, simplifying, be identified with Kitcher's quadruple <L, S, R, Q>, while Kitcher's M would be part (only part) of the Agent's constituency. As we shall see below, we may call the *Framework – Agent* couple our "nuclear scheme" in analyzing mathematical knowledge.

To briefly expand on the idea of *Framework*, let me underscore that it is not meant to denote a theory in the idealized sense of logic (infinite set of statements closed under logical consequence) but rather a theory in the concrete, constructivist sense of our actual practice. Hence by the four elements of the quadruple we mean:

- A given language L (mixing natural language, technical expressions, and symbolic means)
- A given series of statements, propositions, theorems S (actually given and proven, if so; which may or may not include axioms)
- The diverse collection R of forms of reasoning and methods that are linked with those statements and their proofs
- And the series of questions Q (basic or advanced, some of them called conjectures) that are actually emphasized by agents in a community

Notice that by the language L one does not necessarily mean a formal language; my Framework approach is intended to be applicable directly to the analysis of historically given practices, such as Euler's in his *Introductio in Analysin Infinitorum*, and also that of Frege in his *Grundgesetze*.

This becomes possible precisely because we consider the couple *Framework – Agent*. Normal frameworks such as that of Euler (or the one developed by a university professor teaching Algebra or basic Analysis) cannot be made to stand alone. Only formal systems like the ones studied by Mathematical Logic could.

Let us know expand on the idea of the *Agent*. This can be analysed at different levels, e.g.,

- At that of *normal* agents with typical cognitive abilities and basic practices

This is a very important idea for my current purposes. Meant are abilities such as Perception (a high-level cognitive system, based on both visual inputs and motor outputs), Language (meant is oral language), and the practices of working with written language, symbols, and basic diagrams. The viewpoint is simply that those cognitive abilities and some of the basic practices are common to Euclid, Russell, a typical undergraduate student of math today, and myself.

But one can also analyse the Agent at

- A more concrete level close to historical actors, with their specific metamathematical views and research agendas
- Or even the collective level of communities (e.g., research schools) where one can talk of a typical agent of a community

The couple *Framework – Agent* is *not* to be identified with a mathematical practice. Usually, to analyze mathematical practice, one shall need to consider a plurality of agents and a plurality of frameworks. Needless to say, mathematical practice always involves *communities* of agents in interaction, and a diversity of theories which in fact can be linked with *different* frameworks. Hence I say that the couple *Framework – Agent* plays the role of a *nuclear scheme* in the analysis of mathematical knowledge.

A clarification may be in order. The partial, specific analyses offered by contributors to Mancosu (2008) can be regarded as narrower perspectives on facets of the couple *Framework – Agent* and its workings. Normally they tend to emphasize some cognitive elements or mental ingredients, but also some associated practices or techniques. That is, e.g., the case with studies of visual thinking or diagrammatic elements.

Even if we disregard subtler aspects of real mathematical practices, such as the images of mathematics they incorporate, or the values that are being promoted by participants in the practice, we are still left with sufficient material for an interesting analysis of the constitution of mathematical knowledge.

6.2 From Mathematical Practice to the Interplay of Practices

It is a key thesis of my approach that several different levels of knowledge and practice are coexistent, and that their links and interplay are crucial to mathematics. They coexist (or can coexist) both historically during the very same period, and also within an agent, the individual mathematician. Hence, in my view the Kuhnian element in Kitcher's analysis is very misguided, and even makes it impossible for his approach to address the key epistemological issues.

In all cases that I can think of, the analysis of a certain mathematical Framework $<L, S, R, Q>$ requires, at the very least, to consider its connections with another level of knowledge and practice. This is particularly the case, as remarked above, when the framework under analysis is not a formal system. Moreover, in the most simple cases of a mathematical framework, the connections with a technical practice are crucial. Let me indicate some examples in a very schematic and simplified way. The first refers to the way in which counting practices are intertwined with reckoning arithmetic, and this in turn with the structural theory of $\mathbb{N}$:[1]

---

[1] In an individual's development, the first practice will be taught in early childhood, the second in primary school, the third in undergraduate university studies.

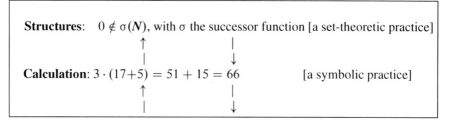

The following two examples are merely sketched:

Measuring practices/fraction arithmetic/theory of proportions
Practical geometry/Euclidean geometry/Cartesian geometry

In the second example for instance, we observe mathematical practices having to do with fraction arithmetic that arise in connection with technical practices of measurement (rods or ropes, the Egyptian cubit, acres, etc.). After an important transition motivated by discovery of the incommensurability of certain homogeneous quantities, there emerged a sophisticated practice based on Eudoxos' theory of proportions (presented in the *Elements*, book V). I have avoided mention of practices having to do with the real numbers, because in this case the web of interrelated practices is particularly dense.

The connections between those different practices are systematic and *mathematiké* in the etymological sense, i.e., can be taught & learnt. All of this means, again, that my analysis of mathematical knowledge is crucially *centered on the agents*. The links between different frameworks are not given abstractly, but rather established concretely by an agent, thanks to his or her (normal) cognitive abilities and mastery of the relevant practices.

The key thesis, then, is that we have *working knowledge of several different practices and of their systematic interconnections*. This causes *links* that restrict the admissible, guiding the formation of new concepts, and the adoption of new principles. Such links also lead to the objectiveness of results, as we shall see.

In connection with the above, let me insist that it is a key element for this approach to avoid reductionism and excessive drive toward systematicity. This (Fregean, Quinean) drive actually gets in the way of understanding mathematical knowledge.[2] At least when the understanding that we seek for is of the cognitive, pragmatic, and historic kind.

Notice also that the approach I am proposing puts an emphasis on interconnections between mathematical practices and other kinds of practice, in such a way that the problem of the "applicability" of mathematics ceases to be posed as external to mathematical knowledge itself, and becomes internal to its analysis.

---

[2] See Ferreirós (2005).

6.3 Elementary and Advanced Mathematics

The contrast between elementary and advanced mathematics is important for an epistemological analysis of mathematical knowledge. It should be obvious that I do not wish to construe this contrast as a dichotomy, but rather as marking noteworthy positions along a spectrum.[3] In my view, when we come to advanced math, it is essential to incorporate into the scheme a Hypothetical Conception: the role of constitutive hypotheses in laying out new practices. This in turn suggests the need to consider the interplay of certainty and hypothesis in mathematics.

6.3.1 From our standpoint, it seems best to identify *elementary mathematics* as those layers of mathematical knowledge that are strongly rooted in our cognitive systems, in the normal agent and her everyday practices. Consider the technical practices of counting, measuring, or drawing geometricals, and the mathematical practices that are linked with them. (A lot has still to be done concerning the cognitive science of these, to be sure; one the aims of my approach is to converge with this kind of studies.)

Notice how reckoning arithmetic links directly with counting practices, although it introduces the symbolic means that are so characteristic of mathematics. With such means, elementary arithmetic introduces methods of calculation, and algorithms. But, in a sense, this mathematical practice remains within the context of the purely elementary (strong cognitive and everyday-practical roots). The jump to advanced math comes with structural theories of $\mathbb{N}$, as I have termed them above.

Notice also how Euclidean geometry can be understood as a theoretical study of practical geometry. In the process, idealizations are introduced in well-known ways (dimensionless points, one-dimensional lines), but Euclidean geometry remains a theory of geometrical drawings, of diagrammatic constructions.[4] However, it seems to me that in the process we start moving from elementary to advanced mathematics. The main reason for this assertion is the role played by the conception of the continuum in geometry. (Continuum assumptions are read into the diagrams, and it is well known that this happens already in Book I, prop. 1, the construction of an equilateral triangle.)

6.3.2 Let me now say a few words about what I call Hypothetical Conception of advanced mathematics. First, when I use this expression, '*hypothesis*' is used in its etymological sense, without the connotation that mathematical hypotheses may represent aspects of physical reality. (To use a more elaborate language, the hypotheses are *constitutive* for a certain body of mathematics, but not representational.) The Hypothetical Conception is obviously linked with what is often called

---

[3] The question, however, has to be left somewhat open, in the sense that the spectrum is not a continuum, and it seems to me that there are significant breaks in it.

[4] See Mander's chapter in Mancosu (2008). A reconstruction of the *Elements* as elementary mathematics goes *against* the tendency of reflections on the subject since Pasch and Hilbert.

"quasi-empiricism," although I wish to avoid the simplistic "reduction" of mathematical methodology to that of the sciences, and even the idea that inductive methods are at play when mathematical hypotheses are established.

There is no need to take the Hypothetical Conception as an all-or-nothing approach; one can (in my view, ought to) accept that certain parts of mathematics are not hypothetical. One could argue that some parts of the theoretical body, especially elementary arithmetic, enjoy a special non-hypothetical status – a status of *certainty*. At the same time, I would argue that the theories of modern mathematics are based on propositions that are neither evident nor certain. Some such principles are the Axiom of Choice (in whatever version we formulate it, set-theoretic or category-theoretic), the Axiom of Continuity or Completeness for R, or say the assumptions of Infinity and Power Sets.[5]

Thus, the question arises as to the interplay between the "certain" (strong cognitive and practical roots) and the "hypothetical". The hypotheses we are talking about can be regarded as *"well-founded fictions:"* in particular, as we shall see, the properties assigned to them are not at all arbitrary.

We should understand modern mathematical theories as systems *based on hypotheses*, and mathematical truths as parts of such systems. (The system can be a foundational axiom system, say *ZFC*, or simply a mathematical theory in the field of algebra, geometry, topology.) Now, if starting from hypotheses one emphasizes the deductivist or inferential aspect of mathematical knowledge, we would develop a *hypothetico-deductive view* that might converge with If-thenism.

A quite different emphasis can be given by placing the idea within the context of an *analysis of mathematical practices* and their history. Particularly important is to emphasize mathematical practi*ces* in the plural, as we have done above, and the links between them. If-thenism, with its deductivist drive, is oriented towards a static, monolithic perspective on mathematics; it is system-driven. My approach, however, aims to remain far from both reductionism and over-systematicity (see above).

Thus, the Hypothetical Conception I am promoting is oriented towards a dynamical analysis of the emergence of new systems, and the constraints they have to satisfy. According to my standpoint, such hypothetical systems are not based on mere convention, but arise out of a richly embedded network of practices, remaining far from arbitrariness.[6]

6.3.3 One of the differences between math & myth lies in the systematic interconnections inside the web of practices. Another key difference can be found in the complexity of the cognitive and practical roots of elementary math. (Metaphorically

---

[5] Let me add a word on the contrast between the basic hypotheses I'm interested in, and conjectures. Many people are interested in how conjectures are formulated and tested in practice, eventually coming to confirm or falsify them, before they are proved; see Mazur (1997). But beyond confirmation or refutation, there's a third possibility for hypotheses: to become fixed and *solidified* as basic principles, as axioms, and this is the phenomenon I am most interested in.

[6] In this process, the interactions between mathematics and the sciences have played (and still play) a crucial role.

speaking, these two aspects can be compared with the inductive step and the basis of a proof by complete induction.) We must always remember that advanced math is linked systematically with the elementary, and thus with its strong roots.

Advanced knowledge is placed in a complex network of intertwined knowledge and practices, and here of course one could redevelop ideas presented by previous authors about thematisation, hybridization, etc. (I mean to refer to the work of Lakatos, Kitcher, Cavaillès, and others.)

In the sequel, I am interested in how new practices and new forms of knowledge are developed in advanced mathematics. The view is not one of analyzing theories once they have been carefully worked out: my interest lies more in the *emergence* of knowledge, the constitution of new "notional champs," and their subsequent systematization, clarification, consolidation, and eventual axiomatization. Examples of "notional champs" before consolidation, on which I have done some work myself, are: Riemann surfaces as of 1860, set theory as of 1880, abstract structures as of 1920.

6.4 On the Objectivity of Mathematical Knowledge

New hypotheses are not simply invented, and the results obtained about the newly 'invented' objects are typically not arbitrary. Let me consider briefly some examples from set theory. The idea is simply that previous mathematical practices, in particular arithmetical ones, have conditioned the admissible principles of set theory, and have led to objective results.

We shall see that the traditional dichotomy invention/discovery is simply inadequate to analyse the conceptual subtleties of the mathematician's handling of "well-founded fictions." But there is no great mystery here: a full analysis of what is going on can be produced, as I hope to show.

6.4.1 Take for instance the Axiom of Infinity and the Axiom of Power Sets. I claim that their introduction as new hypotheses can be *explained* by reference to the web of mathematical practices given (in Western mathematics) around 1850:[7] the state of mathematical practices, the perceived need to rigorize analysis, the ideal of *arithmetisation* (and its accompanying ideal of full autonomy of mathematics), combined all of them with the *core assumption* that the continuum is a point-set.

No constructivist or predicativist approach could bridge the gap between $\mathbb{Q}$ and $\mathbb{R}$; thus, when Cantor and Dedekind confronted the problem of defining $\mathbb{R}$ on the basis of $\mathbb{Q}$, both axioms of Infinity and Power Sets were indispensable. This remains true despite the fact that none of those principles was clearly formulated,

---

[7] This of course, understood as a historical claim, is nothing new: historians and philosophers of math have traditionally thought this way. On the topic of the modern definitions of the real numbers, see Jourdain (1910), Cavaillès (1962), Dauben (1979), Epple (2003), or Ferreirós (1999), especially chap. IV.

let alone argued for, in their works. (In particular, the axiom of Power Sets and the accompanying idea of combinatorialism, or the admission arbitrary subsets, remained implicit for a long time, in spite of being crucial to the new set-theoretic practice.[8]) All of this led to the *introduction of infinite sets* guided by the idea of the continuum as a point-set.

6.4.2 Furthermore, I claim that results such as Cantor's proof of the non-denumerability of $\mathbb{R}$, and others like the existence of bijections from $\mathbb{N}$ to proper subsets of itself, were totally non-arbitrary – modulo the admission that the naturals and the real numbers are to be conceived as infinite sets (i.e., roughly speaking, the hypothesis encapsulated in the Axiom of Infinity).

If we accept the hypothesis of an actually infinite set of all natural numbers, it is inevitable to admit as well the existence of any *definable set* of natural numbers. This means, any subset of $\mathbb{N}$ that can be specified by a formula or a formal predicate, obvious examples being the even numbers, the set of prime numbers, or the set of multiples of 391. But now, arithmetical knowledge has the implication that there are one-to-one correspondences between $\mathbb{N}$ and each of the subsets just mentioned.

For instance, Euclid's proof, that no finite collection of primes contains all the prime numbers, can easily be supplemented to prove that for each n there is an nth prime number (by complete induction). Correspondences like the bijection from $\mathbb{N}$ to the set of even numbers, or to the multiples of 391, can be concretely exhibited. What I aim to emphasize is that, once the step of assuming actually infinite sets of numbers is done, the rest follows by previously available mathematical methods. No new set-theoretic method is needed.

6.4.3 If the first step into set theory is the hypothetical assumption of infinite sets, the second is the even bolder positing of power sets. To remain within the relatively concrete case of the natural numbers, the second crucial hypothesis is the existence of a domain comprising the totality of "all possible" subsets of $\mathbb{N}$. This involves a *denial of the requirement that infinite sets be definable*, or "determined by a concept" (see Bernays 1935).

We are not necessitated to admit combinatorial power sets – a claim which is even proved empirically, by the existence of intuitionist and constructivist mathematicians. But *once* we have adopted the idea of arbitrary subsets and assumed the existence of the corresponding power sets, *it is not arbitrary, but necessary, to conclude* that the power set $\wp(\mathbb{N})$ has greater cardinality than the set $\mathbb{N}$.

Similarly, once we admit (as usual in the nineteenth century) that the real numbers are on a par with the naturals, admission of the set $\mathbb{N}$ calls equally for the admission of an infinite set $\mathbb{R}$ of all real numbers. And given these two sets, previously available mathematical methods lead to the result that no denumerable

---

[8] For a detailed discussion of these issues, and their connection with the "core assumption" that the continuum is a point-set, see my forthcoming paper 'On arbitrary sets and ZFC'. Concerning the crucial role of combinatorialism in set theory, see, e.g. Maddy (1999).

sequence of reals can exhaust $\mathbb{R}$.[9] As a matter of fact, Cantor's first proof of the non-denumerability theorem (1874) merely uses a principle of completeness for the real line that had been employed by Bolzano in 1817, Cauchy in 1821, Weierstrass in his lectures of the 1860s, etc.

Bolzano's principle says that a sequence of nested, closed intervals in $\mathbb{R}$ determines at least one real number (point) r that belongs to all the intervals; i.e., r belongs to the intersection of all the closed intervals. Now, assuming given a denumerable sequence s_n of real numbers, Cantor employs this sequence to define a sequence of nested intervals, in such a way that the Bolzano principle entails the existence of a real number which cannot be in the sequence s_n.[10]

Exactly like before, once the step of assuming actually infinite sets of numbers is done, the rest follows by available mathematical methods, well-established in previous mathematical practice; in particular, no new set-theoretic method is needed. (It could be argued as well that Cantor's diagonal method is not specifically set-theoretic, and it does not go substantially beyond mid-nineteenth century practices in analysis. But this argument is not needed for the conclusion I want to establish.) As one can see, the introduction of new *hypothetical* elements *within the web* of mathematical practices can still give rise to *objective* results, due to systematic links between these practices.

References

Bernays P (1935) Sur le platonisme dans les mathématiques, *L'Enseignement Mathématique* 34. English version in Benacerraf P, Putnam H (eds) Philosophy of mathematics: selected readings. Cambridge University Press, Cambridge, 1983
Cantor G (1874) Über eine Eigenschaft des Inbegriffes aller reellen algebraischen Zahlen, Jour. für reine Math. 77, 258–62. Reprinted in Cantor, Abhandlungen (1932), 115–118. English trans. in W. Ewald, From Kant to Hilbert, vol 2. Oxford University Press, Oxford, 1996
Cavaillès J (1962) Philosophie mathématique. Hermann, Paris
Dauben J (1979) Georg Cantor. His mathematics and philosophy of the infinite. Harvard University Press, Cambridge, MA
Epple M (2003) The end of the science of quantity. In: Jahnke HN (ed) A history of analysis. American Mathematical Society/London Mathematical Society, Providence, RI
Ferreirós J (1999) Labyrinth of thought. A history of set theory and its role in modern mathematics, 2nd edn (2007). Birkhäuser, Basel
Ferreirós J (2005) Dogmas and the changing images of foundations. Philos Sci 5:27–42 (cahier special Heinzmann G, Nabonnand P (ed))
Jourdain P (1910) The development of theories of mathematical logic and the principles of mathematics. Quart J Pure Appl Math 41:324–352; 43:219–314; 44:113–128
Kitcher P (1984) The nature of mathematical knowledge. Oxford University Press, Oxford

---

[9] I formulate matters this way just to avoid use of Cantor's notion of cardinality. With this notion, the above sentence translates into the statement that Card($\mathbb{R}$) > Card($\mathbb{N}$).
[10] Details concerning the proof can be found in Dauben (1979) or Ferreirós (1999).

Kuhn TS (1962) The structure of scientific revolutions. Chicago University Press, Chicago, IL
Maddy (1998) Naturalism in mathematics. Oxford University Press, Oxford
Mancosu P (ed) (2008) The philosophy of mathematical practice. Oxford University Press, Oxford
Mazur B (1997) Conjecture. Synthese 111:197–210

Chapter 7
Einstein, Kant, and the A Priori

Michael Friedman

Kant's original version of transcendental philosophy took both Euclidean geometry and the Newtonian laws of motion to be synthetic a priori constitutive principles – which, from Kant's point of view, function as necessary presuppositions for applying our fundamental concepts of space, time, matter, and motion to our sensible experience of the natural world. Although Kant had very good reasons to view the principles in question as having such a constitutively a priori role, we now know, in the wake of Einstein's work, that they are not in fact a priori in the stronger sense of being fixed necessary conditions for all human experience in general, eternally valid once and for all. And it is for precisely this reason that Kant's original version of transcendental philosophy must now be either rejected entirely or (at least) radically reconceived. Most philosophy of science since Einstein has taken the former route: the dominant view in logical empiricism, for example, was that the Kantian synthetic a priori had to be rejected once and for all in the light of the general theory of relativity.

Yet Hans Reichenbach took the latter route in his first published book: in *Relativitätstheorie und Erkenntnis Apriori* (1920) he proposed instead that Kantian constitutively a priori principles of geometry and mechanics should be *relativized* to a given time in a given theoretical context. Such principles still function, throughout the development from Newton to Einstein, as necessary presuppositions for applying our (changing) conceptions of space, time, and motion to our sensible experience, but they are no longer eternally valid once and for all. For example, while Euclidean geometry and the Newtonian laws of motion are indeed necessary conditions for giving empirical meaning to the Newtonian theory of universal gravitation, the situation in Einstein's general theory relativity is quite different. The crucial mediating role between abstract mathematical theory and concrete sensible experience is now played by the light principle and the principle of equivalence, which together insure that Einstein's revolutionary new description of gravitation by a four-dimensional geometry of variable curvature in fact says something about

---

M. Friedman (✉)
Department of Philosophy, Stanford University, Stanford, CA 94305, USA
e-mail: mlfriedman@stanford.edu

concrete empirical phenomena: namely, the behavior of light and gravitationally interacting bodies.

In my recent book, *Dynamics of Reason* (2001), I have taken up, and further developed, Reichenbach's idea. But my implementation of this idea of relativized constitutively a priori principles (of geometry and mechanics) essentially depends on an historical argument describing the developmental process by which the transition from Newton to Einstein actually took place, as mediated, in my view, by the parallel developments in scientific philosophy involving, especially, Hermann von Helmholtz, Ernst Mach, and Henri Poincaré. However, since this argument depends on the concrete details of the actual historical process in question, it would therefore appear to be entirely contingent. How, then, can it possibly be comprehended within a properly *transcendental* philosophy? Indeed, once we have given up on Kant's original ambition to delineate in advance the a priori structure of all possible scientific theories, it might easily seem that a properly transcendental argument is impossible. We have no way of anticipating a priori the specific constitutive principles of future theories, and so all we can do, it appears, is wait for the historical process to show us what emerges a posteriori as a matter of fact. So how, more generally, can we develop a philosophical understanding of the evolution of modern science that is at once genuinely historical and properly transcendental?

Let us begin by asking how Kant's original transcendental method is supposed to explain the sense in which certain fundamental principles of geometry and mechanics are, in fact, both a priori and necessary. This method, of course, appeals to Kant's conception of the two rational faculties of sensibility and understanding. The answer to the question "how is pure mathematics possible?" appeals to the necessary structure of our pure sensibility, as articulated in the Transcendental Aesthetic of the *Critique of Pure Reason*; the answer to the question "how is pure natural science possible?" appeals to the necessary structure of our pure understanding, as articulated in the Transcendental Analytic. Yet there is an obvious objection to this procedure: how can such proposed transcendental explanations inherit the (assumed) a priori necessity of the sciences whose possibility they purport to explain unless we can also somehow establish that they are the *unique* such explanations? From our present point of view, for example, it does not appear that Kant's explanation of the possibility of pure mathematics is uniquely singled out in any way; on the contrary, our greatly expanded conception of purely logical or analytic truth suggests that an appeal to the faculty of pure sensibility may, after all, be explanatorily superfluous. Indeed, from the point of view of the anti-psychological approach to such questions that dominated much of twentieth-century analytic philosophy, it appears that all consideration of our subjective cognitive faculties is similarly explanatorily superfluous.

In Kant's own intellectual context, however, explanations of scientific knowledge in terms of our cognitive faculties were the norm – for empiricists, rationalists, and (of course) Aristotelians. Everyone agreed, in addition, that the relevant faculties to consider were the senses and the intellect; what was then controversial was the precise nature and relative importance of the two. Empiricist views, which denied the existence of the pure intellect or its importance for scientific knowledge, were, for Kant, simply out of the question, since they make a priori rational knowledge incomprehensible. Moreover, the conception of the pure intellect that was most salient for Kant was that of Leibniz, where the structure of this faculty is delineated, in effect, by the logical forms of traditional Aristotelian syllogistic. But this conception of the pure intellect, Kant rightly saw, is entirely inadequate for representing, say, the assumed infinite extendibility and divisibility of geometrical space, which had recently proven itself to be both indispensable and extremely fruitful in Newtonian mathematical physics. Nevertheless, Newton's own conception of space as the divine sensorium was also unacceptable on theological and metaphysical grounds, and so the only live alternative left to Kant was the one he actually came up with: space is a pure form of our sensibility (as opposed to the divine sensibility), wherein *both* (infinitely iterable) geometrical construction *and* the perception of spatial objects in nature (like the heavenly bodies) then become first possible.

It is of course entirely contingent that Kant operated against the background of precisely these intellectual resources, just as it is entirely contingent that Kant was born in 1724 and died in 1804. Given these resources, however, and given the problems with which Kant was faced, the solution he came up with is not contingent. On the contrary, the intellectual situation in which he found himself had a definite "inner logic" – mathematical, logical, metaphysical, and theological – which allowed him to triangulate, as it were, on a practically unique (and in this sense necessary) solution.

Beginning with this understanding of Kant's transcendental method and its associated rational necessity, we can then see a way forward for extending this method to post-Kantian developments in both the mathematical exact sciences and transcendental philosophy. We can trace out how the "inner logic" of the relevant intellectual situation evolves and changes after Kant in response to both new developments in the mathematical exact sciences themselves and the manifold and intricate ways in which post-Kantian scientific philosophers attempted to reconfigure Kant's original version of transcendental philosophy in light of these developments. That each of these successive new intellectual situations has its own "inner logic" implies that the enterprise does not collapse into total contingency; that, in addition, they successively evolve out of, and in light of, Kant's original system suggests that it may still count as transcendental philosophy.

Hermann von Helmholtz's neo-Kantian scientific epistemology, for example, had deep roots in Kant's original conception. In particular, Helmholtz developed a distinctive conception of space as a "subjective" and *"necessary* form of our external intuition" in the sense of Kant; and, while this conception was certainly developed within Helmholtz's *empirical* program in sensory psychology and psycho-physics, it nevertheless retained important "transcendental" elements. More specifically, space is "transcendental," for Helmholtz, in so far as the principle of free mobility (which allows arbitrary continuous motions of rigid bodies) is a necessary condition for the possibility of spatial measurement – and, indeed, for the very existence of space and spatial objects. Moreover, the condition of free mobility represents a natural generalization of Kant's original (Euclidean) conception of geometrical construction, in the sense that Euclidean constructions with straight-edge and compass, carried out within Kant's form of spatial intuition, are generated by the group of specifically Euclidean rigid motions (translations and rotations). The essential point, however, is that free mobility also holds for the classical non-Euclidean geometries of constant curvature (hyperbolic and elliptic), and so it is no longer a "transcendental" and "necessary" condition of our spatial intuition, for Helmholtz, that the space constructed from our perception of bodily motion obeys the specific laws of Euclidean geometry. Nevertheless, Helmholtz's generalization of the Kantian conception of spatial intuition is, in an important sense, the *minimal* (and in this sense unique) such generalization consistent with the nineteenth-century discovery of non-Euclidean geometries.

The great French mathematician Henri Poincaré then transformed Helmholtz's conception in turn. In particular, Poincaré's use of the principle of free mobility (which plays a central role in his philosophy of geometry) is explicitly framed by a hierarchical conception of the mathematical sciences, beginning with arithmetic and proceeding through analysis, geometry, mechanics, and empirical physics – where, in particular, each lower level of the hierarchy (after arithmetic) *presupposes* that all earlier levels are already in place.

This hierarchical conception of the mathematical sciences underlies Poincaré's fundamental disagreement with Helmholtz. For Helmholtz, as we have seen, the principle of free mobility expresses the necessary structure of our form of external intuition, and, following Kant, Helmholtz views all empirical investigation as necessarily taking place within this already given form. Helmholtz's conception is Kantian, that is, in so far as space has a "necessary form" expressed in the condition of free mobility, but it is also empiricist in so far as which of the three possible geometries of constant curvature obtains is then determined by experience. For Poincaré, by contrast, although the principle of free mobility is still fundamental, our actual perceptual experience of bodily "displacements" arising in accordance with this principle is far too imprecise to yield the empirical determination of a specific mathematical geometry: our only option, at this point, is to *stipulate* Euclidean geometry by convention, as the simplest and most convenient idealization of our actual perceptual experience. In particular, experiments with putatively rigid bodies, for Poincaré, involve essentially physical processes at the level of mechanics and experimental physics, and these sciences, in turn, *presuppose* that the science of geometry is already firmly in place. In the context of Poincaré's hierarchy, therefore, the principle of free mobility expresses our necessary freedom to choose – by a "convention or definition in disguise" – which of the three classical geometries of constant curvature is the most suitable idealization of physical space.

One of the most important applications of Poincaré's hierarchical conception involves his characteristic perspective on the problem of absolute space and the relativity of motion explained in his discussion of the next lower level in the hierarchy: (classical) mechanics. Poincaré's key idea is that what he calls the (physical) "law of relativity" rests squarely on the "relativity and passivity of space" and therefore reflects the circumstance, essential to free mobility, that the space constructed from our experience of bodily displacements is both homogeneous and isotropic: all points in space, and all directions through any given point, are, necessarily, geometrically equivalent. Thus, Poincaré's conception of the relativity' of motion depends on his philosophy of geometry, and this is especially significant, from our present point of view, because Poincarés ideas on the relativity of motion were also inextricably entangled with the deep problems then afflicting the electrodynamics of moving bodies that were eventually solved (according to our current understanding) by Einstein's special theory of relativity.

I shall return to Einstein below, but I first want to emphasize that the connection Poincaré makes between his philosophy of geometry and the relativity of motion represents a continuation of a problematic originally prominent in Kant. Helmholtz, as we have seen, transformed Kant's philosophy of space and geometry, and Ernst Mach, among others, participated in a parallel transformation of Kant's approach to the relativity of motion – which finally eventuated in the modern concept of an inertial frame of reference. Neither Helmholtz nor Mach, however, established any kind of conceptual connection between the foundations of geometry and the relativity of motion – which, at the time, appeared to be entirely independent of one another. On Kant's original approach to transcendental philosophy, by contrast, the two were actually very closely connected. While Kant's answer to the question "how is pure mathematics possible?" essentially involved his distinctive perspective on Euclidean constructive operations, his answer to the question "how is pure natural science possible" involved an analogous constructive procedure by which Newton, from Kant's point of view, arrived at successive approximations to "absolute space" via a definite sequence of rule-governed operations starting with our parochial perspective here on earth and then proceeding to the center of mass of the solar system, the center of mass of the Milky Way galaxy, the center of mass of a system of such galaxies, and so on *ad infinitum*. Indeed, the way in which Kant thereby established a connection between the problem of space and geometry and the problem of the relativity of motion was intimately connected, in turn, with both the overarching conception of the relationship between sensibility and understanding that frames his transcendental method and his characteristic perspective, more generally, on the relationship between constitutive and regulative transcendental principles. ("Absolute space," in particular, is a forever unreachable regulative idea of reason.)

Now it was Mach, as I have suggested, who first forged a connection between Kant's original solution to the problem of "absolute space" and the late nineteenth-century solution based on the concept of an inertial frame of reference. (Kant, from a modern point of view, is constructing a sequence of better and better approximations to what we now call an inertial frame of reference.) And it is clear, moreover, that Poincaré was familiar with this late nineteenth-century solution as well. It is also clear, however, that Poincaré's attempt to base his discussion of the relativity of motion on his philosophy of geometry runs into serious difficulties at precisely this point; for Poincaré is here forced to distinguish his "law of relativity" from what he calls the "principle of relative motion." The latter applies only to inertial frames of reference, moving uniformly and rectilinearly with respect to one another, while the latter applies, as well, to non-inertial frames of reference in a state of uniform rotation: it follows from the "relativity and passivity" of space, for Poincaré, that uniform rotations of our coordinate axes should be just as irrelevant to the motions of a physical system as uniform translations. Therefore, the full "law of relativity," as Poincaré says, "ought to impose itself upon us with the same force" as does the more restricted "principle of relative motion." Poincaré must also admit, however, that the more extended "law of relativity" does not appear to be in accordance with our experiments (e.g., Newton's famous rotating bucket experiment).

It is for this reason that Einstein's appeal to what he calls the "principle of relativity" in his 1905 paper on special relativity is independent of Poincaré's "law of relativity," and it is also independent, accordingly, of Poincaré's "conventionalist" philosophy of geometry. Einstein's principle is limited, from the beginning, to inertial frames of reference (moving relative to one another with constant velocity and no rotation), and his concern is to apply this (limited) principle of relativity to both electro-magnetic and mechanical phenomena. Thus, in particular, whereas Poincaré's "law of relativity" involves very strong a priori motivations deriving from his philosophy of geometry (based on the "relativity and passivity of space"), Einstein's "principle of relativity" rests on the emerging experimental evidence suggesting that electro-magnetic and optical phenomena do not in fact distinguish one inertial frame from another. Einstein "conjectures" that this experimentally suggested law holds rigorously (and for all orders), and he proposes to "elevate" it to the status of a presupposition or postulate upon which a consistent electrodynamics of moving bodies may then be erected. Hence, Einstein's understanding of the principle of relativity is also independent of Poincaré's carefully constructed hierarchy of the mathematical sciences, and it is for precisely this reason, I suggest, that Poincaré himself could never accept Einstein's theory.

Nevertheless, it appears overwhelmingly likely that, although Einstein did not embrace Poincaré's "conventionalist" philosophy of geometry, Einstein's use of the principle of relativity was explicitly inspired by Poincaré's more general methodology described in *Science and Hypothesis* – according to which the fundamental principles of mechanics, in particular, are "conventions or definitions in disguise" arising from "experimental laws" that "have been elevated into principles to which our mind attributes an absolute value." In Einstein's case, the experimental law in question comprises the recent results in electrodynamics and optics, and Einstein now proposes to "elevate" both the principle of relativity and the light principle (which together imply that the velocity of light is invariant in all inertial frames) to the status of "presuppositions" or "postulates." These two postulates together then allow us to "stipulate" a new "definition of simultaneity" (based on the assumed invariance of the velocity of light) implying a radical revision of the classical kinematics of space, time, and motion. In particular, whereas the fundamental kinematical structure of an inertial frame of reference, in classical mechanics, is defined by the Newtonian laws of motion (a revised version of) this same structure, in Einstein's theory, is rather defined by his two postulates.

A central contention of Kant's original version of transcendental philosophy, as we know, is that the Newtonian laws of motion are not mere empirical laws but a priori constitutive principles on the basis of which alone the Newtonian concepts of space, time, and motion can then have empirical application and meaning. What we have just seen is that Einstein's two fundamental "presuppositions" or "postulates" play a precisely parallel role in the context of special relativity. But we have also seen significantly more. For Poincaré's conception of how a mere empirical law can be "elevated" to the status of a "convention or definition in disguise" is a continuation, in turn, of Kant's original conception of the constitutive a priori. Whereas Helmholtz's principle of free mobility generalized and extended Kant's original theory of geometrical construction within our "subjective" and *necessary* form of external intuition," Poincaré's idea that specifically Euclidean geometry is then imposed on this form by a "convention or definition in disguise" represents an extension or continuation of Helmholtz's conception. In particular, specifically Euclidean geometry is applied to our experience by precisely such a process of "elevation," in which the merely empirical fact that this geometry governs, very roughly and approximately, our actual perceptual experience of bodily displacements gives rise to a precise mathematical framework within which alone our properly *physical* theories can subsequently be formulated.

This same process of "elevation," in Einstein's hands, then makes it clear how an extension or continuation of Kant's original conception can also accommodate new and surprising empirical facts – in this case, the very surprising empirical discovery (to one or another degree of approximation) that light has the same constant velocity in every inertial frame. It now turns out, in particular, that we can not only impose already familiar and accepted mathematical frameworks (Euclidean geometry) on our rough and approximate perceptual experience, but, in appropriate circumstances, we can also impose entirely unfamiliar ones (the kinematical framework of special relativity). Einstein's creation of special relativity, from this point of view, thus represents the very first instantiation of a relativized and dynamical conception of the a priori – which, in virtue of precisely its historical origins, has a legitimate claim to be considered as genuinely constitutive in the transcendental sense.

Yet Einstein's creation of the general theory of relativity in 1915 involved an even more striking engagement with Poincaré's "conventionalist" methodology, which, I contend, makes the transcendentally constitutive role of this theory's fundamental postulates (the light principle and the principle of equivalence) even more evident.

The first point to make, in this connection, is that the principle of equivalence (together with the light principle) plays the same role in the context of the general theory that Einstein's two fundamental "presuppositions" or "postulates" played in the context of the special theory: namely, they define a new inertial kinematical structure for describing space, time, and motion. Because Newtonian gravitation theory involves an instantaneous action at a distance (and therefore absolute simultaneity), it was necessary after special relativity to develop a new theory of gravitation where the interactions in question propagate with the velocity of light. And Einstein solved this problem, via the principle of equivalence, by defining a new inertial–kinematical structure wherein the freely falling trajectories in a gravitational field replace the inertial trajectories described by free particles affected by no forces at all. The principle of equivalence, in this sense, replaces the classical law of inertia holding in both Newtonian mechanics and special relativity. But the principle of equivalence itself rests on a well-known empirical fact: that gravitational and inertial mass are equal, so that all bodies, regardless of their mass, fall with exactly the same acceleration in a gravitational field. In using the principle of equivalence to define a new inertial–kinematical structure, therefore, Einstein has "elevated" this merely empirical fact (recently verified to a quite high degree of approximation by Lorand von Eötvös) to the status of a "convention or definition in disguise" – just as he had earlier undertaken a parallel "elevation" in the case of the new concept of simultaneity introduced by the special theory.

Nevertheless, Einstein did not reach this understanding of the principle of equivalence all at once. He first operated, instead, within an essentially three-dimensional understanding of special relativity, and he proceeded (in the years 1907–1912) to develop relativistically acceptable models of the gravitational field by considering the inertial forces (like centrifugal and Coriolis forces) arising in non-inertial frames of reference within this framework. It was in precisely this context, in particular, that Einstein finally (in 1912) came upon the example of the uniformly rotating frame (the rotating disk) – in which, due to a Lorentz contraction in the direction of rotation, circles around the center of rotation obey a non-Euclidean geometry – and it is at this point (and only at this point) that he then arrived at the conclusion that the gravitational field may be represented by a non-Euclidean geometry.

It was in precisely the context of this line of thought, finally, that Einstein found that he now had explicitly to oppose Poincaré's "conventionalist" philosophy of geometry. Yet Einstein's argument – as described in *Geometrie und Erfahrung* (1921) – was far from a simple rejection of Poincaré's methodology in favor of a straightforward "empiricism." For Einstein also famously says, in the same work, that "*sub specie aeterni*" Poincaré is actually correct – so that, in particular, Einstein's reliance on a Helmholtzian conception of "practically rigid bodies" is here merely provisional. I have suggested, therefore, that we can best understand Einstein's procedure as one of delicately situating himself *between* Helmholtz and Poincaré. Whereas Einstein had earlier followed Poincaré's general "conventionalist" methodology in "elevating" the principle of relativity (together with the light principle) to the status of a "presupposition" or "postulate," he here follows Helmholtz's "empiricism" in rejecting Poincaré's more specific philosophy of geometry in favor of "practically rigid bodies." It does not follow, however, that Einstein is also rejecting his earlier embrace of Poincaré's general "conventionalist" (or perhaps we should say "elevationist") methodology. Indeed, Einstein had already side-stepped Poincaré's specific philosophy of geometry in the case of special relativity, and for essentially the same reason he explicitly opposes it here: Poincaré's rigid hierarchy of the sciences, in both cases, stands in the way of the radical new innovations Einstein himself proposes to introduce.

But why was it necessary, after all, for Einstein to engage in this delicate dance between Helmholtz and Poincaré? The crucial point is that Einstein thereby arrived at a radically new conception of the relationship between the foundations of (physical) geometry and the relativity of space and motion. These two problems, as we have seen, were closely connected in Kant, but they then split apart and were pursued independently in Helmholtz and Mach. In Poincaré, as we have also seen, the two were perceptively reconnected once again, in so far as Poincaré's hierarchical conception of the mathematical sciences incorporated both a modification of Helmholtz's philosophy of geometry and a serious engagement with the late nineteenth-century concept of inertial frame. Indeed, it is for precisely this reason, as we now see, that Poincaré's scientific epistemology was so important to Einstein. Einstein could not simply rest content with Helmholtz's "empiricist" conception of geometry, because the most important problem with which he was now faced was to connect the foundations of geometry with the relativity of motion. But Einstein could not rest content with Poincaré's conception either, because his new models of gravitation had suggested that geometry has genuine physical content.

Einstein's radically new way of reconfiguring the relationship between the foundations of geometry and the relativity of motion therefore represents a natural (but also entirely unexpected) extension or continuation of the same conception of dynamical and relativized constitutive a priori principles he had first instantiated in the creation of special relativity. Just as he had earlier shown how an extension or continuation of Kant's original conception could accommodate new and surprising empirical facts (the discovery of the invariance of the velocity of light), Einstein here shows how a further extension of this same tradition can do something very similar in facilitating, for the first time, the application of a non-Euclidean geometry to nature. In this case, however, it is not the relevant empirical fact (the well-known equality of gravitational and inertial mass) that is surprising, but the entirely unforeseen connection between this fact and the new geometry. And what makes this connection itself possible, for Einstein, is precisely the principle of equivalence – which thereby constitutively frames the resulting physical space-time geometry of general relativity in just the same sense that Einstein's two fundamental "presuppositions" or "postulates" had earlier constitutively framed his mathematical description of the electrodynamics of moving bodies in special relativity. Whereas the particular geometry in a given general relativistic space-time is now determined empirically (by the distribution of mass and energy in accordance with Einstein's field equation), the principle of equivalence itself is not empirical in this sense. This principle is instead *presupposed* – as a transcendentally constitutive condition – for any such geometrical description of space-time to have genuine empirical meaning in the first place.

The historicized version of transcendental philosophy I am attempting to exemplify therefore sheds striking new light, I believe, on the truly remarkable depth and fruitfulness of Kant's original version. Kant's particular way of establishing a connection between the foundations of geometry and the relativity of motion – which, as we have seen, lies at the heart of his transcendental method – has not only lead, through the intervening philosophical and scientific work of Helmholtz, Mach, and Poincaré, to a new conception of the relativized a priori first instantiated in Einstein's theories, it has also led, through this same tradition, to a radically new reconfiguration of the connection between geometry and physics in the general theory of relativity itself. There can be no question, of course, of Kant having "anticipated" this theory in any way. The point, rather, is that Kant's own conception of the relationship between geometry and physics (which was limited, of necessity, to Euclidean geometry and Newtonian physics) then set in motion a remarkable series of successive reconceptualizations of this relationship (in light of profound discoveries in both pure mathematics and the empirical basis of mathematical physics) that finally eventuated in Einstein's theory.

Chapter 8
Causal Models and the Asymmetry of State Preparation

Mathias Frisch

8.1 Introduction

It appears to be both natural and intuitive to think of the world as causally evolving. We conceive of events in the present as being caused by events in the past and, in turn, as acting as causes for what happens in the future. But it is also a widespread view—at least among philosophers of physics – that this conception is not part of how mature physics represents the world. According to this view, the notion of cause survives – if at all – as part of a 'folk' scientific conception of the world but has no place in our mature theories of physics. In this paper I will first critically examine considerations in favor of this causal skepticism and then discuss a strategy for defending a role for causal notions in physics, focusing on the asymmetry of the causal relation.

Many recent arguments questioning the legitimacy of causal notions in physics are descendents of Bertrand Russell's famous attack on the notion of cause (Russell 1918). Russell's paper has received a fair amount of attention in the recent literature (see, e.g., Price and Corry 2007), but there is another precursor to the contemporary debate that prefigures many of today's arguments – an exchange between Bas van Fraassen and Nancy Cartwright – and I want to focus on that exchange here.

In the next section I will outline what I take to be van Fraassen's main arguments for the claim that the distinction between causes and non-causes can only be drawn extra-scientifically. In Section 8.3 I will sketch a general framework for incorporating asymmetric causal relations into a theory's models. In Section 8.4 I will discuss one pervasive asymmetry – the asymmetry of state preparation – that suggests that asymmetric causal relations can play a legitimate role even in physics. I will end with a brief conclusion.

M. Frisch (✉)
Department of Philosophy, University of Maryland, College Park, MD 20742, USA
e-mail: mfrisch@umd.edu

8.2 van Fraassen's Challenge

In a critical review of (van Fraassen 1989) Cartwright asks why "van Fraassen does not want to allow causality anywhere inside [the] models," which, according to his version of the semantic view, comprise the content of a physical theory. (Cartwright 1993, 424) van Fraassen responds to this question as follows:

> To me the question is moot. The reason is that, as far as I can see, the models which scientists offer us contain no structure which we can describe as putatively representing causings, or as distinguishing between causings and similar events which are not causings. Cartwright says that if models contain [parts representing] ordinary objects around us (such as cats, and cats lapping milk) then they contain [parts which represent] causes. The question will still be moot if the causes/non-causes distinction is not recoverable from the model. Some models of group theory contain parts representing shovings of kid brothers by big sisters, but group theory does not provide the wherewithal to distinguish those from shovings of big sisters by kid brothers. The distinction is made outside the theory. If Cartwright herself draws, extra-scientifically, a distinction between causes and non-causes, she can describe models furnished by science in terms of that distinction. But it may be a 'hidden variable' description. She may be thinking of the structures scientists use to model data as themselves parts of larger, more articulated structures that carry the distinctions she makes. (van Fraassen 1993, 437, 438)

van Fraassen's answer to Cartwright's question appears to be not only that the models of physical theories as a matter of fact do not draw a distinction between causes and non-causes but, what is more, that it is impossible to interpret these models causally. A theory's models, he says, contain no structures which we "*can* describe as representing causings" (my emphasis). But what is it about the structure of our theories or models that might lead us to such a causally austere view?

van Fraassen's remarks echo a claim by Russell, who maintained in his discussion of Newtonian cosmology as paradigmatic physical theory, that "in the motion of mutually gravitating bodies, there is nothing that can be called a cause and nothing that can be called an effect; *there is merely a formula*." (Russell 1918, 141, my emphasis) While Russell took a theory to be identified with a set of formulas, van Fraassen argues that a theory consists of a set of state-space models. But even though they disagree on whether theories ought to be understood syntactically or semantically, Russell and van Fraassen agree that there is no place for causal notions in physical theorizing. One might think that their argument against causes in physics is simply this:

1. The content of a physical theory is exhausted by a set of state-space models or a set of formulas.
2. Causal relations are not part of the formulas or models of a theory.
3. Therefore, causal relations are not part of the content of physical theories.

As it stands, however, premise (1) is false. Mathematical physics provides us with mathematical models or representations of the world, yet on their own mathematical models do not represent anything. How a given model or class of models represents the world depends on how the model is interpreted. Thus, no theory of physics can be strictly identified with a set of formulas or state-space models, since, minimally, a theory has to contain an interpretation which tells us which bits of the formalism are hooked up with which bits of the world. But once we acknowledge that the radically austere view of theories as consisting solely in a mathematical formalism or set of models is untenable and that an interpretive framework needs to be part of a theory, it is no longer obvious why that framework does not allow us to "describe [certain structures] as putatively representing causings."

This point is also stressed by Cartwright, who says that van Fraassen's state space models

> are models of the equations, not models of the physical systems the equations are supposed to treat. When science constructs a picture of bit of the world, the image is far richer. [...] The scientific image of nature is no more devoid of cause and causings than is our everyday experience. The appearance to the contrary arises from looking only at science's abstract statements of law, and not how those are used to describe the world. (Cartwright 1993, 426)

The question, thus, is how rich an image of the world a theory can present and how rich the interpretive frameworks are within which our theories' state space models are embedded. While Cartwright maintains that causal descriptions can be part of a theory's interpretive framework, and, hence, of the scientific image of the world, van Fraassen thinks that causal notions are an extra-scientific addition to that image.

One extremely quick argument for the claim that causal notions constitute an extra-scientific addition to the interpretation of scientific theories is suggested by Russell's famous remark that the word 'cause' is not used in the advanced sciences. Thus, one might try to argue that causal notions cannot be part of our theories' interpretive framework, since scientists do not use causal discourse in describing the world. However, the premise of this argument can easily be shown to be false. As has been pointed out repeatedly – for example, by Suppes (1970) and more recently by Hitchcock (2007) – the words 'cause', 'causal', and related words are still widely used in contemporary physics. But what is the status of causal discourse in which physicists engage? On the one hand, one might think that when physicists use causal language, they are merely offering an informal commentary on the science using everyday language. This seems to be van Fraassen's view, who agrees with Russell that mature *physics* only provides us with non-causal models embodying functional dependencies but nevertheless acknowledges that *physicists* often describe the world in causal terms. On the other hand, some of the examples of causal discourse in physics certainly seem to suggest a more central role to causal notions than that of an informal gloss. For example, a widely used textbook on classical electrodynamics singles out a principle of causality as "the most sacred tenet in all of physics" (Griffiths 1989, 399).

van Fraassen invokes two different kind of considerations in support of this view, appealing to the asymmetry of the causal relations and to its modal character, respectively. Cartwright's disagreement with van Fraassen focuses largely on the latter issue. According to van Fraassen, scientific theories present us with a Humean picture of the world, free from modal properties. Allowing causings into a theory's models, he maintains, takes us outside of the realm of science proper and into that of "woolly metaphysics," as Jim Woodward has put it in characterizing van Fraassen's worry (in Woodward 2003). By contrast, Cartwright argues that a demodalized Humean picture of the world is incoherent. Many events, such as milk lappings and photon scatterings, are intrinsically causal, according to her, and therefore the idea of a world exactly like ours but stripped of causings is "ridiculous." (Cartwright 1993, 427) Thus, far from it being the case that the models of a theory "contain no structure which we can describe as putatively representing causings", it is not clear that, in a world causal through and through, we can coherently think of the models as not representing causings.

The other considerations to which van Fraassen appeals concern the asymmetry of the causal relation. One reason for why the distinction between causes and non-causes is not recoverable from our theories' models, he suggests, is that these models do not allow us to draw an asymmetric distinction between cause and effect. The causal relation is asymmetric – there is a difference between shoving and being shoved – but the group-theoretic models of our theories do not reflect this asymmetry. Since the causal asymmetry is generally taken to line up with the temporal asymmetry in the sense that effects do not precede their causes – at least in the kind of circumstances with which we are familiar – this point is often expressed by appealing to the time-reversal invariance of the laws of (most of) our mature physical theories. As Russell (1918) put it, "the laws make no difference between past and future." That is, the problem for causal notions is not that we cannot, as van Fraassen says, describe structures as representing causings but rather that our models do not allow us to represent causes *as* causes, since the causal relation is asymmetric and our models do not allow us to represent this asymmetry.

The underlying argument is most directly made in terms of our theories' fundamental equations:

1. The fundamental equations of all mature physical theories are time-reversal invariant.
2. There is no place for an asymmetric notion of cause within a physical theory with time-reversal invariant laws.
3. Therefore, there is no place for an asymmetric notion of cause in mature physical theories.

Of course, it does not follow from the fact that a theory's dynamical equations are time-reversal invariant that all of the theory's state space models will be time-symmetric. In fact 'most' models, in some intuitive sense, will not be time symmetric, reflecting asymmetries between the initial and final conditions characterizing a given model. But the entire class of a theory's models will be time-symmetric in the sense that for each time-asymmetric model M there will be a time-reversed model M^* that also satisfies the theory's dynamical equations. From this symmetry van Fraassen concludes that the distinction between causes and non-causes must be a distinction made outside the theory.

Despite his belief that causal notions play no role in how physics represents the world, van Fraassen agrees with Cartwright that our overall conception of the world ineliminably involves causal notions and that causal discourse is irreducible to non-causal notions. The way to avoid a causal metaphysics and nevertheless acknowledge an important place for an asymmetric notion of cause in our conception of the world, according to him, is to locate causal discourse within psychological and intentional discourse. The ultimate source of causal notions, van Fraassen suggests, is our conception of ourselves as agents and the use of causal notions in physics is a metaphorical and analogical extension of language that has its basic meaning only in the context of folk psychology.

van Fraassen does not make this point explicitly, but his discussion may give the impression that the two types of consideration are closely linked and that the only way in which we can grant an important role to an irreducible asymmetric notion of cause in our conception of the world without embracing a rich causal metaphysics is to locate the source of such notions outside of the image of the world presented to us by physics. One of my aims in this paper is to challenge this impression. I think one can afford a scientifically legitimate role to asymmetric causal notions in physics without embracing a rich causal metaphysics. In the next section I want to suggest how causal notions can be incorporated into a theory's class of models in a way that is metaphysically neutral or non-committal.

8.3 Causal Models

Let us assume van Fraassen's framework of theories as represented by a class of state space models. We can then think of interpreting a theory causally as the equivalent of embedding the theory's state space models into larger model-theoretic structures by introducing asymmetric relations between state space variables. The state of a system $S(t)$ is given by the values of a set of variables $s_1(t), s_2(t), \ldots, s_n(t), \ldots$, which may be finite or infinite. The dynamical laws of a theory define a class of dynamical models specifying dynamically possible sequences of states, which can be represented in terms of state space models. We then define an asymmetric, transitive, and non-circular relation $C = <S(t_i), S(t_j)>$ over the set of states S, which defines a partial ordering over the set of states in a model. C is interpreted as the causal relation: $S(t_2)$ bears C to $S(t_1)$ exactly if $S(t_1)$ is a cause of $S(t_2)$. If two states do not stand in relation C then they are not causally related. The result is a class of what I want to call *potential causal models* of a theory. Depending on the theory in question, we can also introduce more fine-grained causal relations $<s(t_i), s(t_j)>$ defined over individual state variables s_i.

One might object that physicists do not explicitly represent their theories in terms of causal structures of this kind. But the reason for this might be that the causal structures at issue will usually be quite 'boring' and hence not in need of an explicit representation: often their content is exhausted by the claim that the state of a system is a cause of future states of the system.[1] By making the formal framework explicit, however, we see that asymmetric causal notions need not

---

[1] There is, however, at least one field in fundamental physics in which causal relations of this kind are introduced explicitly – the causal set approach to quantum gravity (see, e.g., Rideout and Sorkin 2000).

constitute a vague or imprecise addition to a physical theory, as Suppes (1970) and Hitchcock (2007) have suggested, but can be introduced into a theory in a mathematically well-defined manner. Moreover, embedding a theory's state space models into richer causal structures does not in itself carry any weighty metaphysical commitments with it. As van Fraassen himself has argued, accepting a theory that embeds the theory's observational substructures into a richer set of models does not yet settle the question as to what our metaphysical commitments to the entities posited by our models ought to be. We might be realists, who believe in the unobservable substructures postulated by our theories, or we might be constructive empiricists, who do not think that our grounds for accepting a theory are not also good reasons for believing in its unobservable substructures. Similarly, taking the theory's state space models to be embedded into richer, causal structures in itself is metaphysically non-committal. In particular, adopting causal models carries with it no metaphysical commitment to the existence of anything like a 'causal glue' between causally related events. Thus, there appears to be room for a stance on the issue of causation that "involves no metaphysics" (van Fraassen 1993, 439) yet does not follow van Fraassen in taking causal discourse to be ineliminably psychological.

Causal structures do, however, introduce asymmetric relations that are not part of the non-causal state space models, which van Fraassen identifies with the content of a physical theory. Thus, van Fraassen worries that adding such structures amounts to providing a "'hidden variable' description" and that there can be no empirical or (more generally) scientifically legitimate reason for accepting the richer causal structures instead of non-causal state space models. Yet it seems to me that a closer look at theorizing in physics reveals a far more prominent role for causal notions than van Fraassen wants to allow. First, there are theoretical contexts in which physicists appeal to time-asymmetric causal constraints to restrict the range of physically possible models. In terms of the terminology introduced above, causal constraints are invoked to restrict the class of *potential causal models* to a proper subclass of models that are *causally possible*. For example classical dispersion relations are derived from an explicitly time-asymmetric causal constraint (see Frisch 2009a,b).

Second, there are phenomena that exhibit asymmetries in prevailing initial or final conditions. If actual systems in the domain of a time-reversal invariant theory are best represented by models most of which exhibit the same kind of temporal asymmetry – that is, if there is an asymmetry between the initial and final conditions characterizing models of typical actual systems – then this might be evidence for causal relations among the physical quantities involved. A paradigmatic example of this is the temporal asymmetry characteristic of waves in the presence of wave sources, as I have argued elsewhere.[2]

Both these cases involve an asymmetry in the class of state-space models representing actually occurring phenomena. But, as I will argue in the next section, there can even be scientifically legitimate evidence for time-asymmetric causal structures in cases where the class of state-space models representing the phenomena is time-reversal invariant.

---

[2] See (Frisch 2005), and also (Frisch 2000, 2006, 2008).

8.4 The Asymmetry of State Preparation

In this section I will argue that our experimental interactions with physical systems exhibit a temporal asymmetry even in the case of systems that are best modeled with the help of theories with time-reversal invariant laws and that this asymmetry is best thought of as a causal asymmetry. In explicit premise-conclusion form, my argument is this:

1. There is a temporal asymmetry characterizing experimental interventions into otherwise closed systems.
2. If there is such an asymmetry, it is best explained by appealing to a causal asymmetry.
3. If a concept plays a role in the best explanation of a phenomenon, the concept plays a legitimate role in science.
4. Therefore, asymmetric causal notions play a legitimate role in science.

Why should we accept premise (1)? It is a striking fact about experimental interactions that we can only intervene into a system 'from the past,' as it were. Consider a system S that is governed by both past and future deterministic laws. That is, let us assume that the final state $S_i(t_f)$ of the system is uniquely determined by the initial state $S_i(t_i)$, where $t_i < t_f$, together with the dynamical laws and the boundary conditions; and that the initial state $S_i(t_i)$ is similarly determined by the final state $S_f(t_f)$. Thus, if S is closed between t_i and t_f, then the initial and final states are both dependent on each other. Nevertheless there is an asymmetry of state preparation in the following sense. We can prepare the system in its initial state $S_i(t_i)$ without making use of any knowledge we might have of the system's dynamical evolution between t_i and t_f; and we can subsequently calculate the system's future evolution for times $t > t_i$ from the initial state, the dynamical laws, and the boundary conditions. But we could not similarly first prepare the system's final state at t_f without using our knowledge of the dynamics and then take the final state together with the laws to calculate the system's past evolution for $t < t_f$. (Of course we cannot first prepare the system in S_f and then let it *evolve* into S_i. That is not what the asymmetry consists in. Rather the asymmetry consists in the fact that we cannot first prepare the system in S_f without *making use* of facts about the dynamical evolution and *then calculate* what the system's past evolution from S_i to S_f must have been, given the dynamical laws and the boundary conditions.)

There are two ways in which we can prepare the system in a specific final state at t_f. First, we can make use of our knowledge of the dynamical laws to determine the initial state in which the system has to start out at t_i in order to evolve into the final state in question and prepare the system in the appropriate initial state. In that case our ability to prepare the system in its final state relies crucially on our knowledge of the system's evolution between t_i and t_f. If the system is closed between t_i and t_f, we need to know which state S_i will evolve into the state S_f we are trying to set up, because we can only 'prepare' the system in the state S_f by preparing it in S_i. By contrast, we can prepare the system in an initial state S_i without any knowledge of the dynamical evolution of the system between t_i and

t_f, even though the dynamical laws and boundary condition determine what that evolution is. While, given the dynamical laws and boundary conditions, the initial and final states determine each other, we do not need to make use of that fact if our goal is to set up the system in some specified *initial* state. But we do need to make explicit use of the dynamical evolution between t_i and t_f if our goal is to set up the system in a specified *final* state.

We can imagine, for example, that one experimenter is responsible for preparing a system S in an initial state S_i or a final state S_f and that a different experimenter is responsible for setting up the boundary conditions. If the first experimenter wants to prepare the system in a certain initial state, he can do that without knowing what boundary conditions the second experimenter chooses to set up. But if the first experimenter wants to make sure that the system ends up in a certain final state, he needs to know what the boundary conditions will be in order to make sure he prepares the system in the appropriate initial state.

A second way of preparing the system in a final state S_f is to prepare the system in that state directly by intervening into the system between t_i and t_f. In this case we do not need to make use of our knowledge of the evolution between initial and final times. But then the system will not be closed between t_i and t_f and we cannot use the dynamical laws and boundary conditions governing the closed system to retrodict the initial at t_i. Thus, there is a way for first experimenter to prepare the system in S_f without knowledge of the boundary conditions which the second experimenter tries to set up, but at the cost of having to violate these conditions through his intervention and, thus, by losing any ability to retrodict the evolution of the system with the help of the dynamical equations.

That is, even systems that are governed by both past and future deterministic dynamical equations exhibit an asymmetry of state preparation. In the case of a system that is closed between t_i and t_f and initial and final sates mutually determine each other, given the dynamical equations and boundary conditions, we can only prepare a system in a given final state S_f by making explicit use of the dynamics and prepare the system in the corresponding initial state S_i. If instead we directly intervene on the state of the system at t_f, we can no longer use the state at t_f to retrodict the state at t_i (while we can, of course, directly intervene on the state of the system at the earlier t_i and then predict the system's evolution until some later time t_f). We can only intervene into a system from its past.

My reason for claiming that the asymmetry of state preparation is best understood as a causal asymmetry is that the asymmetry is a paradigm case of the asymmetry characterizing interventions, as understood by interventionist accounts of causation (see, e.g., Woodward 2003). In particular, if S_f is an effect of S_i, then according to an interventionist account of causation there are two ways by which one can intervene on the system to set S_f to a particular value: first, we can intervene on S_i, which in turn will affect the value of S_f; or, second, we can intervene directly on S_f, which 'breaks the causal arrow' from S_i to S_f and, therefore, makes it impossible to retrodict the value of S_i on the basis of the value of S_f. Thus, interventionist accounts of causation predict that experimental systems will exhibit an asymmetry of state preparation, if earlier states of the system are causes of later

8 Causal Models and the Asymmetry of State Preparation

states. Positing an asymmetric causal relation between the states of a system at different times clearly provides *an* explanation of the asymmetry of state preparation. Moreover, it seems to me that there is no other fully worked out and equally as successful non-causal alternative explanation of the asymmetry. Thus, I take it that this asymmetry provides us with empirically justified reasons for embedding non-causal state space models of a system into richer causal models.

Now, van Fraassen recognizes that causal notions such as manipulation and control are an integral part of science but maintains that these notions play a role only in applied science and that their use there can be accounted for entirely as an analogical and metaphorical extension of psychological discourse. Thus, van Fraassen might argue that the asymmetry of state preparation does not point to a genuinely scientific asymmetry, since it reveals itself only in experimental interactions with physical systems and in virtue of being an asymmetry of manipulation and control falls under the domain of our folk-scientific conception of ourselves as agents.

Yet there clearly is no direct argument from the claim that our use of causal discourse in science arises from our experimental interactions with physical systems to the conclusion that causal distinctions are drawn extra-scientifically. There are many scientifically legitimate accounts of phenomena for which our only evidence comes from experimental interactions. Indeed, the asymmetry of state preparation seems to show that the causal asymmetry is a more robustly objective feature of such interactions than van Fraassen's account of causal talk as metaphorical or analogical extension of everyday life allows. If describing experimental interactions with physical systems in causal terms is *only* metaphorical or analogical, then a causal account cannot provide a scientific explanation of the asymmetry of state preparation. As a useful contrast case consider the following. We might metaphorically describe a ball rolling up an inclined plane as 'struggling to reach the top of the plane.' In this case there are no facts about the (macroscopic) physical situation that are not fully accounted for in terms of the initial and boundary conditions and the dynamical equations and the metaphorical description truly cannot serve any genuine scientifically explanatory purpose. By contrast, in the case of the asymmetry of state preparation we do not merely choose to describe in asymmetric causal terms facts about physical systems that also can exhaustively be captured in a non-causal description of that system. Rather, the asymmetry characteristic of our interactions with such systems goes beyond what can be captured in an acausal and time-symmetric description of the system.

In general, two kinds of account of the asymmetry of state preparation seem possible. Either one can appeal to asymmetric causal relations between the states of a system at different times, along the lines I have suggested here. According to this account, the asymmetry is due to an intrinsic asymmetry characterizing the system. Or one might try to argue that the asymmetry is due to what are ultimately non-causal physical features of the kinds environment into which the systems with which we interact are embedded. This second kind of explanation either might appeal to general asymmetries of the physical environment, such as thermodynamic features, or might more narrowly focus on asymmetries characterizing physical agents intervening in experimental systems. Variants of the latter account have been defended by Healey

(1983) and Price (2007). Now, I do not know of a principled argument that can establish that no non-causal account of the asymmetry of state preparation can succeed. But I also do not know of any existing account that offers more details than might be contained in a promissory note. The task for any such account is to show that the asymmetry of state preparation varies with hypothetical changes to the environment. Thus, Healey's and Price's arguments rely crucially on intuitions concerning what the direction of causation would be in hypothetical anti-thermodynamic environments, but it is not clear why we should share their intuitions. That is, to my mind neither has shown convincingly that in a different thermodynamic environment state-preparation would be from the future, as it were. But in the absence of a fully developed convincing alternative, the causal account remains the best explanation of the asymmetry of state preparation.

8.5 Conclusion

I argued that a theory's non-causal state space models can be embedded into richer structures containing asymmetric causal relations in a way that does not carry with it a commitment to a particular causal metaphysics. The use of causal models does not merely constitute the introduction of a 'hidden variable description' but can be empirically justified in various ways. In this paper I presented one such justification that appeals to a pervasive asymmetry characterizing all our interactions with experimental systems: the asymmetry of state preparation.

References

Cartwright N (1993) Defence of 'this worldly' causality: Comments on van Fraassen's laws and symmetry. Philos Phenomenol Res 53(2):423–429
Frisch M (2000) '(Dis-)solving the puzzle of the arrow of radiation. Br J Philos Sci 51:381–410
Frisch M (2005) Inconsistency, asymmetry and non-locality: a philosophical investigation of classical electrodynamics. Oxford University Press, New York
Frisch M (2006) A tale of two arrows. Stud Hist Philos Sci Part B Stud Hist Philos Modern Phys 37(3):542–558
Frisch M (2008) Philosophical issues in electromagnetism. Philos Comp 4/1 (2009): 255–270, 10.1111/j.1747-9991.2008.00192.x
Frisch M (2009a) 'The most sacred tenet?' Causal reasoning in physics. Br J Philos Sci 60:459–474
Frisch M (2009b) Causality and dispersion: a reply to John Norton. Br J Philos Sci 60:487–495
Griffiths D (1989) Introduction to electrodynamics, 2nd edn. Prentice-Hall, Englewood Cliffs, NJ
Healey R (1983) Temporal and causal asymmetry. In: Swinburne R (ed) Space, time, and causality. Reidel, Dordrecht, pp 79–104
Hitchcock C (2007) What Russell got right. In: Price H, Corry R (eds) Causality, physics, and the constitution of reality: Russell's republic revisited. Oxford University Press, Oxford
Price H (2007) Causal perspectivalism. In: Price H, Corry R (eds) (2007)
Price H, Corry R (eds) (2007) Causality, physics, and the constitution of reality: Russell's republic revisited. Oxford University Press, Oxford

Rideout DP, Sorkin RD (2000) A classical sequential growth dynamic for causal sets. Phys Rev D 61:16

Russell B (1918) On the notion of cause. Mysticism and logic and other essays. Longmans, Green, New York

Suppes P (1970) A probabilistic theory of causality. North-Holland, Amsterdam

van Fraassen B (1989). Laws and symmetry. Oxford University Press, Oxford

van Fraassen B (1993) Armstrong, Cartwright, and Earman on laws and symmetry. Philos Phenomenol Res 53(2):431–444

Woodward J (2003) Making things happen: A theory of causal explanation. Oxford University Press. Oxford

Chapter 9
Bell-Type Inequalities from Separate Common Causes

Gerd Graßhoff and Adrian Wüthrich[*]

9.1 Introduction

Bell-type inequalities provide predictions of observable frequencies of measurement outcomes in a particular experiment in quantum mechanics (the EPR-Bohm experiment). Although the actual experiments were conducted with polarized photons, for our purposes we can consider the Bell inequalities as having being tested in the familiar setup proposed by Bohm and Aharonov (1957), which is based on Einstein, Podolsky and Rosen's original formulation of the paradox that bears their names (EPR) (Einstein et al. 1935). We will refer to the experimental setup described by Bohm as the EPRB experiment.

In the EPRB experiment the particles of a pair in the spin singlet state are separated from each other by an arbitrary distance. We assume that one particle flies into the left wing of the experimental setting, the other particle into the right wing. Each particle's spin is measured relative to one of three directions. The individual particle measurements yield an apparently random sequence of results that are either "spin up" or "spin down" relative to the chosen measurement direction. A comparison of measurement results when parallel measurement settings are chosen shows a perfect (anti)correlation between the outcome of the measurement performed on the left particle and the outcome of the measurement performed on the right particle.

The assumption that the correlations require a local causal explanation leads to predictions (a Bell inequality) that contradict quantum mechanics as well as, by current standards, the experimental data. Thus, at least one of the assumptions needed to derive a Bell inequality must be wrong.

The argument has the form of a *reductio ad absurdum*: from the falsity of the conclusion, the falsity of one of the premises is inferred. The strength of the argument rests on the fact that:

- The derivation is deductive.

G. Graßhoff and A. Wüthrich (✉)
History and Philosophy of Science, Sidlerstrasse 5, University of Bern, CH-3012 Bern, Switzerland

[*] Speaker at the conference (EPSA, Madrid 2007). Title of the talk: Minimal Assumption Derivation of a Bell-Type Inequality.

- All the assumptions are explicit.
- The set of assumptions is minimal.

By "minimal" we mean that no subset of the assumptions implies a Bell inequality. Since derivations of Bell-type inequalities are often invoked as arguments against the causal closedness of the physical world, the derivation should above all suppose a notion of causal explanation that is "non-trivial" and "as weak as possible" (van Fraassen 1982, p. 27).

9.2 Common Causes for Correlated Events

The event types that are observed to be correlated in a EPRB experiment are assumed, justifiably, not to stand in a *direct* causal relation. If we are to explain the correlations at all, we have to do so by postulating a *common cause* for the correlated effects. Most derivations, lacking as they do a sound theory of causal relevance, resort to Reichenbach's *common-cause* principle (Reichenbach 1956) providing a seemingly necessary condition for causal relevance (*screening-off*). The principle states that, conditional upon the instantiation of the common cause, the events in question are uncorrelated. This statistical condition is weaker than the demand that the common cause be sufficient for its effects, but it still captures the idea that, with the exception of the common cause, there is no reason for the event types to be correlated.

As Reichenbach explicated the screening-off condition only with reference to one pair of correlated events, it is not clear how one should apply this principle to the EPRB setup (cf. Hofer-Szabó et al. 1999). Traditional derivations assume that there is a *common* common cause for all correlated pairs of events and, therefore, a *common screener-off* C:

$$p(L^+ R^- | M_i N_i C) = p(L^+ | M_i N_i C) p(R^- | M_i N_i C).$$

This, however, is still an unjustifiably strong assumption. Nothing in Reichenbach's notion of common causes dictates that the common cause should be common to all correlated pairs. A more general application of Reichenbach's condition only demands that, for *each* correlated pair, there is a (possibly different) common cause. From the assumption of *separate* common causes, only the following screening-off condition can be justified:

$$p(L^+ R^- | M_i N_i C_{ii}) = p(L^+ | M_i N_i C_{ii}) p(R^- | M_i N_i C_{ii}),$$

where the common causes C are indexed as being the cause of the measurement outcomes of a specific choice of measurement directions.

The C indices express neither a causal nor a statistical dependence between the common causes and the measurement operations. They are simply labels to distinguish between types of events (at the source where the particles are created).

The distinguished event types at the particle source are not mutually exclusive descriptions of a "total state". A common cause responsible for a correlation exhibited by a certain measurement need not occur together with the corresponding measurement. All that is required is that the screening-off conditions hold for the common causes with respect to the measurement results. The common causes may or may not be instantiated in each run of the experiment, irrespective of which measurement event type is chosen to be instantiated. Separate common causes need not be causally relevant to the choice of measurements nor is a causal influence backwards in time from the measurement choices on the common causes required.

9.3 A Common Screener-Off Is Not a Common Common Cause

Hofer-Szabó et al. (1999) claim that without the unjustified strong assumption, made by traditional derivations, of a *common* common cause "Bell's inequality *cannot* be derived" (p. 388, emph. in the original). Graßhoff et al. (2005) proved the contrary, supposing reasonable locality and other independence conditions. This derivation assumes a weaker notion of common cause explanations than traditional derivations and, therefore, provides a stronger *reductio ad absurdum* argument against the possibility of the EPRB correlations having a common cause explanation.

One could surmise, though, that the conjunction of the separate common causes just makes up a *common common cause*. But, in general, this is not the case, since, for instance, from

$$p(A_i B_i | C_i) = p(A_i | C_i) p(B_i | C_i) \tag{9.1}$$

and

$$p(A_j B_j | C_j) = p(A_j | C_j) p(B_j | C_j) \tag{9.2}$$

it does not follow that

$$p(A_i B_i | C_i C_j) = p(A_i | C_i C_j) p(B_i | C_i C_j). \tag{9.3}$$

Only under specific circumstances is the conjunction of separate common causes a common screener-off. For instance, in the case of a perfect correlation between A and B, (9.3) does follow from (9.1), as the conditional probabilities are zero or one. Thus, in the case of perfect correlations, the conjunction of the separate common causes is a common screener-off and, therefore (supposing Reichenbach's notion of common cause), could, but need not, be a common common cause (since the Reichenbach condition is, at most, a necessary condition).

Common screener-offs, however, are not common common causes, since Reichenbach took the screening-off condition to be only a necessary and not a sufficient condition for qualifying as a common cause – and rightly so. In the case of perfect correlations, the screener-offs are sufficient for the correlated events ($p(A_i|C_i) = 1$, for instance). The conjunction of the common causes ($C_i C_j$) is still a sufficient condition for the effects and, therefore, a (trivial) common screener-off.

But it is clear, either intuitively or given the appropriate causal theory (Graßhoff and May 2001), that C_j (the common cause for A_j and B_j), for instance, does not contribute to the correlations of A_i and B_i and hence is not an appropriate part of a common cause for that correlation. Sufficient conditions for events tend to include more than their causes. Only *minimally sufficient* conditions can qualify as causes for events (Graßhoff and May 2001).

9.4 "Genuine" Separate Common Causes

9.4.1 Relative Minimality of Derivations

In Graßhoff et al. (2005) we do not assume a common common cause in deriving a Bell-type inequality. In this respect the derivation is weaker than traditional derivations. However, in a different respect, Clauser et al.'s (1969) derivation, for instance, is weaker: it is not assumed, in the case of parallel measurement settings, that the outcomes are perfectly anticorrelated; on the cost that they require a common common cause. The minimality of the set of assumptions of Graßhoff et al. (2005) is, therefore, only relative to the group of derivations that assume perfect anticorrelations. A derivation of a Bell-type inequality from separate common causes without the assumption of perfect anticorrelations was published in April 2006 as an electronic preprint before being published in print (Portmann and Wüthrich 2007). A similar derivation has been published by Hofer-Szabó (2008).

In Portmann and Wüthrich (2007) a small deviation from perfect anticorrelation was allowed, as for instance: $p(L^+|R^- M_i N_i) = 1 - \varepsilon$. Together with the assumption of separate common causes (and traditional locality and independence assumptions), an inequality was derived that is of the same form as the Clauser-Horne inequality (Clauser and Horne 1974) but less restrictive by essentially the amount ε. Quantum mechanical predictions contradict the predictions of a common-cause model obeying the assumptions of Portmann and Wüthrich (2007) for $\varepsilon \leq 2.689 \cdot 10^{-5}$. It is, however, extremely difficult to obtain such high precision in experiments confirming perfect anticorrelation, but only then is the inequality violated by the quantum mechanical predictions and experimental data. Thus, common-cause models obeying the assumptions of Portmann and Wüthrich (2007) can hardly be ruled out by empirical data.

There are, however, *theoretical* reasons for opposing the possibility of such a common-cause model. If the model is not to be dismissed by the violation of the inequality derived by Portmann and Wüthrich (2007), the deviation from perfect anticorrelation should be greater than $2.689 \cdot 10^{-5}$. Yet, the theoretical predictions of a deviation from perfect anticorrelation by, for instance, quantum gravity effects, are much smaller. Thus, current theories, which, contrary to standard quantum mechanics, predict a slight deviation from perfect anticorrelations, would hardly be able to explain a deviation by the required amount.

A further proviso is in order: it is an open question whether even the original (more restrictive) Clauser-Horne inequality can be derived from the assumptions of Portmann and Wüthrich (2007). In that case common-cause models obeying the assumptions of Portmann and Wüthrich (2007) would be ruled out by an empirical falsification of the Clauser-Horne inequality – as are the models considered by these authors (Clauser and Horne 1974).

9.4.2 "Genuine" Separate Screener-Offs

Only when deviations from the perfect anticorrelations (PCORR) postulated by quantum mechanics are allowed is the conjunction of the separate screener-offs not a common screener-off. For Hofer-Szabó (2008) only then are there no "implicit" common screener-offs (C-SCR), and only such derivations can provide a derivation of Bell's inequality from "genuine" separate screener-offs (S-SCR). But also Graßhoff et al.'s (2005) derivation is a conclusive derivation of a Bell inequality (BELL) from separate screener-offs (and from separate common causes). The fact that, in the case of perfect correlations, the existence of a common screener-off is implied by the existence of separate screener-offs does not invalidate the statement that (together with PCORR and other assumptions X) Bell's inequality follows from the assumption of separate screener-offs. If only separate common causes, the conjunction of which is not a common screener-off, counted as "genuine", then there would be no genuine separate common causes for perfect correlations.

Graßhoff et al. (2005) contains the statement that S-SCR, PCORR and X implies BELL:

$$\text{S-SCR, PCORR, X} \rightarrow \text{BELL}. \tag{9.4}$$

From traditional derivations it is known that

$$\text{C-SCR, PCORR, X} \rightarrow \text{BELL}. \tag{9.5}$$

Hofer-Szabó (2008) grounds his critique on the fact that

$$\text{S-SCR, PCORR} \rightarrow \text{C-SCR}. \tag{9.6}$$

However, the (granted) truth of this statement does not invalidate Graßhoff et al.'s (2005) claimed implication (9.4). Thus, in that respect (quite apart from the non-identity of common causes and screener-offs discussed above) Hofer-Szabó's (2008) critique also does not invalidate the Graßhoff et al. (2005) argument.

9.5 Summary

Traditional derivations of Bell-type inequalities assume, without sound justification, a common cause that is identical for all correlated pairs of events. However, a Bell-type inequality can also be derived from the weaker assumption that, for each correlated pair, there is a (possibly different) common cause. The resulting set of assumptions, one of which by *reductio ad absurdum* must be false, may or may not include the assumption of perfect anticorrelation with parallel measurement settings (Graßhoff et al. 2005; Portmann and Wüthrich 2007). The proof that a Bell-type inequality can be derived from separate common causes has been conjectured to be impossible (Hofer-Szabó et al. 1999, p. 388), but the argument is not invalidated by the mathematical construction of a common screener-off from the separate common causes, contrary to what, we read, is claimed in Hofer-Szabó (2008).

References

Bohm D, Aharonov Y (1957) Discussion of experimental proof for the paradox of Einstein, Rosen, and Podolsky. Phys Rev 108(4):1070–1076
Clauser JF, Horne MA (1974) Experimental consequences of objective local theories. Phys Rev D 10(2):526–535
Clauser JF, Horne MA, Shimony A, Holt RA (1969) Proposed experiment to test local hidden-variable theories. Phys Rev Lett 23(15):880–884
Einstein A, Podolsky B, Rosen N (1935) Can quantum-mechanical description of physical reality be considered complete? Phys Rev 47(10):777–780
Graßhoff G, May M (2001) Causal regularities. In: Spohn W, Ledwig M, Esfeld M (eds) Current issues in causation. Mentis, Paderborn, pp 85–114
Graßhoff G, Portmann S, Wüthrich A (2005) Minimal assumption derivation of a Bell-type inequality. Br J Philos Sci 56(4):663–680
Hofer-Szabó G (2008) Separate-versus common-common-cause-type derivations of the Bell inequalities. Synthese 163(2):199–215
Hofer-Szabó G, Rédei M, Szabó LE (1999) On Reichenbach's common cause principle and Reichenbach's notion of common cause. Br J Philos Sci 50(3):377–399
Portmann S, Wüthrich A (2007) Minimal assumption derivation of a weak Clauser-Horne inequality. Stud Hist Philos Modern Phys 38(4):844–862. Electronic preprint: quant-ph/0604216
Reichenbach H (1956) The direction of time. Dover, New York
van Fraassen BC (1982) The charybdis of realism: Epistemological implications of Bell's inequality. Synthese 52(1):25–38

Chapter 10
Entanglement, Upper Probabilities and Decoherence in Quantum Mechanics

Stephan Hartmann and Patrick Suppes

Quantum mechanical entangled configurations of particles that do not satisfy Bell's inequalities, or equivalently, do not have a joint probability distribution, are familiar in the foundational literature of quantum mechanics. Nonexistence of a joint probability measure for the correlations predicted by quantum mechanics is itself equivalent to the nonexistence of local hidden variables that account for the correlations (for a proof of this equivalence, see Suppes and Zanotti 1981).

From a philosophical standpoint it is natural to ask what sort of concept can be used to provide a "joint" analysis of such quantum correlations. In other areas of application of probability, similar but different problems arise. A typical example is the introduction of upper and lower probabilities in the theory of belief. A person may feel uncomfortable assigning a precise probability to the occurrence of rain tomorrow, but feel comfortable saying the probability should be greater than $1/2$ and less than $7/8$. Rather extensive statistical developments have occurred for this framework. A thorough treatment can be found in Walley (1991) and an earlier measurement-oriented development in Suppes (1974). It is important to note that this focus on beliefs, or related Bayesian ideas, is not concerned, as we are here, with the nonexistence of joint probability distributions. Yet earlier work with no relation to quantum mechanics, but focused on conditions for existence has been published by many people. For some of our own work on this topic, see Suppes and Zanotti (1989).

Still, this earlier work naturally suggested the question of whether or not upper and lower measures could be used in quantum mechanics, as a generalization of probability. To show that an affirmative answer is possible, and, we hope of some philosophical interest, is the general purpose of this paper.

Following Suppes and Zanotti (1991) the initial focus is to construct an upper-probability measure for Bell-type correlations. Such a construction was sketched in the paper just mentioned, but full details are needed here to study the decoherence decay of such systems, our second topic.

S. Hartmann (✉)
Center for Logic and Philosophy of Science, Tilburg University, 5000 LE Tilburg, The Netherlands
e-mail: s.hartmann@uvt.nl

P. Suppes
Stanford University, 220 Panama Street, Stanford, CA 94305-4101, USA

Computation of decoherence times is an important feature of decoherence theories. The literature in fact includes specific results on of whether or not most entangled systems of quantum particles have an expected decoherence time that is much too fast for humans or other animals to make any brain computations that are quantum mechanical. (For a skeptical view of this possibility, see Suppes and de Barros 2007.)

The question of special interest here is whether a computation of decoherence decay of the upper probability measure we construct gives a good approximation of the decay time obtained from direct quantum mechanical calculations of the decoherence decay of the "too active" quantum correlations.

For later use, we give here the definition of upper probability.

Definition 10.1. Let Ω be a nonempty set, F a Boolean algebra on Ω, and P^* a real-valued function on F. Then $\Omega = (\Omega, F, P^*)$ is an *upper probability space* if and only if for every A and B in F

1. $0 \leq P^*(A) \leq 1$;
2. $P^*(\emptyset) = 0$ and $P^*(\Omega) = 1$;
3. If $A \cap B = \emptyset$, then $P^*(A \cup B) \leq P^*(A) + P^*(B)$.

Axiom 3 on finite subadditivity could be strengthened to σ-subadditivity but we are not concerned with that issue here.

10.1 Upper Probabilities in Quantum Mechanics

We use the standard notation familiar in the Bell inequalities which we review very briefly. For definiteness, but not required, we can think of a Bell-type experiment in which we are measuring spin for particle A and for particle B. More generally, we may think of A and B as being the location of measuring equipment and we observe individual particles or a flux of particles at each of the sites. Here we will think of individual particles because the analysis is simpler. The measuring apparatus is such that along the axis connecting A and B we have axial symmetry and consequently we can describe the position of the measuring apparatus just by the angle of the apparatus A or B in the plane perpendicular to the axis. We use the notation w_A and w_B for these angles. The basic form of the locality assumption is shown in terms of the following expectation:

$$E(\mathbf{M}_A | w_A, w_B, \lambda) = E(\mathbf{M}_A | w_A, \lambda) \qquad (10.1)$$

What this means is the expectation of the measurement $\mathbf{M}_A$, of spin of a particle in the apparatus in position A, given the two angles of measurement for apparatus A and B as well as the hidden variable λ, is equal to the expectation without knowledge of the apparatus angle w_B, of B. This is a reasonable causal assumption and is a way of saying that what happens at B should have no direct causal influence on what happens at A. On the other hand, we have the following theoretical result for spin, well confirmed in principle for the case where the measuring apparatuses are both set at the same angle:

$$P(M_A = -1 | w_A = w_B = x \,\&\, M_B = 1) = 1 \tag{10.2}$$

If the angles of the apparatus are set the same, we have a deterministic result in the sense that the observation of an EPR state at B will be the opposite at A, and conversely. Here we are letting 1 correspond to spin $1/2$ and -1 correspond to spin $-1/2$. What Bell showed is that on the assumption there exists a hidden variable, four related inequalities can be derived for settings A and A' and B and B' for the measuring apparatus. We have reduced the notation here in the following way in writing the inequalities. First, instead of writing M_A, we write simply A, and second, instead of writing Cov(A, B) for the covariance, which in this case will be the same as the correlation, of the measurement at A and the measurement at B, we write simply AB. With this understanding about the conventions of the notation, we then have as a consequence of the assumption of a hidden variable the following set of inequalities, which in the exact form given here are due to Clauser et al. (1969):

$$-2 \leq \mathbf{AB} + \mathbf{AB'} + \mathbf{A'B} - \mathbf{A'B'} \leq 2 \tag{10.3a}$$
$$-2 \leq \mathbf{AB} + \mathbf{AB'} - \mathbf{A'B} + \mathbf{A'B'} \leq 2 \tag{10.3b}$$
$$-2 \leq \mathbf{AB} - \mathbf{AB'} + \mathbf{A'B} + \mathbf{A'B'} \leq 2 \tag{10.3c}$$
$$-2 \leq -\mathbf{AB} + \mathbf{AB'} + \mathbf{A'B} + \mathbf{A'B'} \leq 2 \tag{10.3d}$$

Quantum mechanics does not satisfy these inequalities in general. To illustrate ideas, we take as a particular case the following:

$$\mathbf{AB} - \mathbf{AB'} + \mathbf{A'B} + \mathbf{A'B'} < -2$$

We choose

$$\mathbf{AB} = \mathbf{A'B'} = -\cos 30° = -\frac{\sqrt{3}}{2}$$

$$\mathbf{AB'} = -\cos 60° = -\frac{1}{2}$$

$$\mathbf{A'B} = -\cos 0° = -1$$

So the inequality (10.3c) is violated by this example, since from quantum mechanics Cov($\mathbf{AB}$) = $-\cos$ (angle $\mathbf{AB}$) and

$$-\frac{\sqrt{3}}{2} + \frac{1}{2} - 1 - \frac{\sqrt{3}}{2} < -2.$$

First we must compute the probabilities for the pairs with given correlations, using dots for missing arguments. So

$$P(\mathbf{A} = 1) = p(1 \,\cdots\,) = p(-1 \,\cdots\,) = \frac{1}{2}$$

since $E(A) = 0$, and by similar arguments and notation

$$P(\mathbf{B'} = 1) = p(\cdots 1) = p(\cdots -1) = \frac{1}{2}$$

For ease of reading we replace "−1" by "0".

Now the correlation
$$\mathbf{AB} = -\frac{\sqrt{3}}{2}$$

so
$$-\frac{\sqrt{3}}{2} = p(1 \cdot 1 \cdot) + p(0 \cdot 0 \cdot) - p(1 \cdot 0 \cdot) - p(0 \cdot 1 \cdot).$$

But by symmetry
$$p(1 \cdot 1 \cdot) = p(0 \cdot 0 \cdot)$$

and
$$p(1 \cdot 0 \cdot) = p(0 \cdot 1 \cdot).$$

So solving, we obtain
$$4p(1 \cdot 1 \cdot) - 1 = \frac{\sqrt{3}}{2}$$

and
$$p(1 \cdot 1 \cdot) = -\frac{\sqrt{3}}{8} + \frac{1}{4}$$
$$p(1 \cdot 0 \cdot) = \frac{\sqrt{3}}{8} + \frac{1}{4}.$$

Similarly for $\mathbf{A'B'} = -\sqrt{3}/2$:

$$P(\mathbf{A} = 1, \mathbf{B'} = 1) = p(\cdot 1 \cdot 1) = -\frac{\sqrt{3}}{8} + \frac{1}{4}$$
$$p(\cdot 1 \cdot 0) = \frac{\sqrt{3}}{8} + \frac{1}{4}.$$

Next $\mathbf{AB'} = 1/2$, so

$$4p(1 \cdot \cdot 1) - 1 = -\frac{1}{2}$$
$$p(1 \cdot \cdot 1) = \frac{1}{8}$$
$$p(1 \cdot \cdot 0) = \frac{3}{8}.$$

Since $\mathbf{A'B} = -1$

$$4p(\cdot\,1\,1\,\cdot) - 1 = -1$$
$$p(\cdot\,1\,1\,\cdot) = 0$$
$$p(\cdot\,1\,0\,\cdot) = \frac{1}{2}.$$

Since each of the four measurements $\mathbf{A}$, $\mathbf{A'}$, $\mathbf{B}$, and $\mathbf{B'}$ has value ± 1, there are 16 atoms, i.e., atomic events, in our upper probability space Ω. There are not simple elementary probability arguments of the kind we have just been following, to compute the upper probability of these atoms. The reason is simple; the main probabilistic law that must be preserved is the subadditivity of upper probabilities, expressed as Axiom 3 of Definition 10.1. This axiom is, of course, weaker than the standard additivity axiom. If we held onto the standard additivity and use the methods just used for computing probabilities of correlations, we would have atoms with negative probabilities, the sort of thing that happens in quantum mechanics when using the Wigner distribution for position and momentum of a single particle (for details on this, see Suppes 1961).

So, to make what could easily be a longer story short, here are the upper probabilities for the axioms. Since $E(\mathbf{A}) = E(\mathbf{A'}) = E(\mathbf{B}) = E(\mathbf{B'}) = 0$, by symmetry we need find only 8. Here is the list.

$$p^*(1\,1\,1\,1) = p^*(0\,0\,0\,0) = 0$$
$$p^*(1\,1\,1\,0) = p^*(0\,0\,0\,1) = \frac{1}{16}$$
$$p^*(1\,1\,0\,1) = p^*(0\,0\,1\,0) = \frac{1}{8}$$
$$p^*(1\,1\,0\,0) = p^*(0\,0\,1\,1) = \frac{1}{8} + \frac{\sqrt{3}}{8}$$
$$p^*(1\,0\,1\,0) = p^*(0\,1\,0\,1) = \frac{1}{8} - \frac{\sqrt{3}}{8}$$
$$p^*(1\,0\,1\,1) = p^*(0\,1\,0\,0) = \frac{1}{8}$$
$$p^*(1\,0\,0\,0) = p^*(0\,1\,1\,1) = \frac{1}{16}$$
$$p^*(1\,0\,0\,1) = p^*(0\,1\,1\,0) = 0$$

Note that the upper probabilities are non-negative and not greater than 1. What makes them as a whole upper probabilities, not standard probabilities, is that the sum of the 16 is greater than 1:

$$\sum_{i,j,k,l=0}^{1} p^*(i,j,k,l) = 1 + \frac{3}{8}$$

We now verify that each of the correlation probabilities computed earlier satisfy the subadditivity for the four of the 16 atoms that define it as an event. To simplify notation further, in showing these computations, we replace p^* (1 1 0 0), e.g., by 1 1 0 0. So all the following inequalities are really about upper probabilities, but "p^*" has been deleted.

$$\mathbf{AB}: p(1 \cdot 0 \cdot) = \frac{1}{4} + \frac{\sqrt{3}}{8} \leq 1\,1\,0\,1 + 1\,1\,0\,0 + 1\,0\,0\,1 + 1\,0\,0\,0$$

$$\leq \frac{1}{8} + \left(\frac{1}{8} + \frac{\sqrt{3}}{8}\right) + \frac{1}{16} + \frac{1}{16} \leq \frac{3}{8} + \frac{\sqrt{3}}{8}$$

$$p(1 \cdot 1 \cdot) = -\frac{1}{4} - \frac{\sqrt{3}}{8} \leq 1\,1\,1\,1 + 1\,1\,1\,0 + 1\,0\,1\,1 + 1\,0\,1\,0$$

$$\leq 0 + \frac{1}{16} + \frac{1}{8} + \left(\frac{1}{8} - \frac{\sqrt{3}}{8}\right) \leq \frac{5}{16} - \frac{\sqrt{3}}{8}.$$

$$\mathbf{AB'}: p(1 \cdot\cdot 0) = \frac{3}{8} \leq 1\,1\,1\,0 + 1\,1\,0\,0 + 1\,0\,1\,0 + 1\,0\,0\,0$$

$$\leq \frac{1}{16} + \left(\frac{1}{8} + \frac{\sqrt{3}}{8}\right) + \frac{1}{8} - \left(\frac{\sqrt{3}}{8} + \frac{1}{16}\right) \leq \frac{3}{8}$$

$$p(1 \cdot\cdot 1) = \frac{1}{8} \leq 1\,1\,1\,1 + 1\,1\,0\,1 + 1\,0\,1\,1 + 1\,0\,0\,1$$

$$\leq 0 + \frac{1}{8} + \frac{1}{8} + \frac{1}{16} \leq \frac{5}{16}.$$

$$\mathbf{A'B}: p(\cdot 1\,0\,\cdot) = \frac{1}{2} \leq 1\,1\,0\,1 + 1\,1\,0\,0 + 0\,1\,0\,1 + 0\,1\,0\,0$$

$$\leq \frac{1}{8} + \left(\frac{1}{8} + \frac{\sqrt{3}}{8}\right) + \left(\frac{1}{8} - \frac{\sqrt{3}}{8}\right) + \frac{1}{8} \leq \frac{1}{2}$$

$$p(\cdot 1\,1\,\cdot) = 0 \leq 1\,1\,1\,1 + 1\,1\,1\,0 + 0\,1\,1\,1 + 0\,1\,1\,0$$

$$\leq 0 + \frac{1}{16} + \frac{1}{16} + \frac{1}{16} \leq \frac{3}{16}.$$

$$\mathbf{A'B'}: p(\cdot 1 \cdot 0) = \frac{1}{4} + \frac{\sqrt{3}}{8} \leq 1\,1\,1\,0 + 1\,1\,0\,0 + 0\,1\,1\,0 + 0\,1\,0\,0$$

$$\leq \frac{1}{16} + \left(\frac{1}{8} + \frac{\sqrt{3}}{8}\right) + \frac{1}{16} + \frac{1}{8} \leq \frac{3}{8} + \frac{\sqrt{3}}{8}$$

$$p(\cdot 1 \cdot 1) = -\frac{1}{4} - \frac{\sqrt{3}}{8} \leq 1111 + 1101 + 0111 + 0101$$

$$\leq 0 + \frac{1}{8} + \frac{1}{16} + \left(\frac{1}{8} - \frac{\sqrt{3}}{8}\right) \leq \frac{5}{16} - \frac{\sqrt{3}}{8}.$$

10.2 The Decay of the EPR State and the Existence of a Joint Distribution

We calculate the time evolution of the EPR state

$$|EPR> = \frac{1}{\sqrt{2}}(|01> - |10>) \tag{10.4}$$

under the influence of decoherence. The decaying state will not stay pure, but become mixed in the course of time. We therefore calculate the corresponding initial density operator and $P(0) := |EPR><EPR|$ obtain

$$P(0) = \frac{1}{2}(|01><01| + |10><10|) - \frac{1}{2}(|01><10| + |10><01|) \tag{10.5}$$

There are many different ways to model the influence of decoherence on a quantum system described by a quantum state (Schlosshauer 2007). Here we focus on the master equation approach that is popular in quantum optics. According to this approach, a quantum state couples to an environment ("heat bath"), which is modeled as an infinite collection of harmonic oscillators. This way of modeling decoherence takes into account that a quantum system can never be shielded from its environment (Zeh 1973). Note that due to the coupling of the quantum system to the environment, the entanglement of the quantum system in question diffuses into the environment and the reduced state of the system becomes less and less entangled. This reduced state of the quantum system can be obtained by a procedure called "tracing out" the environment variables. This procedure can be justified by noting that nothing is known about the environment and so it is appropriate to take a statistical average. Finally, one obtains a master equation for the reduced state P of our 2-atom system,

$$\frac{\partial}{\partial t} P(t) = -\frac{k}{2} \sum_{i=1}^{2} \left[\sigma_+^{(i)} \sigma_-^{(i)} P(t) + P(t) \sigma_+^{(i)} \sigma_-^{(i)} - 2\sigma_-^{(i)} P(t) \sigma_+^{(i)} \right] =: L\, P(t), \tag{10.6}$$

with the damping constant k. $\sigma_\pm^{(i)}$ are the raising and lowering operator acting on atom i. These operators can be expressed in terms of the Pauli matrices σ_1 and σ_2:

$$\sigma_\pm = \frac{1}{2}(\sigma_1 \pm i\sigma_2)$$

with

$$\sigma_1 = \begin{pmatrix} 0 & 1 \\ 1 & 0 \end{pmatrix}, \quad \sigma_2 = \begin{pmatrix} 0 & -i \\ i & 0 \end{pmatrix}$$

Eq. 10.6 can be formally solved:

$$P(t) = e^{Lt} P(0) \tag{10.7}$$

Using Eqs. 10.5 and 10.6 and after some algebra (see Hartmann 2009), we obtain the time evolution of the EPR state,

$$P(\tau) = e^{-\tau} P(0) + (1 - e^{-\tau}) |00><00| \tag{10.8}$$

with the normalized time parameter $\tau = kt$.

We see that the quantum system under consideration asymptotically reaches the ground state $|00><00|$.

Let us now study the correlation that the decaying quantum state exhibits. In order to connect to the discussion in Section 10.1, we focus on the following four observables:

$$\mathbf{A} = \sigma_1^{(1)} \tag{10.9a}$$
$$\mathbf{A}' = \sigma_1^{(1)} \cos(\alpha) + \sigma_2^{(1)} \sin(\alpha) \tag{10.9b}$$
$$\mathbf{B} = \sigma_1^{(2)} \cos(\alpha) + \sigma_2^{(2)} \sin(\alpha) \tag{10.9c}$$
$$\mathbf{B}' = \sigma_1^{(2)} \cos(\beta) + \sigma_2^{(2)} \sin(\beta) \tag{10.9d}$$

Note that $\mathbf{A}$ and $\mathbf{A}'$ act only on particle 1 and $\mathbf{B}$ and $\mathbf{B}'$ act only on particle 2. Clearly, the expectation values of $\mathbf{A}$, $\mathbf{A}'$, $\mathbf{B}$ and $\mathbf{B}'$ in $P(t)$ all vanish for all times τ:

$$<\mathbf{A}> = <\mathbf{A}'> = <\mathbf{B}> = <\mathbf{B}'> = 0$$

However, the two-particle correlations $<\mathbf{A}\mathbf{B}>$, $<\mathbf{A}\mathbf{B}'>$, $<\mathbf{A}'\mathbf{B}>$ and $<\mathbf{A}'\mathbf{B}'>$ do not vanish. We calculate the expectation values of these operators for the state $P(\tau)$.

$$<\mathbf{A}\mathbf{B}> = -e^{-\tau} \cos(\alpha) \tag{10.10a}$$
$$<\mathbf{A}\mathbf{B}'> = -e^{-\tau} \cos(\beta) \tag{10.10b}$$
$$<\mathbf{A}'\mathbf{B}> = -e^{-\tau} \tag{10.10c}$$
$$<\mathbf{A}'\mathbf{B}'> = -e^{-\tau} \cos(\alpha - \beta) \tag{10.10d}$$

We expect that these correlations can be derived from a joint probability distribution for sufficiently large τ, i.e., when the state is sufficiently decayed. But when precisely is a description of the correlations in terms of a joint probability distribution possible? This question is addressed by the Clauser-Horne-Shimony-Holt inequalities (see Eq. 10.3). To be more specific, let $\alpha = 30°$ and $\beta = 60°$, the example introduced in Section 10.1. It turns out that inequalities (10.3a), (10.3b) and (10.3d) are always satisfied. However, inequality (10.3c) leads to

$$e^{-\tau} \leq \frac{4}{2\sqrt{3}+1}, \qquad (10.11)$$

or $\tau > .1$.

We see that a "classical" description of the correlations is possible already after a very short period of time (in units of k^{-1}).

Instead of the calculation of (10.9) for the decay of the quantum mechanical theoretical correlations, we now compute the upper-probability correlations from the upper-probability values of the 16 atoms. We will label these correlations $<A B>^*$ (superscript for upper).

So, here are the calculations of the four upper correlations.

$<AB>^*: p^*(1 \cdot 1 \cdot) = p^*(1\,1\,1\,1) + p^*(1\,1\,1\,0) + p^*(1\,0\,1\,1) + p^*(1\,0\,1\,0)$

$$= 0 + \frac{1}{16} + \frac{1}{8} + \frac{1}{8} - \frac{\sqrt{3}}{8} = \frac{5}{16} - \frac{\sqrt{3}}{8}$$

$p^*(1 \cdot 0 \cdot) = p^*(1\,1\,0\,1) + p^*(1\,1\,0\,0) + p^*(1\,0\,0\,1) + p^*(1\,0\,0\,0)$

$$= \frac{1}{8} + \frac{1}{8} + \frac{\sqrt{3}}{8} + 0 + \frac{1}{16} = \frac{5}{16} + \frac{\sqrt{3}}{8}$$

so $<AB>^* = 2\left[\frac{5}{16} - \frac{\sqrt{3}}{8} - \left(\frac{5}{16} + \frac{\sqrt{3}}{8}\right)\right] = -\frac{\sqrt{3}}{2}.$

$<AB'>^*: p^*(1 \cdot \cdot 1) = p^*(1\,1\,1\,1) + p^*(1\,1\,0\,1) + p^*(1\,0\,1\,1) + p^*(1\,0\,0\,1)$

$$= 0 + \frac{1}{8} + \frac{1}{8} + 0 = \frac{1}{4}.$$

$p^*(1 \cdot \cdot 0) = p^*(1\,1\,1\,0) + p^*(1\,1\,0\,0) + p^*(1\,0\,1\,0) + p^*(1\,0\,0\,0)$

$$= \frac{1}{16} + \frac{1}{8} + \frac{\sqrt{3}}{8} + \frac{1}{8} - \frac{\sqrt{3}}{8} + \frac{1}{16} = \frac{3}{8}$$

so $<AB'>^* = 2\left[\frac{1}{4} - \frac{3}{8}\right] = -\frac{1}{4}.$

$<A'B>^*: p^*(\cdot\,1\,1\,\cdot) = p^*(1\,1\,1\,1) + p^*(1\,1\,1\,0) + p^*(0\,1\,1\,1) + p^*(0\,1\,1\,0)$

$$= 0 + \frac{1}{16} + \frac{1}{16} + 0 = \frac{1}{8}.$$

$p^*(\cdot\,1\,0\,\cdot) = p^*(1\,1\,0\,1) + p^*(1\,1\,0\,0) + p^*(0\,1\,0\,1) + p^*(0\,1\,0\,0)$

$$= \frac{1}{8} + \frac{1}{8} + \frac{\sqrt{3}}{8} + \frac{1}{8} - \frac{\sqrt{3}}{8} + \frac{1}{8} = \frac{1}{2}$$

so $<A'B>^* = 2\left[\frac{1}{8} - \frac{1}{2}\right] = -\frac{3}{4}.$

$$\begin{aligned}
< A'B' >^* : p^*(\cdot 1 \cdot 1) &= p^*(1\,1\,1\,1) + p^*(1\,1\,0\,1) + p^*(0\,1\,1\,1) + p^*(0\,1\,0\,1) \\
&= 0 + \frac{1}{8} + \frac{1}{16} + \frac{1}{8} - \frac{\sqrt{3}}{8} = \frac{5}{16} - \frac{\sqrt{3}}{8} \\
p^*(\cdot 1 \cdot 0) &= p^*(1\,1\,1\,0) + p^*(1\,1\,0\,0) + p^*(0\,1\,1\,0) + p^*(0\,1\,0\,0) \\
&= \frac{1}{16} + \frac{1}{8} + \frac{\sqrt{3}}{8} + 0 + \frac{1}{8} = \frac{5}{16} + \frac{\sqrt{3}}{8}
\end{aligned}$$

so $< A'B' >^* = 2\left[\frac{5}{16} - \frac{\sqrt{3}}{8} - \left(\frac{5}{16} + \frac{\sqrt{3}}{8}\right)\right] = -\frac{\sqrt{3}}{2}.$

Putting these upper correlations into inequality (10.3c), we get

$$-\frac{\sqrt{3}}{2} + \frac{1}{4} - \frac{3}{4} - \frac{\sqrt{3}}{2} = -\frac{1}{2} - \sqrt{3} < -2.$$

Applying then the same decay rate $e^{-\tau}$, we have

$$(-\frac{1}{2} - \sqrt{3})e^{-\tau} \leq -2,$$

so $e^{-\tau} \leq \dfrac{4}{2\sqrt{3}+1}.$

exactly the same inequality for the decay time as was obtained earlier for the quantum mechanical computation. Yet the two methods are not identical. It is clear that the two proper joint probability distributions at the time $e^{-\tau} = \frac{4}{1+2\sqrt{3}}$ are close but not exactly the same.

References

Clauser JF, Horne MA, Shimony A, Holt RA (1969) Proposed experiment to test local hidden-variable theories. Phys Rev Lett 23:880–884

Hartmann S (2009) Quantum optical master equations for Z atoms: The use of superoperator methods. Unpublished manuscript, Tilburg University

Schlosshauer M (2007) Decoherence and the quantum-to-classical transition. Springer, Berlin/Heidelberg/New York

Suppes P (1961) Probabilistic concepts in quantum mechanics. Philos Sci 28(4):378–389

Suppes P (1974) The measurement of belief. J Roy Stat Soc Ser B 36:160–191

Suppes P, de Barros JA (2007) Quantum mechanics and the brain. Quantum interaction: Papers from the AAAI spring symposium, Technical Report SS-07-08. AAAI Press, Menlo Park, CA, pp 75–82

Suppes P, Zanotti M (1981) When are probabilistic explanations possible? Synthese 48:191–199

Suppes P, Zanotti M (1989) Conditions on upper and lower probabilities to imply probabilities. Erkenntnis 31:323–345
Suppes P, Zanotti M (1991) Existence of hidden variables having only upper probabilities. Found Phys 21:1479–1499
Walley P (1991) Statistical reasoning with imprecise probabilities. Chapman & Hall, London/New York
Zeh HD (1973) Toward a quantum theory of observation. Found Phys 3:109–116

Chapter 11
Gauge Symmetry and the Theta-Vacuum

Richard Healey

11.1 Two Kinds of Symmetry

Abstractly, a symmetry of a structure is an automorphism – a transformation that maps the elements of an object back onto themselves so as to preserve the structure of that object.

A physical theory specifies a set of models – mathematical structures – that may be used to represent various different situations, actual as well as merely possible, and to make claims about them. Any application of a physical theory is to a situation involving some system, actual or merely possible. Only rarely is that system the entire universe: typically, one applies a theory to some subsystem, regarded as a relatively isolated part of its world. The application proceeds by using the theory to model the situation of that subsystem in a way that abstracts from and idealizes the subsystem's own features, and also neglects or idealizes its interactions with the rest of the world.

We can therefore enquire about the symmetries of the class of models of a theory; or we can enquire about the symmetries of a class of situations, whether or not we have in mind a theory intended to model them. The first enquiry may reveal some *theoretical symmetry*: the second may reveal some *empirical symmetry*. An empirical symmetry can be recognized even without a physical theory to account for it. But it does not cease to be empirical if and when such a theory becomes available. A theory may entail an empirical symmetry.

Galilei (1967, pp. 186–7) illustrated his relativity principle by describing a famous empirical symmetry of this kind.

> Shut yourself up with some friend in the main cabin below decks on some large ship, and have with you there some flies, butterflies and other small flying animals... When you have observed all these things carefully..., have the ship proceed with any speed you like, so long as the motion is uniform and not fluctuating this way and that. You will discover not the least change in all the effects named, nor could you tell from any of them whether the ship was moving or standing still.

R. Healey (✉)
Philosophy Department, University of Arizona, 213 Social Sciences, Tucson,
Arizona 85721-0027, USA
e-mail: rhealey@email.Arizona.edu

His implicit claim is that a situation inside the cabin when the ship is in motion is indistinguishable from another situation inside the cabin when the ship is at rest by observations confined to those situations. The claim follows from a principle of the relativity of all uniform horizontal motion. While we know today that an unqualified form of Galileo's claim is false, in a modified form it continues to play an important role in physics.

Galileo's implicit claim is that situations related by a uniform collective horizontal motion are empirically symmetrical. Specifically

> A 1-1 mapping $\varphi : \mathscr{S} \to \mathscr{S}$ of a set of situations onto itself is an *empirical symmetry* if and only if any two situations related by φ are indistinguishable by means of measurements confined to each situation.

A measurement is confined to a situation just in case it is a measurement of intrinsic properties of (one or more objects in) that situation. Note that the reference to measurement is not superfluous here, in so far as a situation may feature unmeasurable intrinsic properties. If every function $\varphi \in \Phi$ is an empirical symmetry of $\mathscr{S}$, then $\mathscr{S}$ is symmetric under Φ-transformations. Note that situations in $\mathscr{S}$ related by a transformation φ may be in the same or different possible worlds: if φ is an empirical symmetry, then $\varphi(s)$ may be in the same world w as s, but only if w is itself sufficiently symmetric.

One may distinguish symmetries of the set of situations to which a theory may be applied from symmetries of the set of the theory's models.

> A 1-1 mapping $f : \mathscr{M} \to \mathscr{M}$ of the set of models of a theory Θ onto itself is a *theoretical symmetry of* Θ if and only if the following condition obtains: For every model m of Θ that may be used to represent (a situation s in) a possible world w, $f(m)$ may also be used to represent (s in) w.

If every function $f \in F$ is a symmetry of Θ, then Θ is symmetric under F-transformations. Theoretical symmetries may be purely formal features of a theory, if all they do is to relate different but equivalent ways the theory has of representing one and the same empirical situation. One model may be more conveniently applied to a given situation than another model related to it by a theoretical symmetry, but the theoretical as well as empirical content of any claim made about that situation will be the same no matter which model is applied. But a theoretical symmetry of a theory may entail a corresponding empirical symmetry, in which case it is not a purely formal feature of the theory.

The empirical symmetry associated with uniform velocity boosts in special relativity is a consequence of a theoretical symmetry of special relativity, if one associates each model of that theory with an inertial frame with respect to which a given situation is represented. For then the empirical symmetry becomes a consequence of the Lorentz invariance of the theory – the fact that the Lorentz transform of any model is also a model of the theory. The Lorentz transform of any model may be used to represent the same situation as the original model (from the perspective of a boosted inertial frame); but it may also be used to represent a boosted *duplicate* of that situation (from the perspective of the original frame). (Here a duplicate of a situation is a situation that shares all its intrinsic properties.) The special theory of relativity entails the empirical symmetry associated with Lorentz invariance by implying that these empirically equivalent situations are not merely empirically indistinguishable by means of measurements confined to those situations, but indistinguishable by reference to any intrinsic properties or relations of entities each involves.

Relativity principles assert empirical symmetries. If local gauge transformations reflect some similar empirical symmetry, then they also represent distinct but indistinguishable situations. But I shall defend the conventional wisdom that the successful employment of Yang-Mills theories warrants the conclusion that local gauge transformations are only theoretical symmetries of these theories that reflect no corresponding non-trivial empirical symmetries among the situations they represent. Local gauge symmetry is a purely formal feature of these theories.

11.2 Warm-Up Exercise: Faraday's Cube

Michael Faraday constructed a hollow cube with sides 12 feet long, covered it with good conducting materials but insulated it carefully from the ground, and electrified its exterior to such an extent that sparks flew from its surface. He made the following entry in his diary in 1836:

"I went into this cube and lived in it, but though I used lighted candles, electrometers, and all other tests of electrical states, I could not find the least influence on them". (Maxwell (1881, p. 53))

Both Faraday and Galileo described observations of symmetries in nature. In each case, different situations are compared, and it is noted that these are indistinguishable with respect to a whole class of phenomena. But while velocity boosts are paradigm empirical symmetries, gauge symmetry is usually taken to be a purely formal feature of a theory. In this case, adding the same constant to all electrical potentials is a symmetry of classical electromagnetism. Why doesn't Faraday's cube provide a perfect analogue of Galileo's ship for local gauge symmetry? (Note that the electric potential transformation $\varphi \to \varphi + a$ is an example of a local gauge transformation $A_\mu \to A_\mu + \partial_\mu \Lambda$ with $A_\mu = (\varphi, -\mathbf{A})$ and $\Lambda = at$.)

There is an important disanalogy between the Lorentz boost symmetry imperfectly illustrated by Galileo's ship and the local gauge symmetry illustrated by Faraday's cube. While both are theoretical symmetries of the relevant theories, only in the former case does this theoretical symmetry imply a corresponding empirical symmetry.

In order that charging the exterior of Galileo's cube should provide an example of the relevant kind of empirical symmetry, two conditions must be satisfied. It must produce a situation inside the cube that differs from its situation when uncharged in a way that corresponds to performing a local gauge transformation on its interior. But despite this difference, the transformed situation must remain internally indistinguishable from the original situation.

To see how it might be possible to meet both conditions, consider the analogous case of a Lorentz-boosted (space!) ship. Even though the situation inside the ship is a perfect duplicate of its situation before boosting, the theory itself implies that these situations are related by a boost transformation: because the only theoretical models that represent *both situations at once* are models in which the two situations are related by a velocity boost.

But classical electromagnetic theory has no analogous implication in the case of Galileo's cube. It contains models, each of which represents the cube both before and after charging, that represent the cube's interior as being in exactly the same state, independent of the charge on its exterior! There is no theoretical or experimental reason to suppose that charging the cube's exterior does anything to alter the electromagnetic state of its interior. Charging the exterior of Faraday's cube is not a way of performing a local gauge transformation on its interior: it is no more effective than painting it blue, or simply waiting for a day! (See Healey 2009, for further discussion of this case.)

11.3 The θ-Vacuum

The ground state of a quantized non-Abelian Yang-Mills gauge theory is usually described by a real-valued parameter θ – a fundamental new constant of nature. The structure of this vacuum state is often said to arise from a degeneracy of the vacuum of the corresponding classical theory. The degeneracy allegedly follows from the fact that "large" (but not "small") local gauge transformations connect physically distinct states of zero field energy. In a classical non-Abelian Yang-Mills gauge theory, "large" gauge transformations apparently connect models of distinct but indistinguishable situations. This seems to show that at least "large" local gauge symmetry is an empirical symmetry.

In clarifying the distinction between "large" and "small" gauge transformations we will be driven to a deeper analysis of the significance of gauge symmetry. But understanding the θ-vacuum will require refining, not abandoning, the thesis that local gauge symmetry is a purely theoretical symmetry.

Before moving to the quantum theory, consider a classical SU(2) Yang-Mills gauge theory with action

$$S = \frac{1}{2g^2} \int Tr(\mathbf{F}_{\mu\nu}\mathbf{F}^{\mu\nu}) d^4x \qquad (11.1)$$

where $\quad \mathbf{F}_{\mu\nu} = \partial_\mu \mathbf{A}_\nu - \partial_\nu \mathbf{A}_\mu + [\mathbf{A}_\mu, \mathbf{A}_\nu]$
and $\quad \mathbf{A}_\mu = A_\mu^j \frac{\sigma_j}{2i} \quad$ transform as

$$\mathbf{A}_\mu \to \mathbf{A}'_\mu = \mathbf{U}\mathbf{A}_\mu \mathbf{U}^\dagger + (\partial_\mu \mathbf{U})\mathbf{U}^\dagger, \ \mathbf{F}_{\mu\nu} \to \mathbf{U}\mathbf{F}_{\mu\nu}\mathbf{U}^\dagger \qquad (11.2)$$

under a local gauge transformation $\mathbf{U}(\mathbf{x}, t)$. (Here σ_j ($j = 1, 2, 3$) are Pauli spin matrices.)

11 Gauge Symmetry and the Theta-Vacuum

The field energy is zero if $\mathbf{F}_{\mu\nu} = 0$: that condition is consistent with $\mathbf{A}_\mu = 0$ and gauge transforms of this. Now restrict attention to those gauge transformations for which $\mathbf{A}'_0 = 0$, $\partial_0 \mathbf{A}'_j = 0$ i.e.,

$$\mathbf{A}_\mu = 0 \to \mathbf{A}'_j(\mathbf{x}) = \{\partial_j \mathbf{U}(\mathbf{x})\} \mathbf{U}^\dagger(\mathbf{x}), \quad \mathbf{A}'_0 = 0 \tag{11.3}$$

These are generated by functions $\mathbf{U} : \mathbb{R}^3 \to SU(2)$.

Those functions that satisfy $\mathbf{U}(\mathbf{x}) \to \mathbf{1}$ for $|\mathbf{x}| \to \infty$ constitute smooth maps $\mathbf{U} : S^3 \to SU(2)$, where S^3 is the 3-sphere. Some of these may be continuously deformed into the identity map $\mathbf{U}(\mathbf{x}) = \mathbf{1}$. But others cannot be. The maps divide into a countable set of equivalence classes, each characterized by an element of the homotopy group $\pi_3(SU(2)) = \mathbb{Z}$ called the *winding number*.

Maps in the same equivalence class as the identity map are said to generate "small" local gauge transformations: these are taken to relate alternative representations of the same classical vacuum. But $\mathbf{A}'_\mu$, $\mathbf{A}''_\mu$ generated from $\mathbf{A}_\mu = 0$ by maps $\mathbf{U}(\mathbf{x})$ from different equivalence classes are often said to represent *distinct* classical vacua, and $\mathbf{A}'_\mu$, $\mathbf{A}''_\mu$ are said to be related by "large" gauge transformations. (It is important to distinguish this claim from the quite different proposition, according to which degenerate *quantum* vacua may be related by a *global* gauge transformation in cases of spontaneous symmetry breaking. We are concerned at this point with a possible degeneracy in the *classical* vacuum of a non-Abelian Yang-Mills gauge theory.)

But if local gauge symmetry is a purely formal feature of a theory, then a gauge transformation cannot connect representations of physically distinct situations, even if it is "large"! And yet, textbook discussions of the *quantum* θ-vacuum typically represent this by a superposition of states, each element of which is said to correspond to a distinct state from the degenerate classical vacuum.

11.4 Two Analogies

Such discussions frequently appeal to a simple analogy from elementary quantum mechanics. Consider a particle moving in a one-dimensional periodic potential of finite height, like a sine wave. Classically, the lowest energy state is infinitely degenerate: the particle just sits at the bottom of one or other of the identical wells in the potential. But quantum mechanics permits tunnelling between neighboring wells, which removes the degeneracy. In the absence of tunnelling, there would be a countably infinite set of degenerate ground states of the form $\psi_n(x) = \psi_0(x - na)$ where a is the period of the potential. These are related by the translation operator $\hat{T}_a$: $\hat{T}_a \psi(x) = \psi(x - a)$. $\hat{T}_a$ is unitary and commutes with the Hamiltonian $\hat{H}$. Hence there are joint eigenstates $|\theta\rangle$ of $\hat{H}$ and $\hat{T}_a$ satisfying $\hat{T}_a |\theta\rangle = \exp(i\theta) |\theta\rangle$.

Such a state has the form

$$|\theta\rangle = \sum_{n=-\infty}^{+\infty} \exp\{-in\theta\} |n\rangle \tag{11.4}$$

where $\psi_n(x)$ is the wave function of state $|n\rangle$. When tunnelling is allowed for, the energy of these states depends on the parameter $\theta \in [0, 2\pi)$. It is as if quantum tunneling between the distinct classical ground states has removed the degeneracy, resulting in a spectrum of states of different energies parametrized by θ, each corresponding to a different superposition of classical ground states.

An alternative analogy is provided by a charged pendulum swinging from a long, thin solenoid whose flux Φ is generating a static Aharonov-Bohm potential $\mathbf{A}$. The Hamiltonian is

$$\hat{H} = \frac{1}{2m}[-i(\nabla - ie\mathbf{A})]^2 + V \tag{11.5}$$

With a natural "tangential" choice of gauge for $\mathbf{A}$ this becomes

$$\hat{H} = -\frac{1}{2ml^2}\left(\frac{d}{d\omega} - ielA\right)^2 + V(\omega) \tag{11.6}$$

where the pendulum has mass m, charge e, length l and angle coordinate ω. If the wave function is transformed according to

$$\psi(\omega) = \exp\left[ie\int_0^\omega lAd\omega'\right]\varphi(\omega) \tag{11.7}$$

then the transformed wave function satisfies the Schrödinger equation with simplified Hamiltonian

$$\hat{H}_\varphi = -\frac{1}{2m}\frac{d^2}{d\omega^2} + V(\omega) \tag{11.8}$$

The boundary condition $\psi(\omega + 2\pi) = \psi(\omega)$ now becomes

$$\varphi(\omega + 2\pi) = \exp\{-ie\Phi\}\varphi(\omega) \tag{11.9}$$

which is of the same form as in the first analogy: $\hat{T}_{2\pi}\varphi = \exp\{i\theta\}\varphi$, with $\theta = -e\Phi$.

Unlike the periodic potential, the charged pendulum features a *unique* classical ground state. The potential barrier that would have to be overcome to "flip" the pendulum over its support can be tunnelled through quantum mechanically, but the tunnel ends up back where it started from! This produces a θ-dependent ground state energy as in the analogy of the periodic potential. But in this case there is a *single* state corresponding to an *external* parameter θ rather than a spectrum of states labeled by an internal parameter θ.

Which is the better analogy? Is the θ-vacuum in a quantized non-Abelian gauge theory more like a quantum state of the periodic potential, or a state of the charged quantum pendulum?

Rubakov (2002) describes both analogies. He notes that vacua of a classical Yang-Mills gauge theory related by a "large" gauge transformation are topologically inequivalent, since their so-called Chern-Simons numbers are different. The Chern-Simons number n_{CS} associated with potential $\mathbf{A}_\mu$ is defined as follows:

$$n_{CS}\left(\mathbf{A}_\mu\right) \equiv \frac{1}{16\pi^2} \int d^3\mathbf{x}\epsilon^{ijk} \left(A_i^a \partial_j A_k^a + \frac{1}{3}\epsilon^{abc} A_i^a A_j^b A_k^c \right) \quad (11.10)$$

and if $\mathbf{A}_\mu'', \mathbf{A}_\mu'$ are related by a "large" gauge transformation of the form (11.3) with winding number n, then $n_{CS}\left(\mathbf{A}_\mu''\right) = n_{CS}\left(\mathbf{A}_\mu'\right) + n$. But in a semi-classical treatment, quantum tunneling between them is possible through quantum tunneling. This suggests that the classical vacua are indeed distinct, and that a "large" gauge transformation represents a change from one physical situation to another. If so, symmetry under "large" gauge transformations is not just a theoretical symmetry but reflects an empirical symmetry of a non-Abelian Yang-Mills gauge theory. This favors the first analogy.

But Rubakov then goes on to offer an alternative (but allegedly equivalent!) perspective, when he says (p. 277):

> From the point of view of gauge-invariant quantities, topologically distinct classical vacua are equivalent, since they differ only by a gauge transformation. Let us identify these vacua. Then the situation becomes analogous to the quantum-mechanical model of the pendulum.

From this perspective, even "large" gauge transformations lead from a single classical vacuum state back into an alternative representation of that same state! Is this perspective legitimate? If it is, how can it be equivalent to a view according to which a "large" gauge transformation represents an empirical transformation between distinct states of a non-Abelian Yang-Mills gauge theory?

11.5 Are "Large" Gauge Transformations Empirical?

Consider first a purely classical non-Abelian Yang-Mills gauge theory. If it has models that represent distinct degenerate classical vacua, what is the physical difference between these vacua? Models related by a "large" gauge transformation are characterized by different Chern-Simons numbers, and one might take these to exhibit a difference in the intrinsic properties of situations they represent. But it is questionable whether the Chern-Simons number of a gauge configuration represents an intrinsic property of that configuration, even if a *difference* in Chern-Simons number represents an intrinsic *difference* between gauge configurations. Perhaps Chern-Simons numbers are like velocities in models of special relativity. The velocity assigned to an object by a model of special relativity does not represent an intrinsic property of that object, even though that theory does distinguish in its models between situations involving objects moving with different *relative* velocities. As we saw, it is this latter distinction that proves critical to establishing that Lorentz boosts are empirical symmetries of situations in a special relativistic world.

So does a *difference* in Chern-Simons number represent an intrinsic *difference* between classical vacua in a purely classical non-Abelian Yang-Mills gauge theory? I see no reason to believe that it does. There might be a reason if the theory included models representing *more than one* vacuum state at once, where the distinct vacua were represented by different Chern-Simons numbers in *every* such model. Such distinct vacua extend over all space. So they could all be represented within a single model only if it represented them as occurring at different times. But topologically distinct vacua are separated by an energy barrier, and in the purely classical theory this cannot be overcome. So there is no representation within a single model of the purely classical theory of vacua with different Chern-Simons numbers. That is why I see no reason to believe that a "large" gauge transformation represents an empirical transformation between distinct vacuum states of a purely classical non-Abelian Yang-Mills gauge theory.

According to a semi-classical theory, vacua with different Chern-Simons numbers *can* be connected by tunnelling through the potential barrier that separates them. So such a theory can model a single situation involving more than one such vacuum state, each obtaining at a different time. Moreover, no model of this theory represents these states as having the *same* Chern-Simons numbers. Perhaps this justifies the conclusion that in a world truly described by such a theory a "large" gauge transformation *would* represent an empirical transformation between distinct vacuum states. But we do not live in such a world.

The θ-vacuum of a fully quantized non-Abelian Yang-Mills gauge theory is non-degenerate and symmetric under "large" as well as "small" gauge transformations. Analogies with the periodic potential and quantum pendulum suggest that it be expressed in the form

$$|\theta\rangle = \sum_{n=-\infty}^{+\infty} \exp\{-in\theta\} |n\rangle \quad (11.11)$$

where state $|n\rangle$ corresponds to a classical state with Chern-Simons number n. But not only the θ-vacuum but the whole theory is symmetric under "large" gauge transformations. So a generator $\hat{U}$ of "large" gauge transformations commutes not only with the Hamiltonian but with all observables. It acts as a so-called "superselection operator" that separates the large Hilbert space of states into distinct superselection sectors, between which no superpositions are possible. Physical states are therefore restricted to those lying in a single superselection sector of the entire Hilbert space. Hence every physical state of the theory, including $|\theta\rangle$, is an eigenstate of $\hat{U}$.

Now there is an operator $\hat{U}_1$ corresponding to a "large" gauge transformation with winding number 1,

$$\hat{U}_1 |n\rangle = |n+1\rangle \quad (11.12)$$

from which it follows that none of the states $|n\rangle$ is a physical state of the theory! This theory cannot model situations involving *any* state corresponding to a classical vacuum with definite Chern-Simons number, still less a situation involving two or more states corresponding to classical vacua with *different* Chern-Simons numbers. Consequently, "large" gauge transformations in a fully quantized non-Abelian Yang-Mills gauge theory do *not* represent physical transformations, and symmetry under "large" gauge transformations is not an empirical symmetry. There is no difference in this respect between "large" and "small" gauge transformations.

11.6 Are They Really Gauge Transformations?

There are several reasons why it remains important to better understand the difference between "large" and "small" gauge transformations. One reason is that doing so will help to resolve the following apparent paradox.

Two beliefs are widely shared. The first belief is that local gauge transformations implement no empirical symmetry and therefore have no direct empirical consequences. The second belief is that global gauge transformations have *indirect* empirical consequences *via* Noether's Theorem, including the conservation of electric charge. The paradox arises when one notes that a global gauge transformation appears as a special case of a local gauge transformation. If local gauge symmetry is a purely formal symmetry, how can (just) this special case of it have even *indirect* empirical consequences?

Another reason is to appreciate why some (e.g., Domenico Giulini) have proposed that we make

> a clear and unambiguous distinction between proper physical symmetries on one hand, and gauge symmetries or mere automorphisms of the mathematical scheme on the other (Giulini 2003, p. 289).

The proposed distinction would classify invariance under "small" gauge symmetries as a gauge symmetry, but invariance under "large" gauge transformations as a proper physical symmetry. It is founded on an analysis of gauge in the framework of constrained Hamiltonian systems.

The guiding principle is to follow Dirac's proposal by identifying gauge symmetries as just those transformations on the classical phase-space representation of the state of such a system that are generated by its first-class constraint functions. In a classical Yang-Mills gauge theory, these are precisely those generated by the so-called Gauss constraint functions, such as the function on the left-hand side of equation

$$\nabla \cdot \mathbf{E} = 0 \tag{11.13}$$

in the case of pure electromagnetism.

Giulini (2003) applies this principle to a quantized Hamiltonian system representing an isolated charge distribution in an electromagnetic field. He concludes that the gauge symmetries of this system consist of all and only those local gauge transformations on the quantized fields that leave unchanged both the asymptotic electromagnetic gauge potential $\hat{A}_\mu$ and the distant charged matter field. A global gauge transformation corresponding to a constant phase rotation in the matter field does *not* count as a gauge symmetry since it is not generated by the Gauss constraint (or any other first-class constraint) function. Rather, global $U(1)$ phase transformations would be associated with what Giulini calls *physical* symmetries. According to Giulini (2003, p. 308)

> This is the basic and crucial difference between local and global gauge transformations.

The formalism represents the charge of the system dynamically by an operator $\hat{Q}$ that generates translations in a coordinate corresponding to an additional degree of freedom on the boundary in the dynamical description. A charge superselection rule, stating that all observables commute with the charge operator, is equivalent to the impossibility of localizing the system in this new coordinate. Consequently, conservation of charge implies that translations in this additional degree of freedom count as physical symmetries for Giulini. So conservation of charge is equivalent both to the existence of these symmetries, and (by Noether's first theorem) to the global gauge symmetry of the Lagrangian. But these physical symmetries do not correspond to gauge symmetries, either global or local, since they affect neither the gauge potential nor the phase of the matter field.

It is hard to argue that these novel physical symmetries are empirical. No operational procedures are specified to permit measurement of the additional degrees of freedom, and these attach on a boundary which is eventually removed arbitrarily far away. But even if such a new physical symmetry were empirical, it would not correspond to any constant phase change. A global gauge symmetry would still not entail any corresponding empirical symmetry.

This delicate relation between global gauge transformations and some other physical symmetry helps to resolve the apparent paradox outlined above. A global gauge transformation is not merely a special case of a local gauge transformation. Indeed, the constrained Hamiltonian approach provides a valuable perspective from which it is not even appropriately classified as a gauge transformation.

This perspective illuminates the distinction between "large" and "small" gauge transformations more generally. As Giulini (1995) put it, in Yang-Mills theories

> it is the Gauss constraint that declares some of the formally present degrees of freedom to be physically nonexistent. But it only generates the identity component of asymptotically trivial transformations, leaving out the long ranging ones which preserve the asymptotic structure imposed by boundary conditions as well as those not in the identity component of the asymptotically trivial ones. These should be considered as proper physical symmetries which act on physically existing degrees of freedom. (p. 2069)

Whether the constrained Hamiltonian approach to gauge symmetry establishes that "large" gauge transformations correspond to empirical symmetries is more sensitive to theoretical context than Giulini's last sentence seems to allow. But the approach certainly shows that not only a global gauge transformation but any "large" gauge transformation not generated by a Gauss constraint is very different from the local gauge symmetries that it does generate.

11.7 The θ-Vacuum in a Loop Representation

The availability of loop representations of quantized Yang-Mills theories has interesting implications for the nature of the θ-vacuum. Recall that when the theory is non-Abelian, "large" gauge transformations with non-zero winding number connect potential states with different Chern-Simons numbers, including different candidates for representing the lowest-energy, or vacuum, state of the field. Requiring that the theory be symmetric under such "large" gauge transformations implies that the actual vacuum state is a superposition of all these candidate states of the form

$$|\theta\rangle = \sum_{n=-\infty}^{+\infty} \exp\{-in\theta\} |n\rangle \qquad (11.14)$$

where θ is an otherwise undetermined parameter – a fundamental constant of nature.

Associated with the θ-vacuum is an additional term proportional to $\epsilon_{\mu\nu\rho\sigma} F^{a\mu\nu} F^{a\rho\sigma}$ that enters the effective Lagrangian density for quantum chromodynamics

$$\mathscr{L}_{QCD} = \overline{\psi}_a (i\gamma^\mu D_\mu - m)\psi^a - \frac{1}{4} F_{a\mu\nu} F^{a\mu\nu} + \frac{\theta}{64\pi^2} \epsilon_{\mu\nu\rho\sigma} F^{a\mu\nu} F^{a\rho\sigma}$$

– unless the value of θ is zero, in which case this term itself becomes zero. It turns out that certain empirical consequences of quantum chromodynamics are sensitive to the presence of this extra term: if it were present, then strong interactions would violate two distinct discrete symmetries, namely parity and charge conjugation symmetry. Experimental tests have shown that $|\theta| \leq 10^{-10}$, making one suspect that in fact $\theta = 0$. This fact – that of all the possible real number values it could take on, θ appears to be zero – is known as the *strong CP problem*. Various solutions have been offered, several of which appeal to some new physical mechanism that intervenes to force θ to equal 0. But from the perspective of a loop representation, there is no need to introduce θ as a parameter in the first place. I quote Fort and Gambini (1991):

> It is interesting to speculate what would happen if from the beginning holonomies were used to describe the physical interactions instead of vector potentials. Probably we would not be discussing the strong CP problem. This would simply be considered as an artifact of an overdescription of nature, by means of gauge potentials, which is still necessary in order to compute quantities by using the powerful perturbative techniques. From this perspective, the strong CP problem is just a matter of how we describe nature rather than being a feature of nature itself. (p. 348)

As Fort and Gambini explain, when a theory is formulated in a loop/path representation, all states and variables are automatically invariant under both "small" and "large" gauge transformations, so there is no possibility of introducing a parameter θ (as in Eq. 11.11) to describe a hypothetical superposition of states that are not so invariant. While the conventional perspective makes one wonder why θ should equal zero, from the loop perspective there is no need to introduce any such parameter in the first place. Once formulated, the loop representation will be equivalent to the usual connection representation with $\theta = 0$.

One can introduce an arbitrary parameter θ into a loop representation of a more complex theory, as Fort and Gambini show. But from the holonomy perspective there would have been no empirical reason to formulate such a more complex theory, and the fact that even more precise experiments do not require it would be a considered a conclusive reason to prefer the simpler theory – the one that never introduced an empirically superfluous θ parameter.

References

Fort H, Gambini R (1991) Lattice QED with light fermions in the P representation. Phys Rev D 44:1257–1262
Galilei G (1632) Dialogue concerning the two chief world systems. Translation by Stillman Drake, 2nd edn (1967) University of California Press, Berkeley
Giulini D (1995) Asymptotic symmetry groups of long-ranged gauge configurations. Modern Phys Lett 10:2059–2070
Giulini D (2003) Superselection rules and symmetries. In: Joos, Zeh, Kieffer, Giulini, Kupsch, Stamatescu (eds) Decoherence and the appearance of a classical world in quantum theory, 2nd edn. Springer, New York
Healey R (2009) Perfect symmetries. Br J Philos Sci 60(4):697–720; doi:10.1093/bjps/axp033
Maxwell JC (1881) An elementary treatise on electricity. In: Garnett W (ed) Clarendon Press, Oxford
Rubakov V (2002) Classical theory of gauge fields. Princeton University Press, Princeton

Chapter 12
The Chemical Bond: Structure, Energy and Explanation

Robin Findlay Hendry

12.1 Introduction

The bond is central to modern chemistry's understanding of the behaviour of matter, figuring in explanations of why chemical reactions happen, what their products are, and how much heat is generated or absorbed in the process. Molecular spectra arise from the vibrations and rotations of bonded groups of atoms. Chemistry provides a wealth of information about the properties and behaviour of individual bonds, and its applications of quantum mechanics offer deep theoretical insights into the structure and bonding of molecules. In this paper I trace the development of classical theories of chemical structure from the nineteenth century to G.N. Lewis. I then develop a *structural* view of the chemical bond within quantum mechanics, which identifies the chemical bond by its explanatory role within classical structure theory.

12.2 Chemical Structure Theory

In the first half of the nineteenth century, organic chemistry emerged from its roots in the study of natural substances derived from plants and animals, and became an experimental discipline focussed on the chemistry of carbon compounds, including artificial ones (Klein 2003, Chapter 2). Chemists analysed the many new substances they had synthesised and isolated, and represented their elemental composition in the new Berzelian formulae, which functioned not merely as repositories of extant experimental knowledge but as 'productive tools on paper or "paper tools" for creating order in the jungle of organic chemistry' (Klein 2003, 2). Chemical formulae integrated both experimental knowledge and theoretical understanding of the composition of chemical substances, allowing chemists to develop an awareness of isomerism, where distinct chemical compounds share the same elemental

---

R.F. Hendry (✉)
Department of Philosophy, Durham University, Durham, DH1 3HN, UK
e-mail: r.f.hendry@durham.ac.uk

composition. Because isomers are different compounds that are alike in their elemental composition, they must differ in the ways in which elements are combined within them: that is, in their internal structure. In the first half of the nineteenth century, however, there was no general agreement about what 'structure' might be. The radical and type theories embodied two quite different conceptions of structure, and sometimes fierce debates raged between their defenders (see Brock 1992, Chapter 6). The situation was complicated by a further lack of consensus on how equivalent or atomic weights should be assigned to elements. The elemental composition of a compound substance concerns the relative proportions in which the elements are combined in it *as equivalents*. For example, the formula 'HCl' does not mean that hydrogen chloride contains equal masses of hydrogen and chlorine. Rather it means that HCl contains different masses of hydrogen and chlorine, which can be scaled to a 1:1 proportion to reflect the power of these elements to combine with fixed quantities of other elements (on an atomistic interpretation 'equivalent' amounts of chlorine and hydrogen contain the same number of atoms). Disagreement over equivalent or atomic weights might mean disagreement over the elemental composition of a particular substance, which could confuse any discussion of its structure. The confusion engendered by disagreement over equivalent or atomic weights was an issue in inorganic chemistry too, but the obscurity of the notion of structure, and the fact that there were proliferating ways of representing structure in diagrams whose import was unclear, meant that in organic chemistry the confusions were ramified and intractable.[1]

Structure theory, as it came to be widely accepted from the 1860s, had its origins in the work of many chemists, but Alan Rocke credits August Kekulé, in two papers published in 1857 and 1858, with forging a scattered set of ideas and insights into 'a clear and methodical elaboration of a unified conception of chemical constitution, with a program for its elucidation' (Rocke 1984, 263). A number of important changes occurred in theoretical organic chemistry in the 1850s, which provided the background to the emergence of structure theory (Rocke 1993). These included a spreading standardisation of atomic weights and molecular formulae, culminating in the Karlsruhe Congress of 1860, and the general acceptance, through Edward Frankland's work on organometallic compounds, that atoms in molecules are linked to fixed numbers of other atoms. This last idea, which came to be known as 'valency,' amounted to a structural application of the Daltonian idea that elements combine in fixed proportions. Kekulé ([1858] 1963, 127) applied fixed valency to carbon, which was 'tetratomic (tetrabasic)' – able to link to four other atoms – and, by allowing it to link to other carbon atoms, he reduced the aliphatic hydrocarbons (methane, ethane, propane, etc.) to a homologous series ([1858] 1963, 126–130). In later papers Kekulé introduced double bonds and extended his treatment to aromatic compounds, producing the famous hexagonal structure for benzene.

Yet Kekulé's presentations of structure theory lacked a clear system of diagrammatic representation. His own 'sausage formulae' were unconducive, and did not

---

[1] In the 1860s August Kekulé highlighted this issue by gathering nineteen competing formulae for acetic acid, nearly all of them inscrutable to the modern eye (see Brock 1992, 253 for a list).

figure prominently in his papers. However, in a paper discussing isomerism among organic acids, Crum Brown ([1864] 1865) published a system of 'graphic notation' that he had developed a few years earlier in his M.D. thesis (see Ritter 2001). Here, structure was shown as linkages between clearly Daltonian atoms (see Fig. 12.1). Crum Brown used his notation to exhibit the isomerism of propylic alcohol and Friedel's alcohol (propan-1-ol and propan-2-ol, respectively: see Fig. 12.2).

Frankland adapted Crum Brown's notation – the double bonds were straightened, the atoms lost their Daltonian circles – and popularised it in successive editions of his *Lecture Notes for Chemical Students*, applying it also to the structure of inorganic compounds (Russell 1971; Ritter 2001). He also introduced the term 'bond' for the linkages between atoms (see Ramberg 2003, 26). James Dewar and August Hofmann developed systems of models corresponding closely to Crum Brown's formulae (Meinel 2004). Dewar's molecules were built from carbon atoms represented by black discs placed at the centre of pairs of copper bands joined at the middle, whose ends represented the valences. In Hofmann's 'glyptic formulae,' atoms were coloured balls (black for carbon, white for hydrogen, red for oxygen, etc.) linked by bonds. Even though they were realised by concrete three-dimensional structures of balls and connecting arms, the three-dimensionality of Hofmann's glyptic formulae was artefactual, the product only of the medium. It had no theoretical basis, and Hofmann retained the

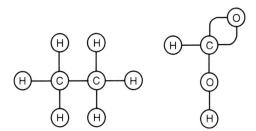

Fig. 12.1 Ethane and formic acid in Crum Brown's graphic notation ([1864] 1865, 232)

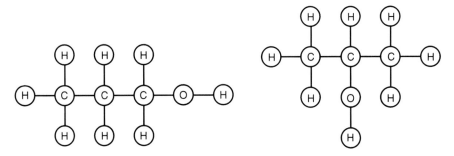

Fig. 12.2 Propylic alcohol and Friedel's alcohol in Crum Brown's graphic notation ([1864] 1865, 237)

'planar symmetrical orientation' of the bonds from Crum Brown and Frankland (Meinel 2004, 252). In short, glyptic formulae no more represented molecules *as* three-dimensional entities than they represented carbon atoms as black, hydrogen atoms as white and oxygen atoms as red.

In the 1870s, mathematician Arthur Cayley related the structures represented in Frankland's diagrams to his work on trees, a kind of undirected graph (Biggs et al. 1976, Chapter 4). Using the analogical connection – atoms being nodes, and bonds vertices – Cayley developed methods for calculating the number of distinct aliphatic isomers with a formula C_nH_{2n+2}. In effect, he was investigating with formal methods a space of structural possibility constrained by the rules of valence, in which atoms are assigned the valence characteristic for their element, and a structural formula should use up or 'saturate' all the available valences. The graphic notation of Crum Brown and Frankland, and the models of Dewar and Hofmann, allowed chemists to investigate this same space through visual reasoning.

How were these formulae and models interpreted? According to Rocke, Kekulé 'established a pattern ... which many other chemists emulated: viz., the distinction between the apparent atomic arrangements deduced from chemical properties ("chemical constitution" or later, "chemical structure"), and the true, actual spatial arrangement of the atoms within a molecule' Rocke (1984, 269). Crum Brown echoed this, cautioning that in his graphical formulae he did not 'mean to indicate the physical, but merely the chemical position of the atoms' (1864, 232). In his *Lecture Notes*, Frankland noted that 'It must carefully be borne in mind that these graphic formulae are intended to represent neither the shape of the molecules, nor the relative position of the constituent atoms' (quoted in Biggs et al. 1976, 59).

One obvious interpretation of these comments is that structural formulae did not represent the 'real' or 'physical' positions of atoms in space but were merely theoretical tools for summarising a compound's chemical behaviour. Perhaps this is part of what is going on, but it can't be the full story for it fails to distinguish between two sets of issues raised by structural formulae considered as representations. One set concerns their content: do they embody any information at all about the microscopic structure of the compound substances they represent? If they do embody structural information, what kind of information is it? The second set of issues concerns truth and inference: could there be any warrant for thinking that the information embodied in structural formulae is true in some absolute sense, and under what epistemic conditions could there be such warrant? It is helpful to set aside the second set of issues: epistemic caution (or at least lip service to it) was common in nineteenth-century chemistry, and understandable given the novelty of structure theory, and the fact that so many prominent figures in chemistry were sceptical of, or downright hostile to, atomistic explanation. Turning to the first set of issues, Kekulé, Crum Brown and Frankland all seem to agree on the limitations of structural formulae, which do not display the spatial positions of atoms within molecules. Yet Frankland also provided something more positive:

> The lines connecting the different atoms of a compound, and which might with equal propriety be drawn in any other direction, provided they connected together the same elements, serve only to show the definite disposal of the bonds: thus the formula for nitric acid

indicates that two of the three constituent atoms of oxygen are combined with nitrogen alone ... whilst the third oxygen atom is combined both with nitrogen and hydrogen. (Quoted in Biggs et al. 1976, 59)

Johannes Wislicenus, who worked on isomerism among the lactic acids in the 1860s, also saw limitations in structural formulae, yet clearly agreed with Frankland's more positive comments:

That the chemical properties of a molecule are most decisively determined by the nature of the atoms that compose it *and* by the sequence of their mutual combination, *the chemical structure* of the molecule, is now a generally shared conviction. (Quoted in Ramberg 2003, 47–8)

If structural formulae displayed the connections between atoms, the 'sequence of their mutual combination,' it is not surprising that Cayley saw them as graphs, embodying topological information only. As such they do fail to provide the spatial positions of atoms, but only because they abstract away from particular spatial arrangements.

Yet in 1873, Wislicenus also saw the possibility that structural formulae might need to be extended into three dimensions, noting that there were still distinct substances that structural formulae could not distinguish. The difference, he thought, must lie in geometrical differences that would be evidenced in 'properties lying on the border areas of physical and chemical relationships, such as solubility, crystal form, water of crystallization, and so forth' (quoted in Ramberg 2003, 48). So it proved: notwithstanding the earlier reservations, structure theory very soon became spatial. In 1874, Jacobus van't Hoff explained why there are two isomers of compounds in which four different groups are attached to a single carbon atom by supposing that the valences are arranged tetrahedrally (the two isomers are conceived of as mirror images of each other). Adolf von Baeyer explained the instability and reactivity of some organic compounds by reference to strain in their molecules (see Ramberg 2003, Chapters 3 and 4). These stereochemical theories were intrinsically spatial, because their explanatory power depended precisely on their describing the arrangement of atoms in space.

12.3 The Electron and the Chemical Bond

In the last section we saw how chemical structure theory, as it emerged in the nineteenth century, developed a theoretical role for the bond: providing the links in the topological structures of molecules and, after van't Hoff and Baeyer, constraining the spatial arrangements of atoms within them. Yet there was no account of what, if any *thing* at all, realised this role. G.N. Lewis was responsible for the first influential theory to fill that gap.[2] The main features of Lewis' account are well known: bonds involve the sharing of paired electrons, in a process in which atoms fill incomplete electron shells. More interesting for this paper is how Lewis (i) views the

---

[2] For the background see Kohler (1971, 1975).

bond proper as intimately connected with the notion of a stable molecular structure that obeys valence rules, and (ii) thereby sets limits on its domain of applicability within chemistry.

In his first published paper on the topic, Lewis argued that 'we must recognize the existence of two types of chemical combination which differ, not merely in degree, but in kind (1913, 1448).' Non-polar compounds are 'immobil,' that is 'unreactive, inert and slow to change into more stable forms, as evidenced by the large number of separable isomers. Inorganic compounds, on the other hand, approach more frequently the ideal polar or mobil type, characterized by extreme reactivity' (1913, 1448–9). The differences are accounted for by different kinds of bonding in the two cases.

> To both types of compounds we should ascribe a sort of molecular structure, but this term doubtless has a very different significance in the two cases. To the immobil compounds we may ascribe a sort of *frame* structure, a fixed arrangement of the atoms within the molecule, which permits us to describe accurately the physical and chemical properties of a substance by a single structural formula. The change from the non-polar to the polar type may be regarded, in a sense, as the collapse of this framework. The non-polar molecule, subjected to changing conditions, maintains essentially a constant arrangement of the atoms; but in the polar molecule the atoms must be regarded as moving freely from one position to another. (Lewis 1913, 1449)

Lewis somewhat overemphasises the contrast, and its alignment with the distinction between polar and non-polar, but one consequence of the mobility of polar compounds is that they exhibit tautomerism, where different molecular structures exist in 'mobil equilibrium' and so 'the compound behaves as if it were a mixture of two different substances' (1913, 1449). By noting that tautomerism is characteristic of polar compounds 'we may account for the signal failure of structural formulae in inorganic chemistry' (1913, 1449). Organic substances are sometimes polar too, so tautomerism arises also in organic chemistry, but it is (he thought) more characteristic of 'immobil' non-polar compounds to exhibit isomerism, where transition between the two forms is restricted, and they can be separated as distinct substances.

This is one respect in which Lewis considers valence formulae to be genuinely representative of structure only for non-polar compounds. Another important respect concerns the directionality of bonding. Lewis considers a proposal to represent the polar bond in potassium chloride with an arrow, as K→Cl, which would signify that 'one electron has passed from K to Cl' but considers it misleading, because even if one could track the electron as it passed from K to Cl, the bond does not arise from this passage, for 'a positive charge does not attract one negative charge only, but all the negative charges in its neighborhood' (1913, 1452). In a non-molecular polar substance like potassium chloride, the bonding is electrostatic and therefore radially symmetrical. Hence an individual ion bears no *special* relationship to any one of its neighbours. Polar bonding is non-directional, and so cannot be represented by the lines connecting atoms in classical structural formulae.

The distinction between polar and non-polar bonding is still present in his next paper we shall consider (Lewis 1916), and 'roughly' but not exactly coextensive with that between inorganic and organic chemistry (1916, 764). But it is now a matter of degree rather than of kind, and relational, because it depends on the environment: a non-polar substance may be polarised by a polar solvent (1916, 765). Bonds arise from atoms with incomplete electron shells filling them either by sharing electrons (in what came to be known as covalent, or shared-electron bonds) or transferring them (the ions and electrostatic bonding of polar compounds). Because shared electrons may not be shared equally, giving rise to partial charges on the bonded atoms, Lewis saw pure covalent and ionic bonds as two ends of a continuum, and offered the pairing of electrons as a unifying explanation of bonding in both polar and non-polar compounds.

In his influential textbook *Valence*, Lewis ([1923] 1966, 20) again presented his theory as a unification, this time of the two great theories of chemical affinity of the nineteenth century. The electrochemical theory saw affinity as arising from the transfer of electricity between atoms: attraction between the resulting opposite charges would explain the stability of the compound. As Lewis ([1923] 1966, 20) pointed out, this account found support in the fact that electrolysis demonstrated an intimate link between electricity and chemical combination, but foundered on the existence of homonuclear species like H_2 and N_2. There were also pairs of analogous compounds like acetic acid (CH_3-COOH) and trichloroacetic acid (CCl_3-COOH) in which positive hydrogen in one compound is substituted by negative chlorine in the other, without any great difference in chemical and physical properties (for fuller details see Brock 1992, Chapter 4).

We met the structural theory already in the last section, but Lewis' comments on it in *Valence* are very interesting. Firstly, he regarded it as so successful that

> No generalization of science, even if we include those capable of exact mathematical statement, has ever achieved a greater success in assembling in simple form a multitude of heterogeneous observations than this group of ideas which we call structural theory (Lewis [1923] 1966, 20–1).

Moreover he recognised the concept of chemical bond as central to the theory:

> The valence of an atom in an organic molecule represents the fixed number of *bonds* which tie this atom to other atoms. Moreover in the mind of the organic chemist the chemical bond is no mere abstraction; it is a definite physical reality, a something which binds atom to atom. Although the nature of the tie remained mysterious, yet the hypothesis of the bond was amply justified by the signal adequacy of the simple theory of molecular structure to which it gave rise. ([1923] 1966, 67)

But in this very versatility and explanatory power there was the danger of overextending structural theory:

> The great success of structural organic chemistry led to attempts to treat inorganic compounds in a similar manner, not always happily. I still have poignant remembrance of the distress which I and many others suffered some thirty years ago in a class in elementary chemistry, where we were obliged to memorize structural formulae of a great number of inorganic compounds. Even such substances as the ferricyanides and ferrocyanides were forced into the system, and bonds were drawn between the several atoms to comply with certain artificial rules, regardless of all chemical evidence. Such formulae are now believed to be almost, if not entirely, devoid of scientific significance. ([1923] 1966, 67)

Some features of Lewis' views on bonding were short-lived, for instance his use of static arrangements of electrons at the corners of cubes to represent the fact that filled outer shells of atoms contain eight electrons. This feature of the theory was dropped by Lewis long before *Valence*, but seemed to exemplify the static and qualitative nature of chemical models of the atom, as opposed to the dynamical and quantitative models favoured by physics (see Arabatzis 2006, Chapter 7). But the electron-pair bond, and his representation of paired electrons as colons between atoms, was taken up with great success by British organic chemists like Christopher Ingold (Brock 1992, Chapter 13), who sought to understand the mechanisms of organic reactions in terms of transfer of electrons (Goodwin 2007). What is most important for this paper, however, is Lewis' subtle understanding of the limited scope of the structural interpretation of the chemical bond. As relatively discrete structural features of molecules, bonds are to be found only where valence formulae are structurally significant.

12.4 Quantum Mechanics and the Chemical Bond

It might be expected that the detailed application of quantum mechanics to chemical bonding, beginning in the second quarter of the twentieth century, would finally answer the question of what bonds are, but the need for approximate methods greatly complicates the explanatory relationship (see Hendry 2004). Quantum mechanics treats molecules as systems of electrons and nuclei interacting electrostatically. This determines a Schrödinger equation for the molecule whose solutions (molecular wavefunctions) correspond to the quantum states available to the system. But Schrödinger equations for chemically interesting molecules are mathematically intractable, and from the 1930s onwards chemists developed *two* approximation schemes that gave workable results for molecules. The successes of these schemes were founded as much on chemical as on mathematical insight. On the one hand, Linus Pauling extended the methods that Walter Heitler and Fritz London had developed for the hydrogen molecule (H_2) and hydrogen molecule-ion, (H_2^+), modelling wavefunctions for molecules as superpositions of states corresponding to classically bonded Lewis structures. The resulting quantum-mechanical states were 'resonance hybrids' of the canonical structures from which they were formed. On the other hand, the molecular-orbital approach of Friedrich Hund and Robert Mulliken built up delocalised molecular orbitals from available atomic orbitals.

Although the approximate solutions could be interpreted in terms of classical bonds, there was some question whether this interpretation amounted to a projection of the bond into a quantum-mechanical reality that is devoid of them. The recovery of bonds, it was suggested, was an artefact of the approximate methods. To this suggestion there are two possible responses. One might regard the great unifying explanatory power of classical chemical structure, to which bonds are central, as an argument for the reality of bonds. If the interpretation of semi-empirical wavefunctions in terms of bonds amounts to projection, this shows only that exact quantum mechanics alone is insufficient to capture the explanatory power of classical molecular structure. This is roughly the position adopted by Linus Pauling (see Hendry 2008 Section 3). Although he recognised the centrality of the bond to chemical explanation, Charles Coulson expressed a more sceptical attitude to bonds:

> From its very nature a bond is a statement about two electrons, so that if the behaviour of these two electrons is significantly dependent upon, or correlated with, other electrons, our idea of a bond separate from, and independent of, other bonds must be modified. In the beautiful density diagrams of today the simple bond has got lost. (Coulson, quoted in Simoes and Gavroglu 2001, 69)

In what remains I will set out a *structural view* of the bond, the core intention of which is to retain the explanatory insights afforded by classical structural formulae. Lewis and Pauling work within this view, but the theoretical role of bonds within structural formulae predated both the discovery of the electron and the advent of quantum mechanics. Also, the view ought to be compatible with discoveries that have come after Lewis and Pauling. Hence it ought not to incorporate too closely the particular conceptions of the material basis of the bond offered either by Lewis (discrete shared electron pairs) or Pauling (hybridisation and resonance). One way to meet these constraints is for the structural view to identify a theoretical function for bonds – continuous with that in the classical formulae – but leave it to empirical and theoretical investigation to identify how it is physically realised within the quantum-mechanical states of molecules. Hence on the structural view, chemical bonds are, at least for molecular substances, material parts of the molecule which are responsible for spatially localised submolecular relationships between individual atomic centres.

There are three sorts of challenge to the structural view:[3] from quantum statistics, electron delocalisation, and substances whose structures cannot informatively be represented by classical valence formulae. The quantum-statistical objection is that electrons are fermions, which is why molecular wavefunctions are anti-symmetrical with respect to electron permutation. So features of molecular wavefunctions cannot depend on the identities of particular electrons. Understood as pairs of electrons with fixed identities, bonds simply cannot be features of molecular wavefunctions. A related problem concerns electron delocalisation: electron density is 'smeared out' over the whole molecule, and so Lewis-style bonds consisting of pairs of electrons held fixed between pairs of atoms are not to be found in real molecules. These two problems rule out the identification of bonds with Lewis-style static electron pairs, but can be addressed within the structural view. The first problem can be addressed by recognising that a bond must be individuated by the atomic centres it links. In so far as electrons participate physically in the bond (as they must) they do so not as individuals, but as the occupancies of non-arbitrary partitions of the full electronic wavefunction that can be associated with the bond. The second problem can be addressed in similar fashion: some part of the total electron density of the molecule is responsible for the features associated with the bond, and it will be a

---

[3] Weisberg (2008) presents further challenges, and Stemwedel (2006) argues for the continued centrality of the classical bond to explanatory mechanisms in organic chemistry.

matter of empirical and theoretical investigation to identify which part. There need be no assumption that it is localised. The structural conception is the heir to the conception of bonds that Lewis inherited from the structure theory of the nineteenth century, but generalised to take account of quantum-mechanical insights.

Turning now to the third kind of challenge, many substances seem to defy informative representation by valence formulae. Some, like BF_3, PF_5 and SF_6, violate the octet rule, according to which the outermost shell of a bonded atom has eight electrons. Such cases were known to Lewis, and constitute objections to the octet rule rather than to the electron-pair bond. For other substances like the boranes (boron hydrides), although Lewis-style valence structures can be drawn (albeit violating the octet rule), these do not reflect known features of their structures like bond lengths and bond angles, which seem to demand distributed multi-centre bonds (see Gillespie and Popellier 2001, Chapter 8). The response to these difficulties is to take a lead from Lewis who, as we saw in Section 12.3, recognised that even where they can be written, valence formulae are not always structurally representative. If we regard the concept of a covalent bond as a theoretical notion associated with structural formulae, the applicability of this notion must be delimited by the applicability of structural formulae themselves. Of course there is *bonding* in paradigm ionic substances like potassium chloride, but their structure and stability is explained electrostatically, without recourse to *directional* relationships between individual atomic centres. The structural view maintains, however, that a proper understanding of the structure and stability of paradigm non-polar substances like methane, to which structurally informative valence structures can be given, must involve bonds, understood as localised and directional submolecular relationships between individual atomic centres.

References

Arabatzis T (2006) Representing electrons: a biographical approach to theoretical entities. Chicago University Press, Chicago, IL
Biggs NL, Keith Lloyd E, Wilson RJ (1976) Graph theory 1736–1936. Clarendon, Oxford
Brock W (1992) The Fontana history of chemistry. Fontana Press, London
Crum Brown A ([1864] 1865) On the theory of isomeric compounds. Trans Roy Soc Edinburgh 23:707–19. Page references are to the reprint in the J Chem Soc 18(1865):230–245
Gillespie R, Paul P (2001) Chemical bonding and molecular geometry: from Lewis to electron densities. Oxford University Press, New York
Goodwin W (2007) Scientific understanding after the Ingold revolution in organic chemistry. Philos Sci 74:386–408
Hendry, RF (2004) The physicists, the chemists and the pragmatics of explanation. Philos Sci 71:1048–1059
Hendry, RF (2008) Two conceptions of the chemical bond. Philos Sci 75:909–920
Kekulé A. ([1858] 1963) Über die Constitution und die Metamorphosen der chemischen Verbindungen und über die chemische Natur des Kohlenstoffs. Ann Chem Pharm 106:129–159. Page references are to the translation as 'The constitution and metamorphoses of chemical compounds and the chemical nature of carbon. In: Theodor Benfey O (ed) Classics in the theory of chemical combination. Dover, New York, pp 109–131

Klein U (2003) Experiments, models and paper tools: Cultures of organic chemistry in the nineteenth century. Stanford University Press, Stanford, CA

Kohler R (1971) The origin of G.N. Lewis's theory of the shared pair bond. Hist Stud Phys Sci 3:343–376

Kohler R (1975) G.N. Lewis's views on bond theory, 1900–16. Br J Hist Sci 8:233–239

Lewis GN (1913) Valence and tautomerism. J Am Chem Soc 35:1448–1455

Lewis GN (1916) The atom and the molecule. J Am Chem Soc 38:762–785

Lewis, GN ([1923] 1966) Valence and the structure of atoms and molecules. Chemical Catalogue Company, Washington, DC. Page references are to the reprint under the same title (Dover, New York, 1966)

Meinel C (2004) Molecules and croquet balls. In: Chadarevian S, Hopwood N (eds) Models: the third dimension of science. Stanford University Press, Stanford, pp 242–275

Ramberg P (2003) Chemical structure, spatial arrangement: the early history of stereochemistry, 1874–1914. Ashgate, Aldershot

Ritter C (2001) An early history of Alexander Crum Brown's graphical formulas. In: Klein U (ed) Tools and modes of representation in the laboratory sciences. Kluwer, Dordrecht, pp 35–46

Rocke AJ (1984) Chemical atomism in the nineteenth century: from Dalton to Cannizzaro. Ohio State University Press, Columbus, OH

Rocke AJ (1993) The quiet revolution of the 1850s: social and empirical sources of scientific theory. In Mauskopf SH (ed) Chemical sciences in the modern world. University of Pennsylvania Press, Philadelphia, PA, pp 87–118

Russell CA (1971) The history of valency. Leicester University Press, Leicester

Simoes A, Gavroglu K (2001) Issues in the history of theoretical and quantum chemistry. In: Reinhardt C (ed) Chemical sciences in the twentieth century: bridging boundaries. Wiley, Weinheim, pp 51–74

Stemwedel J (2008) Chemical bonds in causal explanations of chemical reactions. Paper presented at the 20th biennial meeting of the Philosophy of Science Association, Vancouver, November 2006

Weisberg M (2008) Challenges to the structural conception of bonding. Philos Sci 75(5):932–946

Chapter 13
Randomness, Financial Markets and the Brownian Motion: A Reflection on the Role of Mathematics in Their Interaction with Financial Theory After 1973*

Ghislaine Idabouk

13.1 Introduction

In May of 1973, the Journal of Political Economy publishes an article entitled "The pricing of Options and Corporate Liabilities" by Fischer Black, from the University of Chicago and Myron Scholes from the Massachusetts Institute of Technology (MIT).

The authors address a question that had been a topic of interest among economists since the 1960s: the pricing of financial securities used for speculation and hedging purposes, options.

They derive an option pricing formula which depends on the model's initial parameters (no ad hoc parameter added) and explicitly uses the standard normal distribution. Their article will have a major influence.[1]

First, it is to become the cornerstone of a theoretical stream, which will later be known as Continuous-Time Finance or Mathematical Finance, within the modern financial theory born in the early 1950s.[2] The characteristic of this new stream of theory within the theory is the substantial use of stochastic calculus-a sub-field of modern probability theory which emerged with the work of Japanese mathematician Kiyosi Itô in the 1940s – and more broadly of the theory of stochastic processes. The article will also serve, from a practical standpoint, as the pricing reference. In April of 1973, a month before the article was published, the Chicago Board Options Exchange (CBOE), the first exchange for the standardized trading of

---

G. Idabouk (✉)
Paris Diderot University/Rehseis (CNRS, UMR 7219), Paris, France

* Remarks and comments on this paper are welcome and may be addressesd to the author at: ghislaine.idabouk@gmail.com.

[1] In Section 13.4 of this paper, we will also consider the paper by Robert C. Merton (Merton 1973) as a founding article of Mathematical Finance.

[2] I consider here, as is standard among finance academics, that "Modern Financial Theory" emerged after the publication of Harry Markowitz's paper (Markowitz 1952). This boundary date with previous financial theory might be questioned. Yet, it is beyond the scope of this paper, which addresses mathematical finance after 1973, to do so.

options, was created. In 1975, the CBOE officially adopted the Black and Scholes formula for the pricing of traded options.[3] The annual volume of equity options traded at CBOE was then 14,4 millions, 14 times the amount it was in 1973.

In a broader perspective, the construction of mathematical finance as a theoretical field, since the Black and Scholes article of 1973 and for the following 10 years,[4] is a challenging field of research and analysis for a philosopher of science. It raises many questions.

The first has to do with the mathematization of randomness. Randomness here is randomness of the price processes of the risky primitive financial instruments (for instance stocks) in a financial market. In these articles, it is modeled through the Brownian motion. In traditional neoclassical economic theory, a price is determined through equality of supply and demand that emanate from agents usually assumed to be rational. If one clings to these fundamentals as determinants of a price, what need is there to give a probabilistic representation of the price of a stock?

Another issue that arises from the use of stochastic calculus, and more specifically of the Brownian motion, to model randomness, and from the martingale property which will, soon after the Black Scholes paper, be claimed for the discounted prices of financial securities, is the question of the ideological implications of such models. Indeed, underneath the Brownian motion, there are the ideas of independent identically distributed increments, and the use of the normal distribution to model the random part of the rates of return on financial securities ("log-normal" prices). At this point, let us recall that financial theory is a social science. It therefore is human behaviors and interactions between individuals that are eventually modeled by a normal distribution. The relevance of the normal distribution in many fields of natural sciences is unquestionable. Is it still the case when it comes to social sciences?

Besides, the mathematical martingale property, present in most of the articles of mathematical finance mentioned in this paper, relates to an economic assumption, the "Efficient Market Hypothesis", one of the cornerstones of modern financial theory, first developed by Eugene Fama in his PhD dissertation of 1964[5] to later become a fiercely debated assumption in financial economics through the 1970s and 1980s. This cannot either be regarded as ideologically neutral.

A last interesting feature for a philosopher of mathematics in the development of mathematical finance is the particular role assigned to mathematics here. In 1990, Fisher Black, one of the co-authors of the founding article, stated: "because the formula is so popular, because so many traders and investors use it, option prices

---

[3] See http://www.cboe.com/AboutCBOE/History.aspx

[4] In this article, the analysis is restricted to a limited corpus of academic articles identified in Sections 13.4 and 13.5, which I consider as a first, coherent building block of mathematical finance. The latter has then evolved into several directions to extend the Black and Scholes model (stochastic volatility, jump-diffusion processes) and to address issues of asset pricing in incomplete markets. The analysis of these extensions from a philosophy of science standpoint is left for future research.

[5] Fama's PhD dissertation was published in the January 1965 issue of the Journal of Business (see Fama 1965).

tend to fit the model even when they shouldn't". This sentence, which might seem odd when said by one of the fathers of the formula, is interesting from an epistemological standpoint. So is the fact that the Black and Scholes formula is often referred to by researchers and practitioners as a "self-fulfilling prophecy". Both are an invitation to rethink the relationship between a model, here an economic and mathematical model, and reality. What role do mathematical models play in this story?

This paper aims at shedding some light on the points mentioned above.

13.2 Options: A Brief Overview

An option is a financial instrument that gives its buyer the *right*, but not the obligation, to buy (or sell, depending on the nature of the option) another financial instrument (a stock, a bond or even another option), or a commodity or a currency, at a fixed price (called the *strike* price) at (or until) some future date (called the *maturity* of the option). The other financial instrument (or the commodity or currency) that can be bought or sold is called the *underlying*. An option to *buy* is called a *call* option, an option to *sell* a *put* option. An option that can be exercised only at maturity is a *European* option. An option that can be exercised at any time up to maturity is an *American* option. Consider for instance that you hold one share of the stock of a company. Today, you could sell it for 100 Euros. But you wish to sell it only in 3 months and you fear that until then the price might decrease. Obviously you are not sure of the decrease and wish to benefit from the potential upside movements. So you keep your share of stock and you buy today a put on that stock with exercise price of 100 Euros and maturity 3 months. Obviously, the right to sell at 100 that you buy has a price, called the option *premium*, for instance 5 Euros. If in 3 months, the stock price is 120 you do not exercise your option, you sell your share directly on the market at 120, and you have lost the 5 Euros you paid for the option at time 0. If, conversely, the stock price is 92 in 3 months, you exercise your option and can sell at 100 something worth only 92 on the market. And so you paid 5 Euros at time 0 to gain 100–92 = 8 Euros in 3 months. From this example, we can see that options are insurance instruments (or *hedging* instruments). But they can also be used for *speculative* purposes. Because the value of the option depends on the value of another item, options are called *derivative* instruments.

13.3 The Financial and Economic Context: Options' Markets and Financial Theory

It is clearly not the purpose of this short section to give a detailed history of the development of options' markets in the world over time. This has already been done, in a beautiful way, and not only for options' markets, by other authors (Belze and Spieser 2005). The first purpose is just to make clear that option trading did not start with the opening of the Chicago Board Options Exchange in 1973. The first organized exchanges appeared as early as the fourteenth century. They were initially for the exchange of goods but soon financial instruments were also negotiated. Derivatives were traded as early as the fourteenth century. For instance, markets futures (another type of derivatives) are mentioned in the Verona statutes of 1318.[6] And option trading was practiced in seventeenth century Amsterdam as is mentioned in a book written by a merchant, Jose Penso de la Vega, and published in Amsterdam in 1688.[7] Another point worth stressing is that, however, the particular context of the 1970s had a major impact on the development of financial derivatives, and of modern derivatives' exchanges as we know them now. Indeed, whereas the Bretton Woods agreements had sought, in the aftermath of World War II and with still fresh memories of the Great Depression, to establish a stable international monetary system (in particular through fixed exchange rates between the major currencies and the US dollar, and the creation of the International Monetary Fund, the World Bank and the International Bank for Reconstruction and Development), the 1970s saw a joint process of deregulation, disintermediation and declustering of markets. In August of 1971, in a context of national inflation and balance of payments deficit, Richard Nixon, then President of the United States, decided to abandon the Bretton Woods Agreements and put an end to the dollar-gold convertibility. The dollar is devalued in December 1971 then again in February 1973. In March of 1973, the European central banks let their currencies float against the dollar. Currency risk which had been frozen by Bretton Woods reemerges. On another hand, with the oil shock of 1973–1974 and the deregulation of interest rates in the United States in 1979, risks increase on other fronts. As an answer to this, more and more sophisticated derivative instruments are created to offset exposure to these risks. On April 26th, 1973, the CBOE is created in Chicago, soon followed by other options exchanges: the Philadelphia Stock Exchange in 1975 and the Pacific Stock Exchange in 1976, the Marché des Options Négociables in Paris in 1987 and the London Traded Options Market in 1989.

While markets were undergoing turmoil in the 1970s, and complex financial derivatives were being developed to insure against increasing risks or to trade these risks, financial theory was lagging behind.

Financial theory (or Financial Economics, as it is also often referred to) is a relatively young discipline. Few contributions are earlier than the twentieth century and up to the 1930s financial theory essentially consisted of practical investment rules and statistical techniques. Again, it is beyond the scope of this short section and of this paper to give a detailed history of the construction of financial theory as a field. The focus of this paper is on Mathematical Finance, a sub-field of Financial Economics that emerged in the 1970s after the paper by Black and Scholes was published. However, some elements are worth stressing in order to grasp the theoretical consensus that prevailed in Financial Economics at the end of the 1960s and in

---

[6] See Belze and Spieser (2005, p 198).
[7] See Belze and Spieser (2005, p 241).

the early 1970s, and understand the theoretical background Black and Scholes had when they started thinking about the option pricing problem. There are three main notions at the heart of financial theory's construction and concerns: the notions of value, price and risk of a financial instrument. In 1907 in *The Rate of Interest* (Fisher 1907) then in 1930 in *The Theory of Interest* (Fisher 1930), Irving Fisher develops his theory of impatience and opportunity thus introducing an intertemporal approach in the consumer choice process. Fisher also gives a definition of capital as any asset that produces a stream of income over time. The value of that capital is the present value of the net incomes. In their 1934 book titled Security Analysis (Graham and Dodd 1934), Benjamin Graham and David Dodd introduce the notion of "intrinsic value" or "fundamental value", which they distinguish from the price. John Burr Williams later reasserts this point, distinguishing between "real worth" and market price (Williams 1938). He gives a rule to compute the value (or "real worth") of a financial instrument by discounting the expected future financial cash flows to be paid by that instrument (for instance dividends for a stock). The question of the appropriate discount rate is left unanswered. In 1952, Harry Markowitz introduces his theory of portfolio[8] selection in a static (one-period) model. In his model, prices and returns of financial instruments are modeled as random variables and agents maximize the expected return of their portfolio while minimizing the risk of the portfolio, as measured by the variance of its return. Such agents will later be called mean-variance investors. In the 1952 paper, Markowitz mainly demonstrates that diversification across assets with low correlations reduces the overall risk of the portfolio. He develops his portfolio selection theory in a 1959 book. The works of Markowitz mark the start of Modern Financial Theory (see note 2). In 1958, James Tobin elaborates on Markowitz's framework and introduces for the investors the possibility to invest in a risk-free asset besides the risky assets already present in the Markowitz setting. He gets a theorem known as the two-fund separation theorem: all Markowitz-type investors (mean-variance) invest their wealth in the risk-free asset and in the *same* portfolio of risky assets. What varies from an investor to another is the proportion invested in each of these two funds (Tobin 1958). In 1964, William Sharpe brings a further contribution to this building block of Modern Financial Theory. He develops the *Capital Asset Pricing Model* (CAPM) which is the major contribution in Financial Economics in the 1960s.[9] In a one-period mean-variance framework similar to those of Markowitz and Tobin, Sharpe obtains, under strong assumptions on the investors in his model (they are assumed to have homogeneous anticipations on the expected returns, the standard deviations and the correlations of the returns of the risky assets) and under an equilibrium assumption, that the risky portfolio that Tobin found in his two-fund separation theorem has to contain *all* the risky assets in the model in a proportion equal to their relative market capitalization. This portfolio is called the *market portfolio*. Sharpe then obtains that, for each financial asset in the

---

[8] A portfolio is a combination of financial instruments. The weight of each asset in the portfolio is given in proportion of the total value of the portfolio.
[9] The CAPM was developed independently by Sharpe (1964), Lintner (1969) and Mossin (1966), and also by Jack Treynor, but in an unpublished memorandum (1961).

model, its expected excess return with respect to the risk-free interest rate is equal to a certain asset-specific coefficient times the market premium (which is the expected excess return of the market portfolio with respect to the risk-free interest rate). The multiplying factor is called the *Beta* of the asset and is equal to the covariance of the return of the risky asset with the return of the market portfolio divided by the variance of the return of the latter.[10] Sharpe's contribution will be largely criticized in the 1980s and 1990s. But in the 1960s and 1970s it is central to Financial Economics in at least two respects. First, it changes the notion of risk for a financial asset. Beta becomes the right measure of risk, meaning that it is no longer the variance of the return of an asset that matters but the covariance of its return with the return of the market portfolio as an indicator of systematic risk (instead of overall risk a part of which can be diversified away). Besides, the expected return on a risky financial instrument as given by the CAPM (see footnote 11) becomes the reference discount factor to use when discounting the future cash flows of a financial instrument to assess its present value.

Besides this first building block of Modern Financial Theory, another idea which will be crucial for the construction of Mathematical Finance spreads among economists and financial economists through a paper published by Modigliani and Miller (1958). In this paper, the authors use arbitrage reasoning.[11] In the next two sections, we will see what this reasoning entails and how central this approach will be in Mathematical Finance.

A last idea also largely permeates theoretical considerations in finance in the 1960s and 1970s: the *Efficient Markets Hypothesis* (EMH).[12] It emerged in 1964 in the PhD dissertation of Eugene Fama, a finance student from the University of Chicago (Fama 1965). In his thesis, Fama claims that prices on financial markets reflect all available information. Therefore, a price variation can result only from unpredictable events. A consequence of this first formulation of EMH is that it is not possible to make profits on financial markets. Fama will later refine this idea in a 1970 paper (Fama 1970), distinguishing between three forms of informational efficiency, depending on the nature of the information reflected in the prices on a financial market. In weak form efficiency, today's share price for a stock reflects all past share prices for the stock. In semi-strong form efficiency, today's share price for a stock reflects all publicly available information. In strong form efficiency, both public and private information are reflected in share prices.

---

[10] This can formally be written as $\forall i, E(r_i) = r_f + \beta_i \left(E(r_M) - r_f \right)$, where r_i is the return of the i-th asset, r_M is the return of the market portfolio, r_f is the risk-free rate and $\forall i, \beta_i = \frac{\text{cov}(r_i, r_M)}{V(r_M)}$.

[11] Modigliani and Miller are not the inventors of the notion of arbitrage. It is already present in the works of Irving Fisher. Yet Modigliani and Miller's paper contributed to spreading the use of arbitrage reasoning among financial economists.

[12] Efficiency here is informational efficiency.

13.4 The Founding Articles: Black and Scholes (1973) and Merton (1973)

It is eventually only in its spring of 1973 issue that the Journal of Political Economy publishes the aforementioned Black and Scholes article after first rejecting it in November 1970. This publication also comes one year after the submission of the last revised version by the two authors. Getting the paper published was no easy task. The most likely reasons for the first rejection of the paper are that, when it was first submitted, option pricing was still viewed as a minor topic among academics,[13] that the paper did not look like a standard economics paper of that time, and probably also that Fisher Black was not a full-fledged academic (Mehrling 2005, p 135–136). Yet, the paper was published and shortly afterwards not only were options no longer a minor topic in practice (the CBOE opened just before the Black and Scholes article was published), but the Black and Scholes paper, alongside another paper published the same year by Robert C. Merton (Merton 1973), a young assistant professor and a colleague of Myron Scholes in the Financial Economics group of the MIT Sloan School of Management, revolutionized Modern Financial Theory, leading to the constitution of a radically new sub-field of Financial Economics: Mathematical Finance. Perry Mehrling (Mehrling 2005, p 121–140) beautifully narrates how the three authors got interested in the option pricing problem.[14] Myron Scholes, then an assistant professor at the Sloan School of Management of MIT, started to work on the question around the fall of 1969 while supervising a master's student thesis. Black, on the other hand, "claimed, and no one has ever disputed it, that he achieved the crucial differential equation that characterizes the unique solution to the option pricing problem by June 1969" (Mehrling 2005, p 127). It is worth stressing here that between January 1965 and March 1969, Fisher Black had worked at a consulting firm called Arthur D. Little (ADL) where he often collaborated and exchanged ideas with Treynor, one of the four contributors to the Capital Asset Pricing Model. This element will be crucial to Black's initial approach to the option pricing problem, the one that led him to find that differential equation by June 1969. Black and Scholes start to exchange their ideas on option pricing "in summer or early fall of 1969" (Mehrling 2005, p 129), a collaboration that will lead to the writing of the 1973 seminal paper. In their paper, Black and Scholes seek the price $w(x, t)$ at date t of a European call option with maturity t* and strike price c on an underlying stock with price x. They start by modeling the stock price as a "random walk in *continuous time* with a variance rate proportional to the square of the stock price". More precisely, they model the *return* on the stock as a stochastic process[15] in continuous time, with a deterministic part and a random part, the random part being modeled

---

[13] Although several attempts at option pricing were made, before 1973, by some economists including the father figure of the MIT Economics Department, Paul Samuelson (see Sprenkle 1961; Ayres 1963; Boness 1964; Samuelson 1965; Baumol et al. 1966; Chen 1970).

[14] Mehrling claims though that the early reasons of Black's interest in option pricing are unknown.

[15] A stochastic process is a collection of random variables.

by a Brownian motion.[16] They then solve the option pricing problem with two different approaches both presented in the published article. Each corresponds more or less to the intuitions of one of the two authors on the problem.[17] The first approach historically is actually the one presented in the published paper as the "alternative derivation". It is the one Black used in 1969 to get to the partial differential equation for the option price. But in the perspective of this paper and our analysis of the construction of Mathematical Finance, it is the second approach that is relevant. Black and Scholes start with the intuition that the risk on the option should be offset by the risk on the underlying stock.[18] Thus it should be possible to build a risk-free portfolio[19] made of 1 share of stock and a certain number of units of the option. The number of units of option held in the portfolio changes with time. This approach is called *dynamic hedging*. Using traditional differential calculus intuitions, they get that the portfolio should contain $-\frac{1}{w_1}$ options.[20] They then use again Itô's calculus to write the change in the value[21] of the portfolio over an infinitesimal time interval[22] and find indeed that this change in value is purely deterministic (the stochastic part is canceled). The next crucial step in the reasoning is the *arbitrage* argument. As this portfolio of the stock and the option is risk-free, it should have the same return as the risk-free rate otherwise there would be arbitrage opportunities.[23] By identification of the two expressions, they get the same partial differential equation for the stock price as in the first approach. With a non trivial change of variable, this equation can be transformed into the heat equation of physics and can therefore be solved. The solution is what will become known as the Black Scholes option pricing formula. It depends only on the initial parameters of the model and the cumulative distribution function of the standard normal distribution.

As previously mentioned, Merton was the third protagonist[24] in this story. He got interested in option pricing in 1969 while working with Paul Samuelson (Merton and Samuelson 1969) and he started exchanging ideas with Black and Scholes on the topic in the fall of 1970. But his key contribution for the construction of Mathematical Finance was his 1973 paper: "Theory of Rational Option Pricing".

---

[16] The Brownian motion is a particular stochastic process. It starts at 0 at time 0, has independent and stationary increments (if we consider that the filtration is the natural one), and the increments follow a centered normal distribution with a variance proportional to the time interval.

[17] Although Black acknowledges Merton's contribution for the arbitrage approach.

[18] Black and Scholes do not justify their intuition. They seem to be reasoning as in traditional differential calculus, which in this case is a reasoning error. Yet, the intuition was correct and it can be properly proved using stochastic differential calculus.

[19] Black and Scholes reason in terms of null Beta (CAPM influence throughout).

[20] I used Black and Scholes' notations here. w_1 is the partial derivative of the option price with respect to the stock price.

[21] Value here simply means number of units multiplied by price.

[22] Here, Black and Scholes implicitly use the assumption that the portfolio is "self-financing", a notion which will be introduced later in Mathematical Finance. Otherwise their computation would be wrong.

[23] An arbitrage opportunity is the possibility to make a profit with no cost and no risk.

[24] Merton and Scholes were awarded the Alfred Nobel Memorial Prize the same year for their "new method to determine the value of derivatives" (Black, who died in 1995, could not receive it as it is not awarded posthumously).

Merton's purpose was to give a broader theory of option pricing[25] than the one given by Black and Scholes and also to prove the Black and Scholes formula under less restrictive assumptions than theirs. In particular, he was not satisfied with their use of the CAPM, which he proved to be useless. Merton was also the one who used the full mathematical formalism of stochastic differential equations, largely absent from the Black and Scholes paper. I claim that, in a way, Merton's paper, by "wiping off" all assumptions on investors' preferences (aside from that of a rational pricing theory, see footnote 25) in the Black and Scholes model, is what allowed the latter to become the methodological cornerstone of Mathematical Finance for the pricing of derivatives. And to continue along these lines, the central contribution of the Black and Scholes model, in my opinion, is certainly not the introduction of the Brownian motion and continuous time,[26] or of Itô's calculus. Their essential contribution is to have combined the tools of stochastic processes with the key intuition of dynamic hedging, meaning that the issues of pricing and hedging must be considered simultaneously (they are the same issue actually). The Black Scholes formula for option pricing also gives the hedging strategy. This idea and the no arbitrage assumption are definitely sine qua non ingredients of the Mathematical Finance corpus (or at least of the part that deals with derivatives' pricing), even if we take a look at what has been done after the 1983 boundary of the present article.

13.5 The Consolidation of a Mathematical Finance Corpus: A First Phase (1973–1983)

In the decade that followed the publication of the Black and Scholes and Merton articles, a body of papers consolidated around them, obviously not just papers quoting them, but papers built along the same lines, structured with the same framework: given stochastic processes[27] for the prices (or returns) of the underlying assets, dynamic hedging, Itô's calculus, and the no arbitrage argument to infer the prices of the derivatives.

A first straightforward consolidation was to extend the Black and Scholes paper, which was for European call options on stock, to other underlying assets. For instance, in 1976 Fisher Black himself extended the Black and Scholes formula to options on futures (Black 1976). William Margrabe extended the formula to exchange options (Margrabe 1978) and a few years later, Mark Garman and Steve Kohlhagen applied the Black and Scholes framework for the study of currency options (Garman and Kohlhagen 1983).

---

[25] Initially, "rational option pricing" for Merton is simply such that no asset is dominant nor dominated. Obviously this definition being too generic, he then had to restrict it to be able to get a price for options.

[26] Anyways Black and Scholes were neither the first ones to use the Brownian motion (and not even the geometric Brownian motion) nor the first ones to use continuous time modeling in Financial Economics.

[27] Mostly Brownian motion in this first phase, though not exclusively. The study here is restricted to papers with Brownian motion models, see also footnote 5.

Another consolidation which is more interesting from an epistemological standpoint consisted of three articles published between 1979 and 1981 by three men Harrison, Kreps and Pliska (Harrison and Kreps 1979; Harrison and Pliska 1981; Kreps 1981). All three authors have an undergraduate degree in either engineering or mathematics and a PhD in Operations Research from the same university: Stanford (1970 for Harrison, 1972 for Pliska and 1975 for Kreps). To get an idea of how the contributions of these three authors radically shaped the framework of Mathematical Finance, it is almost enough to look at how Harrison and Kreps approached the "arbitrage theory of option pricing" (Harrison and Kreps 1979). What puzzled them was precisely that the intuition Black and Scholes had had in their second approach turned out to be correct, the fact that in their mathematically non rigorous and highly imperfect model, the risk on the option could indeed be offset by the risk on the stock and that, in the end, the option could be seen as a redundant asset: a combination of the two primitive assets (the stock and the risk-free asset). So they brought in a mathematical way of looking at the problem: namely, given a model of underlying securities prices, what are the derivatives that can actually be priced by arbitrage as combinations of existing assets? This simple anecdote reveals an important aspect, in my opinion, about the construction of Mathematical Finance: financial engineering (meaning the introduction of new derivatives on markets) was a key concern. What the three papers by Harrison, Kreps and Pliska also did was to build bridges between Mathematical Finance concepts and probabilistic theories: stochastic integration theory and martingale theory. For instance, the so called fundamental theorem of arbitrage asset pricing aims at establishing an equivalence, whenever it is possible, between the absence of arbitrage opportunities in a model of a financial market and the fact that there exists a probability measure under which the discounted prices (or the prices expressed in an appropriate numéraire) are martingales. Another example is the obvious link between the notion of a complete market in Mathematical Finance (a market in which any derivative can be replicated) and the martingale representation theorem in probability theory. Another key feature of Mathematical Finance is thus that it strongly builds upon the probabilistic theory of martingales not just by picking a few tools of that theory but rather by using the whole theory as a unifying and robustness-guaranteeing framework.

13.6 Models of Mathematical Finance and the Practice on Financial Markets

Recent works by sociologists of science (MacKenzie and Millo 2003; MacKenzie et al. 2007) have precisely investigated the impact of models like the Black and Scholes model on financial practices in options markets. Among their conclusions, we can find that:

"Option theory [...] is built into the infrastructure of options markets. It helped make those market seem legitimate; it provided a guide to the pricing of options and to hedging the risk they entail; and it has become incorporated into the way market participants talk and think about options" (MacKenzie et al. 2007).

More specifically, Donald MacKenzie introduces the notion of the "performativity" of economics. He distinguishes between four forms of that performativity: generic performativity, effective performativity, Bayesian performativity and counterperformativity and investigates these forms of performativity in the case of option pricing theory. In particular, he shows that up to the mid 1970s, the model fits reality only approximately. Between the mid 1970s and the summer of 1987, reality adjusts to fit the model. After that systematic deviations from the model are observed.

13.7 Conclusion

What about our introductory questions in light of all this?

First, we wondered why we should bother with a probabilistic representation of the price of a stock when neoclassical economic theory gives us supply and demand. Well, simply because supply and demand does not tell us how prices are formed. It would be way too complicated to model the interactions between agents and the feedback effects that actually lead to price formation through that law. Therefore a probabilistic representation through a stochastic differential equation is probably the best that can be done. This, at least partially, answers also the following question about the relevance of the normal distribution in these models of Mathematical Finance. It might not be relevant but it is "performative". Another part of the answer is that the normal distribution is not a necessary assumption of these models.

As for the links between the mathematical martingale property and the Efficient Market Hypothesis, there are indeed links but they are not trivial because the martingale property depends of the probabilistic notion of a filtration, just as the Efficient Market Hypothesis depends on the nature of the information that is reflected in the prices.

The last question was about the role assigned to mathematical models in Mathematical Finance, and more specifically in derivatives' pricing theory. My claim is that they play a normative role. First, in unifying this theory: they are the *only* approach or at least clearly the *dominant* one in Financial Economics when it comes to thinking theoretically about options. Besides, the mere fact that the construction of Mathematical Finance was interwoven with financial engineering concerns stresses the normative role of mathematical models also from the standpoint of financial practice.

References

Ayres HF (1963) Risk aversion in the warrants market. Indus Manag Rev 4:497–505
Baumol WJ, Malkiel BG, Quandt RE (1966) The valuation of convertible securities. Quart J Econ 80:48–59
Belze L, Spieser P (2005) Histoire de la Finance. Le temps, le calcul et les promesses. Vuibert, Paris

Black F (1976) The pricing of commodity contracts. J Financ Econ 3:167–179
Black F, Scholes M (1973) The pricing of options and corporate liabilities. J Pol Econ 81:637–654
Boness J (1964) Elements of a theory of stock-option values. J Pol Econ 72:163–175
Chen A (1970) A model of warrant pricing in a dynamic market. J Financ 25:1041–1060
Fama EF (1965) The behavior of stock market prices. J Business 38:34–105
Fama EF (1970) Efficient capital markets: A review of theory and empirical work. J Financ 25:383–417
Fisher I (1907) The rate of interest. Macmillan, New York
Fisher I (1930) The theory of interest. Macmillan, New York
Garman MB, Kohlhagen SW (1983) Foreign currency options values. J Int Money Financ 2:231–237
Graham B, Dodd DL (1934) Security analysis. McGraw Hill, New York
Harrison JM, Kreps DM (1979) Martingales and arbitrage in multperiod securities markets. J Econ Theory 20:381–408
Harrison JM, Pliska SR (1981) Martingales and stochastic integrals on the theory of continuous trading. Stochast Proces Appl 11:215–260
Kreps DM (1981) Arbitrage and equilibrium in economies with infinitely many commodities. J Math Econ 8:15–35
Lintner J (1965) The valuation of risk assets and the selection of risky investments in stock portfolios and capital budgets. Rev of Econ and Stat 47(1):13–37
MacKenzie D, Millo Y (2003) Construction d'un marché et performation théorique. Sociologie historique d'une bourse de produits dérivés financiers. Réseaux 122:15–61
MacKenzie D, Muniesa F, Siu L (2007) Do economists make markets? Princeton University Press, Princeton
Margrabe W (1978) The value of an option to exchange one asset for another. J Financ 33:177–186
Markowitz HM (1952) Portfolio selection. J Financ 7:77–91
Markowitz HM (1959) Portfolio selection: Efficient diversification of investments. Wiley, New York
Mehrling P (2005) Fischer black and the revolutionary idea of finance. Wiley, Hoboken, NJ
Merton RC (1973) Theory of rational option pricing. Bell J Econ Manage Sci 4(1):141–183
Merton RC, Samuelson PA (1969) A complete model of warrant pricing that maximizes utility. Indus Manag Rev 10:17–46
Modigliani F, Miller M (1958) The cost of capital, corporation finance and the theory of investment. Am Econ Review 48(3):261–297
Mossin J (1966) Equilibrium in a capital asset market. Econometrica 34:768–783
Samuelson PA (1965) Rational theory of warrant pricing. Indus Manag Rev 6:13–31
Sharpe WF (1964) Capital asset prices: A theory of market equilibrium under conditions of risk. J Financ 19:425–442
Sprenkle C (1961) Warrant prices as indications of expectations. Yale Econ Essays 1:179–232
Tobin J (1958) Liquidity preference as behavior towards risk. Rev Econ Stud 25:65–86
Treynor J (1961) Market value, time and risk. Treynor papers
Williams JB (1938) The theory of investment value. Harvard University Press, Cambridge, MA

Chapter 14
Causation Across Levels, Constitution, and Constraint

Max Kistler

14.1 Introduction: Scientific Explanation and Causal Explanation

According to the traditional conception of logical empiricism, all scientific explanations are causal explanations. The deductive-nomological analysis was intended to indicate at the same time what it takes to be a scientific explanation and what it takes to be related as cause and effect. However, it is well known[1] that there are explanations that satisfy the formal requirements of the DN analysis without intuitively being causal: in such explanations, the initial conditions do not appear to refer to a cause of the explanandum. Additional requirements need to be imposed on two facts or events in order for them to be related as cause and effect, requirements that may be alternative or additional to the requirement of playing the logical roles of initial condition and conclusion in a valid DN-argument. One important suggestion is that causation requires the existence of a mechanism linking the cause to the effect. Such a mechanistic conception of causation falls into the wider category of process conceptions of causation according to which: (1) causes and effects are essentially localised in space and time, in other words they are events, and (2) the causal relation between such events is based on a local, intrinsic process the end points of which are the cause and the effect.

14.2 Reducing Causation to Mechanism?

No doubt, mechanistic explanations are causal explanations. It is part of what it means to be a mechanism that it extends from an initial to an end condition, where the former causes the latter. It is clear that initial and end conditions are meant to

---

M. Kistler (✉)
Département de Philosophie, Université Pierre Mendès France, UFR SH, BP 47,
38040 Grenoble cedex 9, France
and
Institut Jean Nicod, Paris
e-mail: kistlerm@upmf-grenoble.fr

[1] See, e.g., Humphreys (1989, p. 300/1), Salmon (1990, pp. 46–50) and Kistler (2002).

bear on different moments in time. Hence there can be no question of a "mechanism" linking two aspects of the same event. As a consequence, a mechanistic analysis avoids the wrong prediction of the DN analysis, that there may be causal relations between different properties of one substance at one time, such as between the temperature and the pressure of a given sample of gas.

But some have made the stronger claim that the concept of causation can be reduced to that of mechanism. According to Stuart Glennan, "events are causally related when there is a mechanism that connects them" (Glennan 1996, p. 49). Glennan himself admits that such a mechanistic account of causation "cannot explain causation in fundamental physics" (Glennan 1996, p. 50). It cannot be true of interactions between elementary particles that the existence of a causal relation is equivalent to the existence of a mechanism. Glennan concludes that there are two fundamentally different kinds of causation and suggests that "there should be a dichotomy in our understanding of causation between the case of fundamental physics and that of other sciences." (Glennan 1996, p. 50).

However, one would need stronger reasons to justify the radical and counterintuitive conclusion that there are two distinct concepts of causation, one for fundamental physical interactions and one for all other causal relations. This consequence is avoided as soon as one abandons the idea that causation can be *reduced* to mechanism. On closer inspection, it appears that the concept of mechanism presupposes that of causation, far from being reducible to it. Providing a mechanistic explanation means to decompose the working of a complex system into a number of simpler subsystems that interact causally with each other. These subsystems can in general themselves be analysed in still simpler subsystems, so that the interactions between the former subsystems can also be mechanistically explained. The crucial point is that each step of the analysis of a mechanism makes essential use of the notion of cause, and thus presupposes it. If one pushes the analysis far enough, one eventually reaches interactions between elementary particles. These however cannot in their turn be given a mechanistic analysis, because elementary particles cannot be decomposed into their parts. It follows that the concept of mechanism cannot be used to analyse the concept of causation and that, quite on the contrary, the concept of causation is among the irreducible conceptual instruments of mechanistic analysis. Mechanist causation rests in the last instance on the causation of fundamental physical processes.

14.3 "Top-Down" and "Bottom-Up" Experiments

Even if the concept of mechanism does not provide the means to reduce the concept of causation, reflection on the mechanistic analysis of complex systems and their experimental investigation may help us answer a major question raised in recent philosophical work on causation. Scientific experiments on mechanisms seem to rely on causal processes crossing the boundary between levels of composition, both in upward and downward direction.

- In "bottom-up" experiments, one manipulates properties ("independent variables") of individual components of a mechanism in order to observe the consequences of this intervention at the level of system properties ("dependent variables"), i.e., properties belonging only to the whole mechanism but to none of its parts.
- In downward or "top-down" experiments, the experimental intervention consists in manipulating system properties and observing its effects on properties of components of the mechanism.

An important category of bottom-up experiments uses the so-called "knockout" technique: organisms are genetically modified in such a way that specific genes are deleted. The observation of the development and behavioural capacities of such animals is taken to license inferences about the causal contribution of the knocked out genes to the development and capacities of the animal.

If there is bottom-up causation, we may expect there also to be top-down causation where a cause consisting in the modification of system properties has effects at the level of the system's microscopic constituents. Indeed, some experimental strategies seem to presuppose its possibility. In techniques of brain-imaging such as fMRI (functional magnetic resonance imaging) and single-cell recording, the experimenter manipulates system properties, e.g., by putting animals in a situation in which they accomplish a specific behavioural task, and observes subsequent modifications of properties at lower levels: fRMI allows to measure nervous activity in specific brain regions; single cell recording allows to observe the activity of individual neurons.[2] Such experiments intervene causally at the level of the organism: one manipulates the behaviour of the whole animal. The measured effect of that intervention lies at the level of the animal's microscopic constituents: one observes modifications of the properties and activities of neurons in the hippocampus.

Are such "interlevel" experiments instances of top-down and bottom-up causation, which means that they are grounded on interlevel causal relations? Scientists' statements suggest an affirmative answer. In Eric Kandel's words, the "biological analysis of learning requires the establishment of a causal relation between specific molecules and learning" (Kandel 2000, p. 1268). More specifically, Kandel acknowledges the existence of downward causation: "Learning produces changes in the effectiveness of neural connections" (p. 1275). Downward causation also seems to be required to make sense of psychotherapy: "Insofar as social intervention works [...] [e.g.] through psychotherapy [...] it must work by acting on the brain" (*ibid.*).

Recent philosophical work on causation also seems to lead to acknowledge bottom-up and top-down causation. According to Woodward (2003), causation can be analysed in terms of manipulability. If a cause of some property or factor E is a factor C such that interventions on C allow to manipulate E, then the bottom-up and top-down manipulations undertaken to understand the working of mechanisms are all cases of causation.

---

[2] See Ludvig et al. (2001).

14.4 The Puzzle of Downward Causation

However, downward causation, through which the evolution of a complex system causally influences the evolution of its own parts, raises considerable conceptual difficulties. Kim (1998) argues that downward causation is conceptually incompatible with two plausible metaphysical principles. The first, suggested by the success of physics in explaining physical phenomena, is the principle of the "causal closure of the domain of physical phenomena". It says that for a given physical event e that takes place at time t, for each time t^* preceding t, there is a complete physical cause c (at t^*) of e.[3]

The second principle used in Kim's argument is that there is no systematic overdetermination of microscopic events by independent micro- and macroscopic events. If event e at t has a complete physical cause c at time t^* (where t^* is earlier than t), then it does not (at least not in the general case) in addition have another complete cause C at the same time t^*, which is independent of c. In particular, if e is a neural event happening in a subject's brain at t, and c is a complete cause of e at the neural level, there will not (at least not in each case) be other complete causes of e that are simultaneous with c; in particular, there will not be a complete cause C at the cognitive level that is independent of c.

Here I can only sketch the argument against the conceivability of downward causation that Kim develops on the basis of these principles.[4] It proceeds in two steps. In the first step, Kim shows that the only way a mental event C could cause a second mental event E, is indirect, by causing, through a process of downward causation, e, the physical basis of E. By causing e, C necessarily brings about E, because e is E's supervenience basis. The supervenience relation entails that every instance of e is necessarily an instance of E. In a second step, Kim argues against the possibility of downward causation, which would, according to the first step, be required for mental causation. Given the causal closure of the physical domain, e has, at the time of C, a complete physical cause c. Now, either C is supervenient on c, in which case C is not an independent cause from c, or C is independent from c, in the sense that one could occur without the other. Then C's causing e is a case of overdetermination of an event, e, by two independent causes, c and C. It is controversial whether overdetermination is possible in exceptional cases, but it is generally taken for granted that it is implausible to suppose that all mental causes are cases of independent overdetermination.[5]

---

[3] Cf. Kim (1998, p. 37/8). See also Lowe (2000a, 2000b, p. 26 ff.).

[4] I have analysed Kim's argument in more detail in Kistler (2005, 1999/2006a, 2006b).

[5] It has been argued, e.g. by Mills (1996) and Walden (2001), that the effects of mental causes are systematically overdetermined by mental and physical causes, and that this overdetermination is not the result of the dependency of the mental causes on the physical causes. Mills makes it clear that "causal overdetermination requires the *distinct, independent* causal sufficiency of P [a physical cause] and of my believing" (Mills 1996, p. 107; italics Mills'). For lack of space, I cannot here examine Mills' and Walden's arguments in detail. Let me just note that Mills' own justification for the causal efficacy of a certain belief, with respect to the fact that his arm raises, contradicts this claim of independence. He justifies it by the truth of a counterfactual according to which

Kim's argument puts us before a dilemma: Either the argument is sound and we must revise our interpretation of interlevel manipulation of complex systems, so that it does not require any downward causation after all, or we abandon one of the two metaphysical principles Kim uses in his argument, so as to open up the logical space for downward causation.

14.5 Analysing Interlevel Causation in Terms of Constitution

Let me begin by the metaphysical notion of constitution, which is used to distinguish a material object from (1) its matter and (2) the set of its parts. In the present context, constitution is used to refer to the latter: the relation between a macroscopic object and the set of its parts. I will use Unger's (1980) example of the relation between a cloud and the droplets it contains, but the same points could be made with any other macroscopic object, such as tables, chairs and living beings. Here are two reasons why the set of tiny drops in a given cloud is not identical with the cloud: first, considering the evolution of the cloud in time, the concept of cloud allows it to persist, i.e., to continue to exist and remain the same cloud, while individual drops enter or leave it. However, each time a drop is added or removed, the set of drops in the cloud changes. Moreover, and this is the second reason for distinguishing the cloud from the set of its drops, even at a given moment of time, it would have been *possible* that the very same cloud contains some more drops or some less. Let us admit Kripke's thesis that all true identity statements of form "A = B", where A and B are rigid designators, are *necessarily* true. It follows from the contrapositive of this thesis that if a statement attributes a contingent relation to A and B, that relation cannot be identity. The fact that there could have been a different set of drops in the cloud, shows that the relation between the set of drops and the cloud is contingent. Therefore it cannot be a relation of identity. Here is where constitution steps in: One can say that the actual set of drops constitutes the cloud although they are not identical.

Three features of constitution will prove important in what follows. First, it is an asymmetric relation: if A constitutes B, it is impossible that B constitutes A. The set of drops constitutes the cloud but the cloud doesn't constitute the set of drops. Second, a given object can be, successively or alternatively, constituted by more than one set of parts. One might express this by saying that some objects allow for "multiple constitution". Third, constitution is a relation of logical and metaphysical, rather than epistemic or nomological type. It is not epistemic because the fact that a given set of drops constitutes the cloud is independent of our knowing or ignoring

---

the belief causes the arm movement in a possible world in which its physical cause is absent. Now, this counterfactual is true only because "worlds in which my belief is accompanied by some physical event that causes the arm-raising preserve actual laws, whereas worlds in which my belief is unaccompanied by any such physical event do not" (Mills 1996, p. 109). This reasoning seems to presuppose that there is a nomic correlation between physical and mental properties, which contradicts their independence.

this fact. The way in which we justify claims of constitution shows that they are not nomological. Hypotheses bearing on laws of nature can only be justified a posteriori, on the basis of observations of facts that are logically independent of each other and of those laws themselves. However, if I know the position and speed of each drop in the cloud, I know and can infer on purely conceptual grounds all properties of the cloud, such as its position, form and density. Therefore, the objects described by the premise (the drops) and the conclusion (the cloud) stand in a logical or metaphysical, rather than a nomological, relation.

Let us now turn to Craver and Bechtel's analysis of apparent cases of downward causation. Take their example of the process that begins with a person's decision to start a tennis game and leads to appropriate tennis-playing behaviour. The latter requires a raise of glucose consumption in the person's muscle cells. The decision, a system property of the person, seems to have effects at the cellular and molecular levels. However, Craver and Bechtel argue that this appearance is misleading, and disappears at closer inspection. "The case can be described without remainder by appeal only to intra-level causes and to constitutive relations" (Craver and Bechtel 2007, p. 559). If this is correct, downward causation can be analysed according to one of two patterns. In scenario 1, C (the decision) determines c (the brain state underlying the decision), which then causes e (enhanced consumption of glucose in muscle cells) by intra-level causation.

In scenario 2, C (the decision) causes E (appropriate behaviour at the level of the organism), which then determines e (enhanced glucose consumption) in a non-causal way.

The first scenario is inadequate if, as is generally assumed, mental events such as decisions to play tennis are multirealisable by many different brain states. Which particular brain state c realises C depends on the person's history and the circumstances. At any rate, C does not by itself determine c. Furthermore, even if it did (in other words, if we abstract away from multiple realisation), the downward determination of a brain event by a mental event could not possibly be construed as a relation of constitution, because constitution is a bottom-up relation.

The same reasons seem to make scenario 2 inadequate: First, E does not in itself determine e because tennis-playing behaviour, and even a given detailed bodily move, can be realised at the molecular level in many ways. Second, E does not constitute e: Parts can be constitutive of wholes but wholes cannot be constitutive of their parts.

14.6 Downward Causation and Downward Constraints

However, it is possible to reinterpret scenario 2 in such a way that it may represent the situation correctly. I suggest modifying Craver and Bechtel's proposal in two respects. First, the downward relation by which E determines e is a relation of *constraint* not of constitution. Second, the constraint imposed on e by E is not complete but partial.

Let me say a few words on the notion of constraint. A constraint limits the possibilities of evolution or change accessible to a system. In a system of equations with n variables, each equation imposes a constraint on the variables, in the sense of limiting the values the variables can take to satisfy the equations. If the variables represent the degrees of freedom of a physical system, i.e., the dimensions within which the state of the system can evolve, the notion of constraint acquires a physical meaning. Each equation expressing a link between the variables expresses a limit imposed on the possibilities of evolution of the system. Each constraint on a macroscopic system diminishes the number of possible states of its constituents. However, as long as there are less constraints than degrees of freedom, the constraints on a system determine its state only partially and not completely.

Contrary to constitution, constraint is not an asymmetric relation. One can say that the state of the parts of a system constrains the state of the whole; but it can also be correct to say that the state of the whole constrains the states of the parts, as when the position of a solid limits the degrees of freedom of the atoms constituting it.

The notion of degree of freedom, and thus the notion of a constraint limiting those degrees of freedom, can be generalized to all determinable properties of a system that can take different values. An animal's body temperature corresponds to a degree of freedom subject to the constraint of remaining within limits imposed by a regulatory mechanism at the level of the organism. However, this temperature constitutes itself a constraint imposed on the possible states of motion of the molecules composing the organism. The overall temperature imposed on the body by the regulatory mechanism limits the space of possible states of motion of the body's constitutive molecules, by fixing the mean kinetic energy of their states of motion. In the same sense, the fact that a given cognitive system is at a given moment in some cognitive state, e.g., of consciously perceiving an approaching tennis ball, imposes a constraint on the possible states of its parts, and first of all on the state of its neurons. It is incompatible with many neuronal states, such as states corresponding to closed eyes or the contemplation of an immobile scene. However, it is only partial and compatible with a great many microscopic states of neurons and molecules.

The process leading from the decision (C) of a person to her playing tennis (E) is an intra-level causal process at the level of the organism. I suggest that the concept of *partial constraint* helps us understand the relation between tennis playing and the underlying microscopic events e taking place in the body, such as enhanced glucose uptake in muscle cells. The state of organism E exerts a constraint on its parts, in the sense that the fact that the organism is in state E limits the space of possible states of its muscle cells. However, the detailed evolution of each muscle cell is also constrained at the cellular and molecular level, by the physical state of the cell and its surrounding.

The notions of constitution and constraint, which are both forms of non-causal determination, make causal relations crossing levels of composition conceivable. It is after all conceptually possible that a change occurring at the level of the parts of a system causes changes at the level of its systemic properties, and that a change of systemic properties causally influences the states of its parts.

With this analysis in mind, let us return to Kim's argument against the possibility of downward causation. According to Kim, the idea that a change in system properties might exercise a causal influence on the properties of the system's parts is incompatible with the principles of the causal closure of the physical domain and of explanatory exclusion. The controversial premise is the principle of the causal closure of the physical domain. Downward causation is possible if there can be microscopic events in complex systems that are not completely determined, in the long run, by same-level events. Cellular or molecular changes in a living organism may, e.g., not be completely determined over long time intervals by other cellular or molecular events. The brain may exhibit "deterministic chaos".[6] The possibility to make predictions about the evolution of a chaotic system is limited to a short time span. In other words, one cannot (deductively) explain a molecular event in a living organism (such as the transformation of an ATP in an ADP molecule in order to release the energy necessary for muscle contraction), on the basis of other molecular events that have occurred much earlier.

One can only draw a metaphysical conclusion – that the state e of the set of parts of the system at t is not causally determined by the state c of the set of parts of the system at t^* – from an epistemic premise – that it is impossible to make long term predictions in some chaotic systems – if one accepts the following two presuppositions. The first concerns the interpretation of the notion of causal determination. Causal relations can be analysed at two levels: they can be construed as relations between particular events, where a "particular" is a concrete object or event having many properties. At that level, it may be hypothesized that causation rests on the transmission of some quantity of energy (or some other conserved quantity) from one event to the other[7]. However, when one is interested in causal explanation, it is in general not sufficient to point to causal relations at the level of events in this sense. One does not only want to know which event made the billiard ball move at time t, but also what it is about the cause event that makes the effect event one in which the ball moves with a speed of 1 m/sec. In other words, the search for a causal explanation aims at establishing a *fact* about the cause event that is responsible for a fact about the effect event. What is causally responsible for the *fact* that the ball moves with 1 m/sec, is a fact bearing on the masses and speeds of the relevant billiard balls at some time earlier than t, say t^*. This "responsibility" of facts bearing on events happening at t^* for facts bearing on events happening at some later time t rests on laws linking the properties that are constitutive of those facts: laws link speeds and masses at t^* to speeds and masses at t. There is an ontological interpretation of this nomic determination: the dependence of the state of the billiard balls at t depends on their state at t^*, independently of our knowledge and description of

---

[6] Cf. Skarda and Freeman (1990), Lehnertz and Elger (2000) and Newman (2001).
[7] This thesis has been defended in Kistler (1998, 1999/2006a, 2006b).

these facts. In a realist framework, true deductive-nomological explanations of facts at t on the basis of facts at t^* have a truth-maker: the causal dependence, or causal responsibility of the latter for the former.

The second presupposition is that the indeterminacy of the state of a chaotic system is not only epistemic but also ontological. No empirical sense can be attached to the hypothesis that a determinable property of a physical system with a continuous value pattern, possesses at time t an absolutely precise value. There are absolute limits to the possible precision of measures that appear in the so-called uncertainty relations of quantum mechanics. Even if the state of a chaotic system has been determined with the absolutely maximal precision at time t^*, that state does not completely determine the state of the system at times t that are sufficiently distant from t^*. In such a chaotic system, the "horizontal" determination of physical events at the physical level is objectively incomplete. This throws doubt on the "principle of closure of the physical domain". In such a system, for a given physical fact at time t, and for times t^* sufficiently earlier than t, there is no physical fact at t^* that completely determines e. This does not mean that such a fact is completely indeterminate. The success of ethology and psychology in explaining numerous animal and human behaviours shows that animals and humans obey to "system laws"[8] constraining their evolution at the level of systemic properties, such as cognitive laws determining actions on the basis of reasoning and decision making. The fact that an organism obeys to such laws means that its evolution obeys constraints at a psychological level. The constraints exercised on the organism by laws at different levels, at the level of the organism as a whole and at various lower levels corresponding to its parts, create no conflict. If the determination of a molecular event is incomplete at its own level, it may nevertheless be completely determined jointly by laws at molecular and system levels. A given molecular event happening in an organism may be partly determined by constraints at the molecular level and partly by downward constraints from the psychological level, insofar as the organism obeys to psychological system laws[9].

The possibility of this scenario shows that, contrary to what Kim's first principle says, present-day scientific knowledge does not exclude the hypothesis that the domain of physical phenomena is not closed. The microphysical state of a complex system at t^* may not completely determine its microphysical state at a much later time t. In such a system, the microphysical state at t may be partially determined in a downward direction by the constraint that the system must, at t, be in a global state compatible with system level laws, such as cognitive laws. The determination of state e is completed by the physical circumstances occurring immediately before e.

---

[8] Cognitive laws linking actions to reasoning and decision are one case of what Schurz (2002) calls "system laws". Insofar as an organism exhibits regularities at the level of the organism, it is what Cartwright (1999) calls a "nomological machine".
[9] I have justified this sketch in a little more detail in Kistler (2006b).

14.7 Conclusion

Mechanisms are causal processes, and their analysis shows that they contain other more elementary causal processes. At the bottom level, there are fundamental physical causal processes that cannot, for lack of parts, themselves be given a mechanistic analysis. Therefore the concept of mechanism cannot be used to provide a noncircular analysis of the concept of causation.

Nevertheless, the analysis of mechanistic explanation can help us decide whether the mind can influence matter, and in particular, whether our decisions to behave can be considered as causes of microscopic changes in our body. Many philosophers take such "downward causation" to be mysterious and incompatible with general metaphysical principles abstracted away from science, such as the principle of the causal closure of the domain of physical events. I have tried to show that partial downward determination of microphysical states of a complex system is conceivable and does not violate any plausible scientific or metaphysical principles.[10]

References

Cartwright N (1999) The dappled world. A study of the boundaries of science. Cambridge University Press, Cambridge
Craver CF, Bechtel W (2007) Top-down causation without top-down causes. Biol Philos, 22:547–63, doi: 10.1007/s10539–006–9028–8
Humphreys PW (1989) The causes, some of the causes, and nothing but the causes. In Kitcher P, Salmon WC (eds.), Minnesota studies in the philosophy of science, vol XII: Scientific Explanation. University of Minnesota Press, Minneapolis, pp 283–306
Glennan S (1996) Mechanisms and the nature of causation. Erkenntnis 44:49–71
Kandel R (2000) Cellular mechanisms of learning and the biological basis of individuality. In: Kandel ER, Schwartz JH, Jessell TM (eds) Principles of neural science, Chap. 63. McGraw-Hill, New York, pp 1247–1279
Kim J (1998) Mind in a physical world. MIT Press, Cambridge, MA
Kistler M (1998) Reducing causality to transmission. Erkenntnis 48:1–24
Kistler M (1999/2006a). Causalité et lois de la nature. Vrin, Paris (1999), Causation and laws of nature. Routledge, London (2006)
Kistler M (2002) Causation in contemporary analytical philosophy. In: Esposito C, Porro P (eds) Quaestio-Annuario di storia della metafisica, vol 2, Brepols, Turnhout (Belgium), pp 635–668
Kistler M (2005) Is functional reduction logical reduction? Croatian J Philos 5:219–234
Kistler M (2006b) The mental, the macroscopic, and their effects. Epistemologia (Genova, Italy), 29:79–102
Lehnertz K, Elger CE (eds) (2000) Chaos in brain? World Scientific, Singapore
Lowe EJ (2000a) Causal closure principles and emergentism. Philosophy 75:571–585
Lowe EJ (2000b) An introduction to the philosophy of mind. Cambridge University Press, Cambridge
Ludvig N, Botero JM, Tang HM, Gohil B, Kral JG (2001) Single-cell recording from the brain of free moving monkeys. J Neurosci Methods 106(2):179–187
Mills E (1996) Interactionism and overdetermination. Am Philos Quart 33:105–17

---

[10] I thank my auditors in Madrid and Reinaldo Bernal for helpful criticism and discussion.

Newman DV (2001) Chaos, emergence, and the mind–body problem. Austr J Philos 79:180–196
Salmon W (1990) Four decades of scientific explanation. University of Minnesota Press, Minneapolis
Schurz G (2002) *Ceteris paribus* Laws: Classification and deconstruction. Erkenntnis 57:351–372
Skarda C, Freeman WJ (1990) Chaos and the new science of the brain. Concept Neurosci 1: 275–285
Unger P (1980) The problem of the many. In: French PA, Uehling TE, Wettstein HK (eds) Midwest studies in philosophy, vol 5. University of Minnesota Press, Minneapolis, pp 411–467
Walden S (2001) Kim's causal efficacy. Southern J Philos 39:441–460
Woodward J (2003) Making things happen. Oxford University Press, Oxford

Chapter 15
Epistemic Consequences of Two Different Strategies for Decomposing Biological Networks

Ulrich Krohs

15.1 Introduction

It is the mission of systems biology to investigate large biological molecular networks, paradigmatic of cellular extension. It accounts for the network in terms of mathematical models of various kinds. An *E. coli* cell contains several thousand species of molecular components (the number being subject to frequent revision) that are engaged in almost twice as many interactions (Keseler et al. 2005; Su et al. 2008). Networks of eukaryotic cells are even larger. Modeling a whole network of such dimensions requires mathematical tools that differ from the usual procedures of pathway modeling. And not only do the tools differ, the kind of results that can be read from models of this new kind also differ dramatically from what biologists learn from the familiar models of metabolic pathways (Westerhoff and Palsson 2004; O'Malley and Dupré 2005; Krohs and Callebaut 2007).

My paper first describes some modeling techniques on which systems biology relies (Section 15.2). I next describe shortly the two main strategies used to decompose molecular networks into modules (Section 15.3). I will then show that following any of the modularization strategies restricts the explanatory goals that might be followed by modeling the modules (Section 15.4). I conclude with a closer look at the claim that decomposition of a network according to structural criteria is neutral, while functional decomposition gives a biased picture of the network. Structural criteria do not emerge out of nowhere but have to be chosen. I demonstrate that the choice of particular structural criteria introduces a bias (Section 15.5).

U. Krohs (✉)
Department of Philosophy, University of Hamburg, Von-Melle-Park 6, 20146 Hamburg, Germany
and
Center for Philosophy of Science, University of Pittsburgh, 817 Cathedral of Learning, Pittsburgh, PA 15260, USA
e-mail: ulrich.krohs@uni-hamburg.de

15.2 Modeling Strategies in Systems Biology

Systems biologists aim at the complete picture of gene regulatory and metabolic networks. In the end, a model of gene regulatory processes or of the interactions among the proteins would include the roles of any nucleic acid sequence of the whole genome, or the interactions between any pair of proteins potentially expressed in the cell. This poses high demands on experimentation and on model building. As two European protagonists of the field have put it, the experimental challenge is the following:

> A complete systems biological approach requires: (i) a (complete) characterization of an organism in terms of what its molecular constituents are, with which molecules they interact, and how these interactions lead to cell function; (ii) a spatio-temporal molecular characterization of a cell (e.g., component dynamics, compartmentalization, vesicle transport); and (iii) a thorough systems analysis of the 'molecular response' of a cell to external and internal perturbations (Bruggeman and Westerhoff 2006, 46).

Modeling such a completely characterized system shall then serve the following goals:

> In addition, information from (i) and (ii) must be integrated into mathematical models to enable knowledge-testing by formulating predictions (hypotheses), the discovery of new biological mechanisms, calculation of the system behavior obtained under (iii), and finally, development of rational strategies for control and manipulation of cells. (ibid.)

These quotes should not be mistaken as the dreams of some data collection freaks or data mining nerds unaware of the epistemic challenge of biological inquiry or of the biological framework which embeds their work. Bruggeman and Westerhoff, also involved in and fostering debates about the philosophy of their field (e.g., Boogerd et al. 2007), are certainly aware of the problems of the gigantic long-term goals they are posing in the second quote, to which any particular study can at best deliver a small contribution. What Bruggeman and Westerhoff envisage is a synthesis of two strategies that are generally viewed as opposing systems biological approaches. The first quote may best be read as the request to integrate both strategies rather then to fight about the question which might be the better one. The first strategy, labeled (i) and (ii) in the first quotation, is the approach of a complete characterization of the components of a cell with the methods of enzyme kinetics; the system is modeled in a bottom-up way by integration of all these pieces of knowledge into one model. The second strategy, being the background of network perturbance studies (iii), is the top down-approach of a characterization of the network as a whole. Top-down modeling usually states the topology of a network without further characterizing the components in any other way than by describing their place within the network (as a "node" of the network) and the interactions they are engaged in ("edges" in the terminology of network analysis). In particular, top down-modeling does not require a kinetic characterization of the components of the network. Bruggeman and Westerhoff's integrative view envisages a modification of the top-down strategy so that complete knowledge about the components of a network is integrated in a topological model of the network.

While the reconciliation of both strategies seems to be a decent goal of systems biological research, it is confronted with the obstacle that the current epistemic goals of bottom-up and top-down systems biology differ. Each of the methodologies constrains the epistemic goals that could be followed, so the goals can hardly be integrated. To understand why this is so, we first need to look closer at the modeling strategies and epistemic goals of both branches, and then go back to strategies of network decomposition (Section 15.3) that must be applied before modeling can start.

The main kinds of models used in systems biology are kinetic models on the one hand (see, e.g., Bruggeman and Westerhoff 2006),[1] and discrete models like Boolean networks on the other (e.g., Chavez et al. 2005; Albert 2005).

Kinetic models, also called continuous state models, are well known from the older research program of metabolic pathway analysis. These models conceive biochemical pathways and networks as systems of enzyme-catalyzed chemical reactions that bring certain functions or physiological effects about. The enzymatic properties that go into such models are investigated by the methods of enzyme kinetics and described mathematically by rate equations in which the kinetic parameters characterize the properties of the enzymes. The reaction system is then formulated as a set of ordinary differential equations. Though such equation systems are not usually solvable analytically, some of their properties can be determined by mathematical analysis. Other results can be obtained by numerical integration, i.e., by running computer simulations. The epistemic goal that is followed by means of kinetic models is to explain how, i.e., by which molecular mechanism, some function, phenomenon or physiological effect is brought about. The mechanism is explanatory *of* that function, phenomenon or effect (Machamer et al. 2000; Machamer 2004; Bechtel and Abrahamsen 2005). For example, a kinetic model of protein biosynthesis explains mechanistically how proteins are produced in a cell (Darden and Craver 2002); the model of the mechanism of G-protein action explains how a molecular signal is transmitted and amplified (Krohs 2004, Chapter 10.1). Kinetic models usually comprise up to twenty or thirty components. Available simulation methods do not allow to model large networks this way. Application of the method in systems biology therefore requires first to break down networks into modules of manageable size.

Things are somewhat different with discrete models, though, to some degree, the need to break down the network persists (Albert 2005). Discrete models are built from an inventory of the network components and of the occurring interactions. They depict the topology of the network. In contrast to kinetic models, neither the components nor their interactions are further characterized, the only exception being the direction of an interaction (e.g., in cases of material flow), which is considered

---

[1] Stochastic models may serve as a more realistic substitute for kinetic models (which counterfactually assume continuity of matter). In these, reaction events, e.g., the transition of a particular molecule of a given molecular species, replace the reaction rates (Rao et al. 2002).

in the model: if node *a* influences node *b*, the reverse is not necessarily the case. Discrete models are based on the assumption that each component can exist in a limited number of states. In the most popular case of Boolean network models, only two states are admitted, "active" and "inactive." With gene regulatory networks, the two states are "expressed" and "not expressed," respectively (Chavez et al. 2005). A Boolean network is analyzed, usually without even looking at actual activity patterns of the involved components, by checking the possible states and sequences of states that a network of the particular topology under investigation may assume. The general setting is similar to a cellular automaton insofar as the activity pattern of the network is calculated in discrete time steps, following updating rules. An active node is considered to activate a node to which it is directly linked. An inactive node conversely deactivates another node if it is linked directly to it. Activation of an already active node has no effect; neither does inactivation of an inactive node.

The epistemic goal pursued by means of Boolean networks is to find "possibilities" of molecular networks. (Of particular interest are states that turn out to be stable, and sequences of activity patterns that happen to be periodical.) The hope is that these possibilities help explaining activity patterns of the cell. However, one must be aware that not all topologically possible states may represent possible states of the particular system under investigation, which consists of components with specific properties. These properties are not recognized in discrete models and impose constraints on the activity pattern of a network that go far beyond those captured by the updating rules of a Boolean network. The advantage of discrete models over kinetic models lies in the less demanding lab-work required to gather the data on which they are built. The tedious characterization of components can be omitted. Instead, only genomic or proteomic data are required, which can be gathered quickly and almost automatically by the high throughput methods of genomics and the other omic disciplines. So the experimental advantage of an approach that aims at discrete models is obvious.

15.3 Delineating Modules

As it was said before, all modeling strategies – even Boolean network modeling – require breaking down the network under consideration into smaller subnetworks. The aim is to delineate modules within the network that are close to being independent networks in themselves, not much influenced by other parts of the network they are embedded in. The two basic options are breaking down a network according to functional grouping of the components or to structural properties of the network. To inquire whether or not each modularization method is compatible with the different explanatory goals, we first need to look at the criteria both methods apply in network decomposition.

Functional modularization is based on the physiological view of a system: the system is regarded as a whole that has certain general capacities, which can be analyzed into more particular capacities and functional contributions of the single components to the capacities of the system. Well-known functional modules of the metabolic network of a cell are the citric acid cycle, pathways of amino acid metabolism, and β-oxidation of fatty acids. Without first identifying the functional hierarchy of cellular metabolism, no function-modular picture of the network would emerge. So functional modularization is biased by the assumption that the organization of an organism *is* functional (Rohwer et al. 1996; Koza et al. 2002; Friedman 2004; Papin et al. 2004; see Krohs and Callebaut 2007 for a discussion).

Structural modularization, on the other hand, is based on the view that the network is a dynamic structure which is characterized by its topology, i.e., the particular arrangement of nodes and edges. According to this view, whether or not the structure of the network is modular does not depend on any set of functional interrelations that may be found in a physiological analysis of the cell, but exclusively on the criterion of how strong, in a purely numerical sense, the components of the network are interconnected. The network is considered having a modular organization only insofar as subnetworks can be isolated that have strong internal and weak external connections. Philosophers as well as biologists often regard the seemingly neutral way of structural decomposition as delivering an authentic picture of the network (Bechtel and Richardson 1993; Schaffner 1998; Onami et al. 2002; Papin et al. 2004; Palsson 2006).

The mere plurality of a functional and a physicalistic approach, as it is mirrored in the two modularization strategies, does not pose in itself a problem. Biologists are generally working with these two different pictures of living nature. Both supplement each other since the physicalistic one describes the mechanisms that realize biological functions. To the extend that both pictures can be mapped onto each other this is the basis for explaining phenomena described in functional language by the mechanisms that bring them about (Craver 2001; Krohs 2004).[2] A mismatch between both accounts, however, questions the unity of the epistemic goals of biology. Exactly this seems to happen in the case of systems biology. As it turns out, structural decomposition of large molecular networks does not reduplicate the functional delineations – which is the very reason for the dispute about functional and structural delineation. The borders of functional units usually crisscross the structural borders.[3] This finding is where problems for the epistemic endeavor of systems biology start.

---

[2] See Krohs (2009b) for a reconstruction of theory structure in such cases.

[3] The citric acid cycle may serve as an example. It is delineated functionally (Krohs 2004: 173) and consequently forms a functional module. But each metabolic intermediate of the cycle is, besides its two edges within the cycle, also involved in many reactions that do not belong to the cycle but link it to other functional modules and help regulate the size of the pools of each of the intermediates (Kornberg 1965; Owen et al. 2002). By an analysis of a discrete model of the network, the external interactions are judged to be stronger than the internal ones. The functional module of the citric acid cycle is therefore not a structural module (Krohs 2009a).

15.4 Compatibility of Explanatory Goals with Delineation Methods

Which epistemic goals are compatible with a particular modularization method? This question can be answered by inquiring the compatibility of modularization methods with modeling strategies. As we have seen in Section 15.2, the two preeminent epistemic goals are mechanistic explanation and finding the possible states of a network. Each of the goals was shown to be accomplishable by one of the modeling strategies. Kinetic models are used in pursuing the goal of a mechanistic explanation of cellular functions. Once a mechanism is modeled, the model can be used to formulate predictions about the behavior of the system under standard or perturbed conditions, which are two other goals formulated by systems biologists (see the second quotation from Bruggeman and Westerhoff 2006 as given in Section 15.2). Discrete models, on the other hand, are chosen when the goal is to obtain results about the principle possibilities of a network. The question may for example be how many steady states or limit cycles the network may maximally assume.[4] This goal, while not included in Bruggeman and Westerhoff's list, is put forward by top down-systems biologists (Chavez 2005; Albert 2005).

I want to look first whether restrictions with respect to modeling may result from functional modularization. Kinetic models describe functional subunits of a network. Since these are singled out by a functional delineation of modules, this delineation method is clearly compatible with mechanistic explanation, which is the goal of kinetic modeling. The question is whether functional modularization is also compatible with discrete models, which depict the structure of the network. Functional modules group together some of the nodes and edges of a network in a way that is not likely to conserve structural delineations; other edges might link these particular nodes strongly to nodes outside the functional module under investigation. However, no other nodes or edges are stated than those present in the network, and the existence of edges to nodes outside the functional module is not denied. The functional module is the result of a different grouping of nodes and edges than in structural modularization. This nevertheless corrupts the application of discrete modeling strategies. Cutting off the connections with the rest of the network, though it leaves the combinatorial possibilities of the subnetwork unaffected, may influence the Boolean behavior of the module as far as the stability of the possible states is at stake: Whether or not a state or sequence of states is stable may be sensitive to the external connections of the module. Boolean models of functionally delineated modules must be expected to behave different from the behavior of the same subnetwork embedded in a structural module. For this reason, functional delineation predetermines further inquiry against Boolean network models and does

---

[4] As mentioned in Section 15.2, it is often the other way around, namely that only a discrete model can be based on a given data set and that it is difficult to find goals that are compatible with this modeling strategy (see also Krohs and Callebaut 2007).

not generally allow for achieving the goal of a reliable count of the stable states of a network. It predetermines scientific inquiry to follow other goals, e.g., those connected to kinetic modeling.

But also, structural delineation is not neutral with respect to achievable epistemic goals. Its pairing with discrete modeling and thus with the goal of an analysis of the possible states of a network is uncontroversial. But we have to inquire about its compatibility with kinetic modeling[5] and the goal of mechanistic explanation. Certainly, a structural module, including the processes its components undergo, can be conceptualized as a molecular mechanism. However, merely depicting a mechanism is not what scientists are aiming at. As we have seen in Section 15.2, the epistemically interesting aspect of models of mechanisms is that they are explanatory: a mechanism that brings some function, phenomenon or physiological effect about is explanatory *of* that function, phenomenon or effect. The phenomenon *of* which a mechanistic model is explanatory is individuated physiologically, i.e., functionally. Since no function (or functional contribution to some capacity) is ascribed to a structural module, a kinetic model of a structural module is no mechanistic explanation of anything of physiological relevance.[6] Structural modularization is therefore incompatible with the epistemic goals that are followed by kinetic modeling.

15.5 Epistemic Preconceptions: How Unbiased Is "Neutral"?

As we have seen, the criteria used for delineating modules of large metabolic or gene regulatory networks restrict the choice of models that can legitimately be used to describe the network and by this also predefine the explanatory goals that can be achieved. It was shown that only functional modularization allows for mechanistic explanation of cellular physiology, while models of structural modules mainly help to clarify why cells may exist in different stable states. I shall not raise an objection against the use of different research methodologies in the pursuit of different explanatory goals. However, as mentioned in Section 15.3, a debate is going on in literature about the soundness of the different modularization strategies in which proponents of structural modularization claim that their method is neutral while they tend to disqualify functional modularization as being biased. The debate seems to be ended by the following knock-down argument: the natural structure of a network can be established by purely mathematical and therefore neutral methods; a functional approach carves a network biased to the structural borders; thus functional decomposition, in contrast to structural methods, yields a picture that does not re-

---

[5] Including stochastic approaches as mentioned in note 1.
[6] It may, however, count as a mechanistic explanation of the *dynamic* of the structural module. But as long as this dynamic is not itself interpreted physiologically – and thus the structural framework is given up in favor of a functional one – it is hard to see the biological relevance of such an explanation.

produce the natural borders. The claim that structural modularization is neutral since it relies on purely mathematical criteria has some initial appeal. Its tenability must nevertheless be challenged.

So what are the grounds for the claim about the neutrality of structural modularization? The application of a mathematical criterion seems to be neutral once it is selected. In the case of modularization, the criterion refers to the strength of interactions between nodes of the network (a module has strong internal and weak external interactions). Even though the criterion can be applied with mathematical strength and precision, it carves modules out of a network by cutting edges to other modules. The criterion helps minimizing the distorting effect of such unavoidable cuts but can not eliminate it. As it was admonished already at the very beginning of the modularity debate: decomposition *inevitably* distorts the picture of a network (Simon 1969). One can at best argue that structural delineation distorts the picture to a lesser extent than functional modularization. Moreover, application of a given criterion is not the whole story. We have to discern between the application of a given mathematical criterion, and the decision for one particular criterion. Graph theory provides several other criteria that can be used to decompose a discrete network. Just to give one example: one could isolate recurring patterns within the network and regard these as its basic components. Using this criterion, so-called significant interaction motifs (Milo et al. 2002; Shen-Orr et al. 2002) rather than modules would show up as the building blocks of a network. This decomposition procedure, which also relies on the application of a purely mathematical criterion, is as "neutral" as the one for structural modularization. The decision for one or the other strategy can therefore not be based on a neutrality claim. Selection of a criterion must be based upon independent reasons, which will often be related to the epistemic goals. It thus introduces an – unavoidable – bias. So selection of a particular decomposition criterion and application of the criterion both introduce a bias or are the source of a distortion of the picture of the network. The very notion of a neutral decomposition strategy turns out to be dubious.

One final remark about bias: if systems biology aims at an unbiased methodology, the first step has to be made on a completely different field. Nothing seems to influence systems biological modeling more than the availability of particular kinds of data sets. The field of top down-systems biology and consequently discrete modeling is largely driven by the availability of cheap (and dirty, i.e., unreliable)[7] structural data that are available courtesy to the high throughput methods developed in the omic disciplines. At the same time, it suffers from a lack of kinetic data (Krohs and Callebaut 2007). The bias toward structural over kinetic data is introduced neither by any decision about epistemic goals, nor by methodological considerations, but simply by convenience. Being driven by easily available data rather than by scientific goals is no small source of bias. I do not want to argue against the attempt of systems biology to gain the best possible insight into molecular networks by use of the available data. However, I wish to emphasize that a picture of systems biological

---

[7] On of the main problems being a high percentage of false positives in the analysis of interactions (Deane et al. 2002; Albert 2005).

networks that would be based on network kinetics might yield yet other criteria for the delineation of modules or other substructures of large molecular networks. There is no reason to regard the topological picture drawn from structural top level-data as superior in any sense to a more elaborate, dynamic picture. Systems biologists of any camp should therefore aim, with Albert (2005) and Bruggeman and Westerhoff (2006), for an inclusion of bottom level, kinetic data, i.e., for the sound way of often tedious biochemical research.

Acknowledgment Support during the later stages of this work by the Deutsche Forschungsgemeinschaft (DFG), grant KR3662/1–1, is gratefully acknowledged.

References

Albert R (2005) Scale-free networks in cell biology. J Cell Sci 118:4947–4957
Bechtel W, Abrahamsen Å (2005) Explanation: A mechanist alternative. Stud Hist Philos Biol Biomed Sci 33:421–441
Bechtel W, Richardson RC (1993) Discovering complexity: decomposition and localization as strategies in scientific research. Princeton University Press, Princeton, NJ
Boogerd FC, Bruggeman FJ, Hofmeyr J-HS, Westerhoff HV (eds) (2007) Systems biology: Philosophical foundations. Elsevier, Amsterdam
Bruggeman, FJ, Westerhoff HV (2006) The nature of systems biology. Trends Microbiol 15:45–50
Chavez M, Albert R, Sontag ED (2005) Robustness and fragility of Boolean models for genetic regulatory networks. J Theoret Biol 235:431–449
Craver CF (2001) Role functions, mechanisms, and hierarchy. Philos Sci 68:53–74
Darden L, Craver CF (2002) Strategies in the interfield discovery of the mechanism of protein synthesis. Stud Hist Philos Biol Biomed Sci 33:1–28
Deane CM, Salwiński Ł, Xenarios I, Eisenberg D (2002) Protein interactions: Two methods for assessment of the reliability of high throughput observations. Mol Cell Proteom 1:349–356
Friedman N (2004) Inferring cellular networks using probabilistic graphical models. Science 303:799–805
Keseler IM, Collado-Vides J, Gama-Castro S, Ingraham J, Paley S, Paulsen IT, Peralta-Gil M, Karp PD (2005) EcoCyc: A comprehensive database resource for *Escherichia coli*. Nucl Acid Res 33(Database Issue):D334–D337
Kornberg HL (1965) Anaplerotic sequences in microbial metabolism. Angewandte Chem Int Edn 4:558–565
Koza JR, Mydlowec W, Lanza G, Yu J, Keane MA (2002) Automated reverse engineering of metabolic pathways from observed data by means of genetic programming. In Kitano H (ed) Foundations of systems biology. MIT Press, Cambridge, MA, pp 95–121
Krohs U (2004) Eine Theorie biologischer Theorien: Status und Gehalt von Funktionsaussagen und informationstheoretischen Modellen. Springer, Berlin
Krohs, U (2009a) The cost of modularity. In: Krohs U, Kroes P (eds) Functions in biological and artificial worlds: Comparative philosophical perspectives. MIT Press, Cambridge MA, pp 259–276
Krohs U (2009b). Structure and coherence of two-model-descriptions of technical artefacts. Technē: Res Philos Technol 13:150–161
Krohs U, Callebaut W (2007) Data without models merging with models without data. In: Boogerd FC, Bruggeman FJ, Hofmeyr J-HS, Westerhoff HV (eds) Systems biology: Philosophical foundations. Elsevier, Amsterdam, pp 181–213

Machamer P (2004) Activities and causation: The metaphysics and epistemology of mechanisms. Int Stud Philos Sci 18:27–39

Machamer P, Darden L, Craver CF (2000) Thinking about mechanisms. Philos Sci 67:1–25

Milo R, Shen-Orr S, Itzkovitz S, Kashtan N, Chklovskii D, Alon U (2002) Network motifs: Simple building blocks of complex networks. Science 298:824–827

O'Malley M, Dupré J (2005) Fundamental issues in systems biology. BioEssays 27:1270–1276

Onami S, Kyoda KM, Morohashi M, Kitano H (2002) The DBRF method for inferring a gene network from large-scale steady-state gene expression data. In Kitano H (ed) Foundations of systems biology. MIT Press, Cambridge, MA, pp 59–75

Owen OE, Kalhan SC, Hanson RW (2002) The key role of anaplerosis and cataplerosis for citric acid cycle function. J Biol Chem 277:30409–30412

Palsson B (2006) Systems biology. Cambridge University Press, Cambridge, MA

Papin JA, Reed JL, Palsson BO (2004). Hierarchical thinking in network biology: The unbiased modularization of biochemical networks. Trends Biochem Sci 29:641–647

Rao CV, Wolf DM, Arkin AP (2002) Control, exploitation and tolerance of intracellular noise. Nature 420:231–237

Rohwer JM, Schuster S, Westerhoff HV (1996) How to recognize monofunctional units in a metabolic system. J Theoret Biol 179:213–228

Schaffner KF (1998) Genes, behavior, and developmental emergentism: One process, indivisible? Philos Sci 65:209–252

Shen-Orr SS, Milo R, Mangan S, Alon U (2002) Network motifs in the transcriptional regulation network of *Escherichia coli*. Nat Genet 31:64–68

Simon HA (1969) The sciences of the artificial, 3rd edn. MIT Press, Cambridge, MA

Westerhoff HV, Palsson BO (2004) The evolution of molecular biology into systems biology. Nat Biotechnol 22:1249–1252

Su C, Peregrin-Alvarez JM, Butland G, Phanse S, Fong V, Emili A, Parkinson J (2008) Bacteriome.org – An integrated protein interaction database for *E. coli*. Nucl Acid Res 36(Database issue): D632–D636

Chapter 16
Matter(s) in Relativity Theory

Dennis Lehmkuhl

16.1 Introduction

In the Stanford Encyclopaedia for Philosophy, in the entry "Intrinsic and Extrinsic properties", Weatherson (2007) writes:

> I have some of my properties purely in virtue of the way I am. (My mass is an example.) I have other properties in virtue of the way I interact with the world. (My weight is an example.) The former are the intrinsic properties, the latter are the extrinsic properties.

The claim that mass is intrinsic is initially plausible. But concepts like 'mass' and 'weight' are surely not theory-independent. I will argue that mass–stress–energy momentum density in relativity theory is *not* an intrinsic property of material systems, that indeed even for Newtonian mass *density* a case can be made to this effect.[1]

Before I make this more precise, let us see how $T_{\mu\nu}$ enters the theory of general relativity (GR). It is right at the core of the theory, in the Einstein (field) equations

$$R_{\mu\nu} - \frac{1}{2}g_{\mu\nu}R = \kappa_E T_{\mu\nu}, \qquad (16.1)$$

which should be compared to the Poisson equation of Newtonian gravitational physics:

$$\nabla^2 \varphi = \kappa_N \rho. \qquad (16.2)$$

Within their respective theories, both the Poisson equation and the Einstein equations are supposed to describe how gravity, represented by the left side of the

---

D. Lehmkuhl (✉)
Oriel College, Oxford University, Oxford OX4 1EW, UK
e-mail: dennis.lehmkuhl@uni-wuppertal.de

[1] For simplicity, I will often just speak of the 'energy–momentum tensor' or even just of the 'energy tensor' of a material system, rather than of a mass–stress–energy–momentum density tensor. Note that $T_{\mu\nu}$ is *not* a tensor density in the mathematical sense: like the scalar field ρ in Newtonian theory, it is a tensor that represents a *physical* density, rather than a mathematical object that transforms as a tensor density.

respective equation, and matter, represented by the right side, interact with each other.[2]

In the Poisson Eq. 16.2, we have the gravitational potential φ on the left-hand side and the mass density ρ on the right-hand side. For the Einstein Eq. 16.1, the role of the mass density ρ is taken over by the mass–energy–momentum tensor $T_{\mu\nu}$, while the left-hand side of the equation is formed by the Ricci curvature tensor $R_{\mu\nu}$, the Ricci scalar R and the metric tensor $g_{\mu\nu}$; the former two being defined in terms of the latter.[3]

I will show that the energy tensor $T_{\mu\nu}$ is in important ways less fundamental than the metric field $g_{\mu\nu}$. Historically, Mach's principle, the idea that the reverse was the case, was very important for Einstein up until 1921. In Einstein (1918, p. 38), he expresses the principle in the following way:

> Mach's principle: The $[g_{\mu\nu}]$-field is *fully* determined by the masses of bodies. Since according to the results of the special theory of relativity mass and energy are the same, and since energy is formally described by the symmetric energy tensor $(T_{\mu\nu})$, Mach's principle says that the $[g_{\mu\nu}]$-field is constrained and determined by the energy tensor.

Einstein's formulation of Mach's principle is often taken as indicating his commitment to a Leibnizian/relationalist programme: spacetime was supposed to be secondary to material objects. Famously, GR does not fulfil Mach's principle as defined above; for example, the original gravitational field Eq. 16.1 allow empty Minkowski spacetime as a solution, among many other matter-free solutions. Even the modified field equations, in which Einstein introduced the cosmological constant λ in order for them to accord with Mach's principle, turned out to allow for non-trivial solutions even if $T_{\mu\nu} = 0$.[4] Furthermore, the left-hand side of the Einstein equations represents only part of the geometric structure – the Ricci curvature $R_{\mu\nu}$ – whereas the Weyl curvature $C_{\mu\nu\sigma\omega}$ is only constrained but not determined by the energy tensor $T_{\mu\nu}$.

We will see that *even if* knowing the energy tensor did uniquely determine the geometric structure, and hence even if Einstein's formulation of Mach's principle was fulfilled, there would still be no reason to regard matter as more fundamental than spacetime in the theory. For the only thing that would really suggest that either matter or geometry was more fundamental would be if the existence of one was

---

[2] The custom in relativity theory is to count radiation like the electromagnetic field as 'matter'. The left-hand side of the Einstein equations is often claimed to describe *both* the geometry of spacetime and the gravitational field. The main issues of the paper do not depend on whether one sees the metric field $g_{\mu\nu}$ as representing the geometry of physical spacetime, as 'just another field, not intrinsically different from the electromagnetic field', or as both at once. I will sometimes call $g_{\mu\nu}$ 'the geometry of spacetime', but people who do not like that and the ontological flavour of this choice of words should just substitute for it 'the gravitational field' or 'the metric field', without this altering the points made in this article. In Lehmkuhl (2008), I discuss the ways in which this alleged double role can be understood.

[3] Both equations also contain coupling constants, $\kappa_N = -4\pi G$ and $\kappa_E = \frac{8\pi G}{c^4}$, where the latter is obtained by demanding that the Einstein equations should go over into the Poisson equation in the non-relativistic limit.

[4] See Hoefer (1994) for details.

a requirement for the existence of the other but not vice versa. We will see that spacetime *can* exist without matter, whereas systems *cannot* possess mass–energy–momentum density without spacetime structure being in place.

I will start out in Section 16.2 by arguing that it does not matter whether there is a functional dependence of energy tensors on the metric tensor. In Section 16.3, I then show that instead we have a variety of definitional dependencies of $T_{\mu\nu}$ on $g_{\mu\nu}$. Hence, *the* main property of matter depends on the relations that hold between material systems and spacetime structure; so that Section 16.4 argues that mass–energy–momentum is a relational property of matter.

All this is a theme touched upon by Einstein in a virtually unknown letter to Felix Pirani, written in 1954, which shows how much Einstein had changed his mind on the issue; he writes that Mach's principle "is tricky, for the T_{ik}, which are supposed to represent "matter", always presuppose the g_{ik} field. [...] one should not speak of Mach's principle anymore."[5] Making the first sentence precise is exactly the point of this paper.

16.2 Explicit Metric Dependence

The fundamental mathematical objects in relativistic field theory are tensor fields like $F_{\mu\nu}$, ϕ and $g_{\mu\nu}$. We often speak of $F_{\mu\nu}$ as 'an electromagnetic field', of ϕ as 'a scalar field' and of $g_{\mu\nu}$ as 'a metric field', and indeed one can interpret the tensor fields as referring to the physical fields in a rather direct way. But one can also argue that $F_{\mu\nu}$ does not represent the electromagnetic field 'as such', but a certain property, set of properties or even trope of the physical electromagnetic field: its amplitude, or electromagnetic *field strength*. Nothing I am going to say depends on this choice; although I favour the interpretation that regards $F_{\mu\nu}$ as representing the fundamental properties of the material system described by $F_{\mu\nu}$.

If one endorses this interpretation, one should note the following (and corresponding facts if one makes a different interpretational choice): the fundamental properties of a material system do not need to be describable by only one tensor field. The fundamental properties of a perfect fluid for example are described by a triple of matter fields, (ρ, v^μ, p), where ρ is the proper density of the particles the fluid consists of, v^μ the velocity field describing the movement of every particle, and p the pressure field, giving the force an arbitrary fluid volume element 'feels' due to the movement of the rest of the fluid.

I will follow common custom and call a tensor field Φ a 'matter field' if it describes a material system. But whereas mass density was represented by a simple scalar field in Newtonian physics, relativistic mass–energy–momentum density $T_{\mu\nu}$ is defined *in terms of* the fundamental matter fields associated with the material system. However, this is not enough: energy tensors also depend on the metric field $g_{\mu\nu}$!

---

[5] See Call No. 17447 of the Einstein Arcives at the University of Jerusalem, also found at the Einstein Papers project at the California Institute of Technology.

One could think that this is obvious, for in their coordinate-independent representation the energy tensors of almost all paradigm material systems contain the metric field explicitly.

For example, the energy tensor of an electromagnetic field is

$$T_{\mu\nu} = \frac{1}{4\pi}(F_\mu{}^\lambda F_{\lambda\nu} + \frac{1}{4}g_{\mu\nu} F^{\sigma\lambda} F_{\sigma\lambda}), \qquad (16.3)$$

while the energy tensor for a perfect fluid is given by

$$T_{\mu\nu} = (\rho + p)v_\mu v_\nu - p g_{\mu\nu}. \qquad (16.4)$$

However, this *functional dependence* on the metric does not tell us much. First of all, it is not a feature that *all* energy tensors share: there are two particularly simple systems whose energy tensor does not depend on the metric explicitly. One of them is a specialisation of a perfect fluid, namely a perfect fluid without pressure, also called *dust* for short. The energy tensor of this system is

$$T_{\mu\nu} = \rho v_\mu v_\nu. \qquad (16.5)$$

One might think that a material system whose energy tensor depends on the metric explicitly might be 'more' dependent on the metric field $g_{\mu\nu}$, and in particular more sensitive to changes in the metric from point to point. But this is not the case. For if a metric is defined we can write Eq. 16.5 equally well as

$$T_{\mu\nu} = \rho g_{\alpha\mu} g_{\beta\nu} v^\alpha v^\beta, \qquad (16.6)$$

and it will still represent the same physics. In particular, curvature will not influence a system represented by Eq. 16.6 any more than a system represented by Eq. 16.5.

The next section will show that the relevant dependence of energy tensors on the metric field is *not a functional* dependence, but a variety of other kinds of dependencies: *definitional*, *representational* and *interpretational* dependence.[6]

16.3 Definitional Dependence on the Metric

16.3.1 Definitional Dependence at the Level of the Matter Fields

In some cases, the matter fields themselves need to have certain properties in order to allow them to play their role in forming the energy tensor of the material system in question. This is particularly true for all kinds of fluids, where we need the velocity

---

[6] The following section rests on discussions with Robert Geroch, Erik Curiel, Stephen Lyle, John Norton and David Malament, to whom I am very grateful for their help and patience. Needless to say, any remaining unclarities or misconceptions are surely mine.

vectors appearing in the respective energy tensor to be of a certain kind. The two simplest fluid systems are 'normal dust' and 'null dust'. The former represents a collection of particles which do not collide with each other (and is hence identical with a pressureless perfect fluid, cf. Eq. 16.4), the latter represents a collection of not directly interacting light rays. The energy tensor of both systems has the form

$$T_{\mu\nu} = \rho v_\mu v_\nu. \tag{16.7}$$

In the case of 'normal dust', the velocity vector v^μ has to be time-like, in the case of 'null dust' it has to be light-like. It is these properties that enables the two versions of Eq. 16.7 to represent normal or null dust respectively, for a time-like vector field represents an object that moves with less than the speed of light (i.e., a material particle), whereas a light-like vector field represents an object that does move with the speed of light (i.e., a light ray).

We will now see that their definition demands spacetime structure to be in place.

In the case of normal dust, one could argue that we need full-blown metrical structure, because an essential part of the model is that the velocity vectors are normalised, $v_\mu v^\mu = -1$, and in order to normalise vectors, indeed in order to assign to them any definite length, we need a metric. In the case of null dust though, we do not need this condition to be fulfilled in order for the velocity vectors to have the properties needed to play their part in forming an adequate energy tensor for the system.

But for both kinds of dust we need conformal structure, i.e., an equivalence class of metrics related by a conformal transformation. More precisely, we need an equivalence class $[g_{\mu\nu}]$ such that for any two elements of the class, say $\tilde{g}_{\mu\nu}$ and $\bar{g}_{\mu\nu}$, we have $\tilde{g}_{\mu\nu} = \omega \bar{g}_{\mu\nu}$, where ω is a smooth, strictly positive function.

If we did not have such a structure in place, there would be no way of saying whether a vector is time-like, space-like or light-like. Hence, neither the energy tensor of normal nor the energy tensor of null dust would be definable. Does this mean that we need spacetime to have only conformal rather than metric structure in order to define the energy tensor of the two dust systems? Indeed, it would not be too surprising if we did not need full-blown metrical structure even for the simplest material systems – after all, even some gravitational phenomena can be described without recurrence to the metric. But we will see below that even the energy tensor of null dust depends on the full metric structure of spacetime in other ways.

For the energy tensors of more complex material systems (and, arguably, for normal dust), we often need the full metric in order for the matter fields contained in them to have the properties they need in order to define the energy tensor. For example, any energy tensor which contains matter fields of the 'same name' but with differing numbers of upper and lower indices (e.g., $F_\nu{}^\lambda$ and $F_{\sigma\lambda}$ in the electromagnetic energy tensor (16.3)) presupposes a unique isomorphism between the tangent space of the manifold and its dual. And we only have such an isomorphism if we have a metric tensor field $g_{\mu\nu}$ defined on the manifold.[7]

---

[7] See Wald (1984, pp. 22–25).

I will call the above dependence of energy tensors on the metric tensor, namely the need for the metric in order for the matter fields appearing in the energy tensor to have certain properties, a *definitional dependence on the level of the matter fields*. We will now see a different kind of dependence on the metric, namely the need for the metric in order for the energy tensor itself to have two crucial properties. In fact, a $T_{\mu\nu}$ that lacked these two properties would be judged unphysical.

16.3.2 Constraint Dependence

In Newtonian physics, the minimum requirement for a mass density tensor to be regarded as representing a physically possible mass density is that it is positive. Similarly, in relativistic field theory it is demanded that the mass–energy density is positive in any frame of reference, for any observer. This constraint is known as *the weak energy condition*. Malament (2007, p. 240) defines it in the following way:

Weak Energy Condition Given any future-directed unit time-like vector ζ^a at any point in M, $T_{ab}\zeta^a\zeta^b \geq 0$.

This condition is not even formalisable without a metric, for we need the metric in order to have unit time-like vector fields (cf. above). A system for which the weak energy condition does not hold could decay towards a lower and lower energy state without end.

But *even if* we could weaken the weak energy condition (further) in a way that would need only conformal structure to be in place, the second property that we demand of an energy tensor in order to be a physical energy tensor needs a full-blown metric tensor field in order to hold. It is the condition that mass–energy–momentum is *covariantly conserved*, i.e., that the conservation equation

$$\nabla^\mu T_{\mu\nu} = 0 \qquad (16.8)$$

holds. This equation would not in general hold for a connection that is derived from a merely conformal metric $\tilde{g}_{\mu\nu} = \omega \bar{g}_{\mu\nu}$. Furthermore, if we assume that Eq. 16.8 holds, the conformal factor ω is determined, and conformal structure is extended to metric structure.[8]

Note that Eq. 16.8 is an automatic consequence of the Einstein field equations, and indeed in a Lagrangian formulation it is a consequence of both the gravitational and the matter field equations.[9] But also note that *this* dependence of $T_{\mu\nu}$ on $g_{\mu\nu}$ is of a rather different kind than the one we encountered in the last subsection. There we found a definitional dependence of energy tensors on the metric tensor field on the level of the matter fields: we need the metric in order for the matter fields in the energy tensor to have certain properties. Here, we have what might be called a definitional dependence on the level of the constraints: we need the metric in order

---

[8] Cf. Hawking and Ellis (1973, p. 63).
[9] See for example Brown and Brading (2002). For a discussion of the role energy conservation has for the substantivalist/relationalist debate, see Hoefer (2000).

for a $T_{\mu\nu}$ to fulfil certain constraints, a failure of which would give us good reason to regard the energy tensor in question as unphysical, as not representational.

16.3.3 Abstract Definitional Dependence

A third kind of dependence on the metric becomes evident if we look at $T_{\mu\nu}$ in the most abstract manner: as a map. Indeed, we can see $T_{\mu\nu}$ as a machine that always gives us the mass–energy density of a given system at a given point p relative to the state of motion of an arbitrary observer. Every such observer (or, differently speaking, every frame of reference at the point for which we want to know the mass–energy density relative to it) is represented by a future-directed, *unit-timelike* vector w^μ (at p). The mass–energy–momentum tensor is then a map $T : T_pM \times T_pM \to \Theta$, where T_pM is the tangent vector space w^μ lives in, and Θ is the set of all possible mass–energy densities ρ. But again, this map can only be defined if a metric $g_{\mu\nu}$ is defined, for otherwise, as we have seen at the beginning of this section, we could not have unit timelike vectors in the first place! We might call this kind of dependence on the metric an *abstract definitional dependence*.[10]

16.3.4 Interpretational Dependence

The fourth kind of dependence of energy tensors on the metric tensor can even be seen to apply to a Newtonian mass density. The latter is represented by a scalar field ρ, whose definition does of course not depend on the metric. But one could argue that one needs some spacetime structure in order to make sense of ρ as representing a physical *density*. In particular, one could argue that one needs a volume-element, with the help of which we can speak of the density in a particular volume; and the natural volume element of spacetime is defined in terms of the metric tensor. Naturally, this reasoning applies equally well to the mass–energy–momentum *density* tensor $T_{\mu\nu}$, which is supposed to encode mass/energy *density*, momentum *density*, momentum current *density* and mass/energy current *density*.

Of course, one can make the volume element arbitrarily small so that even in curved space the Euclidean/Minkowskian metric would be a good enough approximation, but the fact remains that we need a relation to spacetime structure in order to make sense of densities. The *total* mass of an extended Newtonian object can only be properly defined via integrating the mass density over a (Euclidean) volume element anyway; this is even true for the proper definition of the total mass of a point particle, where the integration process involves use of an appropriate delta function.

We have seen that energy tensors depend on the metric in a variety of ways: we have a frequent, although not general, *definitional dependence on the level of the*

---

[10] For the above way of defining the energy–momentum tensor see Malament (2007, p. 240).

matter fields making up specific energy tensors, a *constraint dependence* of every $T_{\mu\nu}$ on $g_{\mu\nu}$, an *abstract definitional dependence* in the sense that we need the metric in order to define $T_{\mu\nu}$ as a map from pairs of unit timelike 4-vectors to energy densities, and finally even an *interpretational dependence* that demands a metric field in order to interpret tensor fields as representing physical densities in the first place. Note that in all these cases the situation did not depend on whether the metric was a static or a dynamical field.

But what follows from all this?

16.4 Mass–Energy–Momentum as a Relational Property

In Section 16.1, I pointed out that the only thing that would really speak to either matter or geometry being more fundamental would be if one was a requirement for the existence of the other but not vice versa. But the tensor field that is supposed to represent *the* main properties of matter, the mass–energy–momentum density tensor $T_{\mu\nu}$, is a non-fundamental field, and it *does* require *both* the metric field $g_{\mu\nu}$ and the matter fields Φ in order to be defined, and in order to have some of its crucial properties.

So can we really say that $T_{\mu\nu}$ is an intrinsic property of material systems?

Lewis (1983, pp. 111, 112) introduces the distinction between intrinsic and extrinsic properties in the following way:

> A sentence or statement or proposition that ascribes intrinsic properties to something is entirely about that thing; whereas an ascription of extrinsic properties to something is not entirely about that thing, though it may well be about some larger whole which includes that thing as part. A thing has its intrinsic properties in virtue of the way that thing itself, and nothing else, is. Not so for extrinsic properties, though a thing may well have these in virtue of the way some larger whole is. The intrinsic properties of something depend only on that thing; whereas the extrinsic properties of something may depend, wholly or partly, on something else.

We have seen in Section 16.3 that energy tensors $T_{\mu\nu}$ depend not only on the matter fields Φ, but also on the metric field $g_{\mu\nu}$; indeed we have seen that energy tensors depend on the metric in a number of ways.

It seems to follow that the energy tensor $T_{\mu\nu}$ must be seen as corresponding to an *extrinsic* property of material systems. For matter only has an energy tensor $T_{\mu\nu}$ associated with it *in virtue* of the relations holding between the material system (whose fundamental properties are represented by the matter fields Φ) and the geometric structure of spacetime (whose fundamental properties are represented by the metric field $g_{\mu\nu}$).

We should now distinguish between relations and relational properties, in order to find out what kind of extrinsic property *exactly* we should see $T_{\mu\nu}$ to be, and to see what it is a property *of*.

An *n*-place relation $G(x_1, \ldots, x_n)$ depends on *n* entries. An example of a 2-place relation is '$Rxy := (x$ is the father of $y)$'. A relation gives rise to a set of relational properties. For example, Hermann Einstein and Albert Einstein stand in certain relations to each other, which give rise to the relational property of Hermann that he is a father, and to Albert's relational property of being a son.

We can make this distinction more precise in the following way: let us call an n-place predicate, where $n \geq 2$, a *relation* if and only if it contains more than one free variable x, y, z, while it is allowed to contain an arbitrary number of designators a, b, c. Let us speak of a *relational property* if we have an n-place predicate that contains at most one free variable, and at least one designator.

Then we can think of Hermann and Albert Einstein in the following way. We can think of them as two individuals (or two systems each of which consists of only one individual) who stand in certain relations to each other and have hence various relational properties. For example, if we denote the (asymmetric) relation 'x is a son of y' as Sxy, Albert as a and Hermann as h, then the fact that Albert possesses the relational property of 'being the son of *someone*' can be expressed by the sentence $\exists y Say$, while the fact that Albert possesses the relational property of 'being the son of Hermann Einstein' can be expressed by the sentence Sah. Both properties are properties of Albert *only*, even though they are due to his standing in a certain relation to Hermann.

Rather than looking at the system consisting only of Albert Einstein and wondering about the relational properties he has in virtue of standing in a certain relation to his father, we could also look at the bigger system (Albert, Hermann) $= (a, h) =: s$ and wonder about the properties the system s has in virtue of the relations its parts stand in. One *intrinsic* property s has (a property that only depends on s itself, but not on the relation s has to anything else) is that its elements stand in a father-son relationship. Let us denote this 1–place property of systems by Fx, and the fact that s possesses this intrinsic property as being expressed by the sentence Fs.

But not much seems to depend on whether we choose the first or the second perspective. In both cases we can say everything we want to say about Hermann and Albert Einstein in this context, and in both cases the important fact is that the relation between the two leads to certain systems possessing certain properties. Still, it is an interesting question whether the energy tensor $T_{\mu\nu}$ is more like Albert's relational property of having a father, $\exists y Fay$ (case 1), or whether it is more alike the property of a system like (Albert, Hermann), that its parts stand in a father-son relationship, $\exists x \exists y Fxy$ (case 2).

If the latter is the case, we would regard the predicate $\mathcal{T}_1(x)$, where the domain of x is 'systems in which both a metric field $g_{\mu\nu}$ and at least one matter field Φ is defined', as expressing the intrinsic property of such systems to have *a* energy tensor (rather than one with particular components associated with them), a property they have because of the relation(s) that hold between the material system and spacetime structure.

If the former is the case, we would say that possession of some mass–energy–momentum density $T_{\mu\nu}$ is a property of only the material parts of the total system, which are represented by the matter fields Φ, a property that the material system has in virtue of its relations to the metric field $g_{\mu\nu}$. We would then say that although $T_{\mu\nu}$ *depends* on the relations the matter fields Φ has to the metric field $g_{\mu\nu}$ (like Albert's property of being the son of Hermann depends on his relations to Hermann), the property 'possessing mass–energy–momentum' is still a property *only of* matter, just as it is a (relational) property of only Albert that he is the son of Hermann. We can reformulate this by calling $\mathscr{T}_2(x, y)$, where the domain of x is 'metric fields' and the domain of y is 'matter fields', the relation which allows us to express the fact that a material system represented by Φ has the relational property of having an energy tensor associated with it. The trouble with this is only that nothing in the representation $\mathscr{T}_2 g_{\mu\nu} \Phi$ gives away that $\mathscr{T}_2$ is a property of the part of the total system represented by Φ only – but that is also the case for using Sah in order to represent system Albert's property of being system Hermann's son.

As in the case of Hermann and Albert, I do not think that much depends on the perspective we choose. Indeed, one could arguably represent both $\mathscr{T}_1$ and $\mathscr{T}_2$ as a map from the set of pairs $(g_{\mu\nu}, \Phi)$ to the set of all energy–momentum tensors, $\mathscr{T} : (g_{\mu\nu}, \Phi) \longrightarrow T_{\mu\nu}(g_{\mu\nu}, \Phi)$. The distinction would merely lie in seeing the set $(g_{\mu\nu}, \Phi)$ as an 'object' or as a pair of objects. Anyhow, we find that material systems possessing (non-vanishing) mass–energy–momentum tensors depends on a *relation* between the material system and the structure of spacetime.

16.5 Conclusion

I started out by describing how Newton's defining property of *material* systems – the property of possessing mass – was generalised by relativistic field theory to the requirement that such systems need to have a mass–energy–momentum density tensor $T_{\mu\nu}$ associated with them.

We have then seen that the definition of energy tensors depends on the metric field in a variety of ways. Hence, in relativistic field theory, mass–energy–momentum cannot be regarded as an intrinsic property of matter, but must be seen as a relational property of matter (or a property of systems containing matter) that it only has because of its relation to spacetime structure.

Is it a surprising result that $T_{\mu\nu}$ does not describe an intrinsic property of matter? After all, nobody would have claimed that the momentum of a particle in Newtonian mechanics was an intrinsic property of this particle; it is a property that only makes sense if we describe that particle as changing its position with respect to something else. Kinetic energy too might be regarded as a relational property in Newtonian mechanics, for it depends on the velocity of the particle, which is a relational property for the same reason as the momentum of the particle. And even mass *density* can be argued to not be an intrinsic property of material systems, although the dependence of mass density on spacetime structure is surely of a weaker kind than that of relativistic mass–energy–momentum density tensors. Arguably, all this is grist to the mill of certain structural realists, who think that there are no intrinsic properties whatsoever.

In the introduction, I mentioned that Einstein himself was strongly motivated by a version of Mach's principle when he created GR, a version in which he claimed the geometric field $g_{\mu\nu}$ should be determined by the energy tensor $T_{\mu\nu}$. Einstein's idea was that spacetime structure should be derived from the properties of material systems. We have seen that the very definition of energy tensors depends on spacetime structure, and that hence even a unique determination of $g_{\mu\nu}$ by $T_{\mu\nu}$ would not be sufficient for spacetime geometry to be reduced to the properties of matter.

But nor does GR accord to what might be called an Anti-Machian principle: even though the matter fields Φ do not determine the spacetime structure $g_{\mu\nu}$, spacetime does not determine the material structures either: both sides only constrain each other. But matter needs spacetime in order to have some of its key properties defined, while spacetime does not depend on matter in this way. Still, in order to get a truly Anti-Machian theory, we would need not only the energy tensor to depend on the metric field, but the matter fields themselves would need to be derivable from the structure of spacetime. A candidate for such a theory is Kaluza-Klein theory, in which the electromagnetic vector potential forms part of the 5-dimensional metric tensor, and hence leads to the electromagnetic field $F_{\mu\nu}$ itself to be derivable from the geometric properties of spacetime.

But this leads us too far afield. The energy–momentum tensor of matter depends on, but is not determined by, the structure of spacetime – and that is enough for today.

Funding

I wish to thank the British Society for the Philosophy of Science for a Doctoral Scholarship which made the work contained in this article possible, together with help from the Arts and Humanities Research Council and Oriel College, Oxford University. I also wish to thank the Einstein Papers Project at the California Institute of Technology for a grant and wonderful hospitality in the Fall of 2008, during which part of the paper was written.

Acknowledgements I am very grateful to Oliver Pooley, Harvey Brown, Jeremy Butterfield, Eleanor Knox, Robert Geroch, Stephen Lyle, Tilman Sauer, Edward Slowik and Stephen Tiley. Each of them read one or more than one version of this paper, and helped improve it significantly with their comments and suggestions.

I also thank Robert Geroch, Erik Curiel, Stephen Lyle and John Norton for very helpful discussions, in particular about the energy tensors of different fluid systems and their relation to spacetime structure, and David Malament and Eric Curiel for enlightening discussions about the more subtle dependencies of energy tensors on spacetime structure.

References

Brown HR, Brading KA (2002) General covariance from the perspective of Noether's theorems. Diálogos 79:59–86
Einstein A (1918) Prinzipielles zur allgemeinen Relativitätstheorie. Ann Phys 55:241–244

Hawking S, Ellis GFR (1973) The large scale structure of space-time. Cambridge monographs on mathematical physics. Cambridge University Press, Cambridge

Hoefer C (1994) Einstein's struggle for a Machian gravitation theory. Stud Hist Philos Sci 25(3): 287–335

Hoefer C (2000) Energy conservation in GTR. Stud Hist Philos Modern Phys 31:187–199

Lehmkuhl D (2008) Is spacetime a gravitational field? In: Dieks D (ed) The ontology of spacetime, vol 2. Elsevier, Amsterdam

Lewis D (1983) Extrinsic properties. Philos Stud 44:197–200

Malament D (2007) Classical relativity theory. In: Butterfield J, Earman J (eds) Philosophy of physics, vol A. Elsevier, Amsterdam

Wald RM (1984) General relativity. The University of Chicago Press, Chicago

Weatherson B (2007) Intrinsic vs. extrinsic properties. In: Zalta EN (ed) The Stanford encyclopedia of philosophy. URL http://plato.stanford.edu/archives/spr2007/entries/intrinsic-extrinsic/

Chapter 17
Individual Particles, Properties and Quantum Statistics

Matteo Morganti

17.1 Introduction

A long-standing debate in the philosophy of quantum mechanics concerns whether or not particles are – or at least can consistently be said to be – individuals, that is, entities that are determinately self-identical and numerically distinct from other entities. Those who favour an affirmative answer must deal with two alleged difficulties:

- The fact that the Principle of the Identity of the Indiscernibles is violated in quantum mechanics, but it is a plausible criterion on the basis of which to attribute individuality to things.
- The 'non-classicality' of quantum statistics, whose peculiarities are readily explained by assuming that quantum systems are composed of non-individuals.

The former problem can be overcome by postulating some form of 'primitive thisness' that individuates the particles independently of their qualities; or, alternatively, by showing that the Identity of the Indiscernibles does after all hold in the quantum domain.[1] This paper assumes that at least one of these strategies is viable, and deals with the latter difficulty.

The claim that since quantum particles obey a non-classical statistics they should be regarded as non-individuals can be found as early as (Born 1926). This position is regarded by many as the 'Received View' on the nature of quantum entities (see, for instance, the historical reconstruction in (French and Krause 2006, Chapter 3). In what follows, the Received View is critically discussed. Section 17.1 outlines the differences between classical and quantum statistics, and the reasoning that leads from these to the conclusion that quantum particles are not individuals. Section 17.2 briefly overviews some existing strategies for avoiding the conclusion. Section 17.3

M. Morganti
Zukunftskolleg and Philosophy Department, University of Konstanz, Universitätsstraße 10, 78464 Konstanz, Germany
e-mail: Matteo.morganti@uni-konstanz.de

[1] For example, by allowing relations to individuate, as in (Saunders 2006a) and (Muller and Saunders 2008).

presents a new proposal, based on a specific and novel understanding of the relevant quantum properties. Section 17.4 adds some comments, and a concluding summary follows.

17.2 Classical and Quantum Statistics

Suppose one has N particles distributed over M possible single-particle microstates, and is interested in knowing the number of physically possible combinations. In classical mechanics, Maxwell-Boltzmann statistics holds. According to it, the number of possible distributions is

$$W = M^N \tag{MB}$$

In the case of quantum particles, fewer arrangements are available. Bose-Einstein statistics (which applies to the particles known as bosons) has it that

$$W = (N+M-1)!/N!(M-1)! \tag{BE}$$

In the case of fermions, the *Exclusion Principle* (dictating that no two fermions can be in the same state) holds and further reduces the number of possibilities, which becomes equal to

$$W = M!/N!(M-N)! \tag{FD}$$

The latter equality expresses so-called Fermi-Dirac statistics.

On the basis of the above, one can calculate the probability of a specific configuration being realised. Assuming equiprobability, such probability is, obviously enough, given by

$$\text{Prob}(s) = T/W$$

with s being the arrangement in question, and T the number of ways in which s can be actualised (and W being calculated, of course, on the basis of the statistics appropriate for the case at hand).

For example, consider a physical system composed of two particles to each one of which two equally probable states are available (that is, for which $M = N = 2$). Classically, one applies MB and obtains $2^2 = 4$ possible arrangements, each one with probability 1/4 of being realised. In the quantum case, there are instead only $(2+2-1)!/2!(2-1)! = 3$ possibilities (BE), or even just $2!/2!(2-2)! = 1$ (FD), and the probabilities for each possible state are 1/3 and 1, respectively.

In more detail, the arrangements available in the situation being considered are the following (x and y being the available states, and the subscripts denoting the – alleged – particle identities):

$$|x>_1 |x>_2 \tag{C1-Q1}$$

$|y>_1 |y>_2$ (C2–Q2)

$|x>_1 |y>_2$ (C3)

$|y>_1 |x>_2$ (C4)

$1/\sqrt{2}(|x>_1 |y>_2 + |y>_1 |x>_2)$ (Q3)

$1/\sqrt{2}(|x>_1 |y>_2 - |y>_1 |x>_2)$ (Q4)

C1–C4 are the possible arrangements in the classical case, Q1–Q4 the configurations available in the quantum case. In particular, Q1, Q2 and Q3 are symmetric states, accessible to bosonic systems; and Q4 the unique possible state for fermions - which is anti-symmetric (Q3 and Q4 describe the *entangled* states typical of quantum mechanics).

In quantum mechanics, then, only (anti-)symmetric states are possible.[2] For such states, particle exchanges do not make a difference: simply put, there is only one way for two 'quantum coins' to be one heads and one tails (quantum particles are said to be *indistinguishable*). Moreover, non-symmetric states are not allowed, that is, analogues of C3 and C4 for quantum systems are never realised. This, the canonical argument goes, suffices for drawing the conclusion that quantum particles are non-individuals of some sort, and consequently lack definite identity conditions. First, particle permutations cannot in principle make a difference in the quantum case because there are no identities that can be permuted. Similarly, non-symmetric states are ruled out because it is impossible for *a specific* particle to have a certain value for an observable and for *another specific one* to have a different value for that observable, as the particles do not have determinate identities allowing for such property-attributions.

17.3 Attempts to Avoid the Conclusion

One well-known argument aiming to block the inference from quantum statistics to particle non-individuality is based on the idea that indistinguishability is insufficient for non-individuality, as classical particles are also indistinguishable but are, of course, treated as individuals. Historically, this view arises from the observation that Maxwell-Boltzmann statistics incorrectly predicts that mixing similar gases at the same pressure and temperature one experiences a change in entropy (this is the so-called 'Gibbs' paradox'). Since, as a consequence of this, in order to make entropy correctly extensive one has to introduce an N! factor excluding permutations already at the classical level, it can be concluded that classical many-particle systems are also indistinguishable. This, in turn, suggests that the differences between quantum and classical statistics have nothing to do with the particles' identity conditions and just mirror the different behaviours of different types of physical systems.

---

[2] And, crucially, transitions from symmetric to anti-symmetric states (or vice versa) are ruled out.

This idea was recently developed by Saunders (Saunders 2006b), who maintained that such differences are due to the fact that the probability measure is continuous in the classical case and discrete in the quantum case.

However, it can be argued that, historically, the failure of extensiveness for entropy in the context of classical statistics counted as one of the reasons for the shift to quantum mechanics as a radically new description of reality. That is, that by eliminating permutations when faced with Gibbs' paradox one is in fact switching to a non-classical *ontological* setting. The demand for an ontological explanation seems legitimate also with respect to Saunders' specific idea about probability measures. What is it that determines the difference Saunders points at, given that – on this construal – it cannot be (in)distinguishability?

A different strategy is developed by Huggett in a series of works (see Huggett 1997, 1999). Huggett argues that the idea that particle permutations should make a difference if particles were individuals depends on a supposition as to the truth of *haecceitism*: namely, the metaphysical doctrine according to which possible worlds can differ exclusively with respect to the identities of the entities inhabiting them (i.e., be distinct in spite of the fact that they are qualitatively identical). However, claims Huggett, haecceitism is by no means necessary for individuality: it is just one possible thesis regarding conditions of trans-world identity, and certainly not an indispensable element when it comes to defining conditions of identity within a world. Hence, it is possible to interpret the difference between classical and quantum particles in terms of haecceitism (respectively, failure of) and not in terms of (non-)individuality.

Even granting this, however, also in this case an ontological account of the quantum domain must still be specified.

In yet another attempt to block the derivation of non-individuality from quantum statistics, Belousek (2000) argues that whether quantum systems truly are permutation invariant depends on whether it is correct to assume the *Fundamental Postulate of Statistical Mechanics* (FPSM) – according to which every distinct equilibrium configuration must be assigned the same statistical weight – in the framework of quantum mechanics. Such an assumption, Belousek argues, is by no means inescapable, as it is usually derived from a postulate which is commonly held to be controversial: the *Principle of Indifference*, imposing uniform probabilities whenever there is no information justifying a different distribution.

Teller and Redhead (Teller and Redhead 2000) raise the objection that it is in any event impossible to attribute non-uniform probabilities absolutely generally: for at least in some cases information is available and interference terms arise that make the assignment of uniform priors necessary for a correct treatment of actual quantum systems.

Perhaps the most 'radical', but at the same time customary, move made by the supporters of the individuality of quantum particles in order to overcome the difficulty with quantum statistics is to postulate certain primitive and non-further-explicable state-accessibility restrictions. Systems of indistinguishable particles are, on this construal, never found in non-symmetric states (and in symmetries other than those allowed for them) just because this is a fundamental feature of reality. Perhaps, this proposal suggests, it is explanatory enough to claim that non-symmetric states simply are not in the symmetrised Hilbert space that correctly represents the actual world. (As for permutation invariance, it is simply an integral characteristic of the allowed (anti-)symmetric states).

Teller (1998) objects that this is not, in fact, a satisfactory explanation. He contrasts the idea of inexplicable state-accessibility restriction with what could plausibly be regarded as a 'proper' explanation of equally basic physical facts. Statistical mechanics, for example, can explain *why* a state of affairs in which a cold cup of tea spontaneously starts to boil is never observed. And this clearly goes to support the plausibility of the theory, even though the fact being pointed at derives from basic truths of thermodynamics and might just be presented as primitive. Furthermore, according to Redhead and Teller (Redhead and Teller 1991, 1992), an additional, independent difficulty exists. Their working presupposition is that when some meaningful part of a theory does not seem to represent anything, one should elaborate on it further and eventually find the real-world counterpart of the bit of formalism apparently devoid of content. Otherwise, in Redhead's (1975) terminology, one has *surplus structure* that should eventually be eliminated. In the case under consideration, not only are non-symmetric states never experienced; nature would be entirely different if they were realized. Therefore, non-symmetric states in quantum theory indeed appear to constitute *in principle* useless surplus structure one had better get rid of. The only way to eliminate such surplus structure, Redhead and Teller continue, is by opting for the Fock space formalism of quantum field theory, a language without 'particle labels' putatively referring to the particles' identities (as in the case of the subscripts in C1–C4 and Q1–Q4 above) which conveys information about 'how many' entities are in a certain state but not about 'which entity is what'. The Fock space formalism, though, appears to dispense not only with the labels, but with what these express at the ontological level too: namely, particle identities.

Redhead and Teller's reasoning relies on assumptions concerning the significance of surplus structure, the relationship between quantum mechanics and quantum field theory and the ontological import of a theory's formalism that are certainly open to discussion. In any event, it seems plausible to say that the mere positing of primitive state-accessibility restrictions is not very adequate as a basis for insisting on the individuality of particles, and appears in fact quite ad hoc in the present context. Why is it that there are such restrictions in the quantum domain if it is true that quantum particles are not significantly different from classical particles?

Given the foregoing discussion, that of the supporters of individuality in quantum mechanics appears to be a rather weak stance, and that quantum particles are non-individuals a natural conclusion to draw. The rest of this paper, nonetheless, suggests a strategy to think otherwise.

17.4 A New Suggestion

In what follows, a precise ontological claim will be articulated: that *particles in quantum many-particle systems of identical particles never possess state-dependent properties intrinsically*, as such properties are always *irreducible properties of the whole*.

More precisely, the claim is that for any multi-particle system of identical quantum particles, the system exhibits both 'canonical' monadic properties (e.g., the total spin of the system, or the mass of one of the particles) and more complex properties that only belong to the system in its entirety and are not reducible to more basic properties of the component particles; and that this is sufficient for explaining quantum statistics without giving up particle individuality.[3]

Suppose a property of a certain type, R is a property exhibited by a whole (call it S) that 'encodes' information about the components of S but is not reducible to the specific properties of those components (as, say, the mass of two tennis balls together is reducible to the mass of the first ball plus the mass of the second).

Next, think about two fair coins: of course, since these are classical objects, a property of the whole such as, for instance, 'one heads and one tails' is always reducible to two monadic intrinsic properties ('heads' and 'tails') possessed by the coins separately. But imagine now the 'one heads and one tails' property being a property of type R, in the sense that there literally is no 'heads' or 'tails' property for each separate coin before the toss; and then generalize to all states available to the coin pair. On such a construal, *despite the coins being individuals*, a complete statistical description in terms of 'heads' and 'tails' would not say anything about any specific coin because *no such information exists in the system*. Crucially, it follows that, in this scenario, switching the coins would not give rise to a new total state – whatever the state is. And yet, it makes perfect sense to regard the coins as individuals (this has, in fact, been *assumed* to be the case). For analogous reasons, non-symmetric states would be impossible.

Something like this, it is claimed here, is exactly what happens for quantum many-particle systems of identical particles.

For these systems, it is indeed the case that one only has information about the particles in the form (assuming again a two-particle and two-value system) '1 has the same value as 2 for property P, namely, x', '1 has the same value as 2 for property P, namely, y' or '1 has opposite value to 2 for property P'.[4] But these are exactly all the arrangements that can possibly be exhibited by the system without the information being reducible to more fundamental facts about the monadic properties of the separate individuals composing it. By analogy with the hypothetical (quasi-)classical scenario just considered, it can thus be conjectured that in the quantum domain – for all many-particle systems and state-dependent properties – particle exchanges

---

[3] It could be argued that these non-reducible properties are dispositions, but this is immaterial to the treatment of quantum statistics that I wish to illustrate here.

[4] The 'has' should, of course, not be understood in the sense that there are definite properties before measurement, but just as a generic description. For present purposes, the issue concerning the exact interpretation of quantum properties can be left open.

do not give rise to new arrangements (i.e., the identities of the particles are not statistically relevant) not because particles are not individuals and consequently do not have well-defined identities but, rather, because the particles' identities do not play a role in the determination of the states that are described by the statistics. As in the case of our two imaginary coins above, that is, all the relevant properties are R-properties and, consequently, permutations do not affect the 'content' of the total state.

A closely related consequence is that one should not expect 'quantum analogues' of classical states such as C4 (that is, non-symmetric quantum states) to exist, because these would require a property-structure different from the one that – it is being maintained – is exhibited by quantum systems. That is, they would require individual particles possessing well-defined values for their observables separately from each other, which is exactly what is ruled out in the present framework. All this also means that the correspondence between states C1 and C2 on the one hand and states Q1 and Q2 on the other is only an appearance due to the formalism. While the former two effectively are states in which each particle is in a determinate state (that is, possesses a value for the property under consideration the latter two are, instead, states in which there is a correlation but no determinate states for the individual particles, exactly in the same way as in the states described by Q3 and Q4. The impossibility of non-symmetric states thus finds a natural explanation.

What has just been conjectured can, clearly, be said to hold for all systems, independently of the number of their individual components. To see this, one just needs to conceive of the right properties. For instance, considering three fermions and two available states, one has $(3 + 2 - 1)!/3!(2 - 1)!$ possible arrangements, namely 4. These are readily described by two 'same value' correlations of the sort already encountered, plus two 'different values' correlations: 'two particles have the same value, namely x, and one particle has the other value, namely y, for property P'; and 'two particles have the same value, namely y, and one particle has the other value, namely x, for property P'. In fact, if one thinks about it, one can see that the explanation of quantum statistics suggested here *must* be deemed satisfactory if an account based on non-individuality is, because the former differs from the latter only with respect to 'where identity is taken out of the picture', so to speak: property-types rather than property-bearers.

17.5 Further Remarks

Let us now consider some possible reactions, and add a few remarks.

(i) The idea that all statistically relevant properties of quantum systems are irreducible R-type properties of the whole is not as 'exotic' as it may seem at first: it essentially consists of an extension to other quantum states of certain existing, and in fact quite widespread, views regarding entangled states. It is commonly claimed that quantum entanglement consists of some form of holism, coinciding with the existence of properties that belong to the entire system and are not reducible to those of the system's component particles. Paul Teller, in particular (Teller 1986, 1989) formulated what he called 'relational holism' as a way to understand EPR correlations and the violations of Bell's inequalities on the basis on inherent relations. Indeed, the view being proposed here essentially consists of the claim that quantum relational holism concerns not only entangled but also non-entangled systems; and that, as a consequence, the claim that the total system's properties are independent of the identities of its components (as individuals) and irreducible to the monadic properties of the latter generalises to all state-dependent properties and states.

(ii) One may object that this proposal retains particle identity in name only, as properties do not, strictly speaking, belong to individuals anymore. To this, it can be replied, first, that individuals still possess their state-independent properties, that is, *specific instances* of the essential properties that make them individuals of a certain kind. Moreover, if it is correct to claim (with Muller and Saunders 2008) that quantum particles are weakly discernible thanks to irreducible and non-supervenient irreflexive relations holding between them, then, far from making individuality empty, this proposal appears in line with the most recent 'empiricist-oriented' attempts to show that quantum mechanics does not in fact force us to give up particle individuality. Therefore, only by insisting that it is monadic state-dependent properties that ought to individuate their bearers can one pursue this line of criticism; but such an insistence appears difficult to justify.

(iii) One may dislike an ontology according to which non-supervenient properties invariably emerge in quantum many-particle systems out of particles that clearly possess monadic properties when they do not belong to the same system. Relatedly, one might be sceptical on the basis of the suspicion that on the present construal the mere *description* of two or more identical particles in the tensor product formalism appears to be able to change the ontological nature of their state-dependent properties. However, first, as already mentioned the kind of non-supervenience being pointed at is something peculiar about the quantum domain in general, and the present proposal simply extends to other systems claims that are already widely accepted for certain physical composites (i.e., entangled systems) under most interpretations of the theory. Secondly, it is certainly *not* the case that description determines what counts as a many-particle system. The latter is assumed to be an objective matter, from which it follows that what has been proposed here should be understood as applying only to tensor product states that describe systems that truly are 'unitary' in the relevant sense, and not just represented as such in the language.[5]

---

[5] When exactly interaction gives rise to such many-particle systems, on the other hand, does not seem to be an issue that should be settled here. As a matter of fact it appears that it *cannot* be settled here, if only because ultimately connected to the infamous measurement problem. The connected question whether reality ultimately consists of a unique 'universe-whole' can also be left open for the time being.

(iv) One might insist on the presence of in principle meaningless surplus structure in the formalism of quantum mechanics. However, it could then equally be maintained that classical mechanics is inadequate as a description of the objects in its domain because it is possible to describe the latter entities as belonging to entangled systems but such systems are never realised in the classical world. In general, given any physical theory and its formalism, it appears always possible to 'cook up' some form of surplus structure. In fact, it seems correct to claim that what counts as surplus structure is not immediately determined given a theory, and ontological presuppositions are fundamental for interpreting the latter (this is essentially the reason why the ontological perspective provided in this paper succeeds where talk of primitive state-accessibility restrictions fails).

(v) Usually, the Eigenstate-Eigenvalue Link is employed when interpreting the quantum formalism. According to it, a physical system actually possesses a specific value for an observable if it is in an eigenstate for that observable corresponding to that value. This licenses inferences such as the following:
[Prob(particle x has property P with value v) = 1] → [(Particle x actually has property P with value v)]
However, it was denied earlier that in states such as, for instance, Q1 above one has two particles each actually possessing a specific value for the given observable as an intrinsic property: the consequent in the above conditional must thus be deemed false. But in such states, the component particles have probability 1 of being detected as having that property (as they are in an eigenstate for that observable): the antecedent is true. Therefore, the Eigenstate-Eigenvalue Link seems to be made invalid by the present proposal. A response to this supposed difficulty could go in two directions. Either operators corresponding to single-particle properties are excluded at the outset on the basis of the holistic nature of the relevant physical systems. Or, alternatively, an amendment to the Eigenstate-Eigenvalue Link is made so that it only applies to the total system. In neither case are the fundamental postulates of quantum theory affected.

(vi) It could be maintained that the picture delineated in this paper essentially amounts to an endorsement of Bohmian mechanics: the attribution of state-dependent properties to the 'whole system', that is, could be regarded as basically the same as the attribution of them to a 'guiding wave'. Similarities indeed exist. But the important difference also exists that no assumption has been made here about uniqueness of positions and initial particle distribution, which are two distinguishing features of Bohmian mechanics. Also, crucially, unlike in Bohmian mechanics the notion of collapse of the wavefunction is retained in the present framework. Therefore, the analogy is only superficial. A closely related objection could be formulated according to which the suggested proposal aims to achieve something that is already obtained by Bohmian mechanics, and consequently turns out to be superfluous. This criticism, however, can easily be turned on its head: the proposed picture of quantum reality, one could argue, achieves some of the allegedly important results of Bohmian mechanics without requiring one to depart from what many see as the correct theory of the quantum world, i.e., standard quantum mechanics.

17.6 Conclusions

Contrary to a widespread belief, it is possible to formulate an explanation of quantum statistics from a perspective according to which quantum particles are individuals. Such an explanation has been formulated here on the basis of an application of the concept of holism (in particular, in the form of Paul Teller's relational holism) to all state-dependent properties of quantum many-particle systems of identical particles. The emerging ontological picture is one in which the peculiarities of quantum statistics are accounted for by modifying our 'classical' intuitions not with respect to the identity conditions of objects but, rather, with respect to the properties exhibited by physical systems.[6]

References

Belousek DW (2000) Statistics, symmetry, and the conventionality of indistinguishability in quantum mechanics. Found Phys 30:1–34
Born M (1926) Quantenmechanik der stoßvorgänge. Zeitschrift für Physik 38:803–827
French S, Krause D (2006) Identity in physics: a historical, philosophical, and formal analysis. Oxford University Press, USA
Huggett N (1997) Identity, quantum mechanics and common sense. Monist 80:118–130
Huggett N (1999) Atomic metaphysics. J Philos 96:5–24
Morganti M (2009) Inherent Properties and Statistics with Individual Particles in Quantum Mechanics. Studies in History and Philosophy of Modern Physics 40:223–231
Muller FA, Saunders S (2008) Discerning fermions. Br J Philos Sci 59:499–548
Redhead M (1975) Symmetry in intertheory relations. Synthèse 32:77–112
Redhead M, Teller P (1991) Particles, particle labels, and quanta. The toll of unacknowledged metaphysics. Found Phys 21:43–62
Redhead M, Teller P (1992) Particle-labels and the theory of indistinguishable particles in quantum mechanics. Br J Philos Sci 43:201–218
Saunders S (2006a) Are quantum particles objects? Analysis 66:52–63
Saunders S (2006b) On the explanation for quantum statistics. Stud Hist Philos Modern Phys 37:192–211
Teller P (1986) Relational holism and quantum mechanics. Br J Philos Sci 37:71–81
Teller P (1989) Relativity, relational holism, and the Bell inequalities. In: Cushing J, McMullin E (eds) Philosophical consequences of quantum theory: reflections on Bell's theorem. University of Notre Dame Press, Notre Dame, IN, pp 208–223

---

[6] I wish to thank the audience in Madrid for their useful questions and comments. An expanded version of this paper can be found in Morganti (2009).

Teller P (1998) Quantum mechanics and haecceities. In: Castellani E (ed) Interpreting bodies: classical and quantum objects in modern physics. Princeton University Press, Princeton, NJ, pp 114–141

Teller P, Redhead M (2000) Is indistinguishability in quantum mechanics conventional? Found Phys 30:951–957

Chapter 18
Evolution and Directionality: Lessons from Fisher's Fundamental Theorem

Samir Okasha

As is well-known, the second law of thermodynamics has the property of being 'time asymmetric', unlike for example the laws of Newtonian mechanics, which are time reversal-invariant. The second law tells us that in a closed thermal system, the entropy can never decrease, and attains its maximal value at equilibrium. This confers a kind of *directionality* on thermodynamical processes. If you were given a sequence of state-descriptions of a closed thermal system, i.e., a complete specification of the system's physical state (including its entropy) at consecutive points in time, you would be able to deduce which direction time is running in. For the second law tells you that that 'lower entropy' corresponds to 'earlier than'.

An interesting question is whether the evolutionary process is similarly directional. Given a sequence of state-descriptions of an evolving biological population, could one deduce the direction of time? This question, relativized as it is to a single biological population, is different from the question of whether one could deduce the direction of time from a sequence of complete descriptions of all the biota present on earth. The answer to the latter question is clearly yes; for as Maynard Smith (1988) points out, some evolutionary changes were pre-conditions for others. For example, the existence of cells was a pre-condition for the evolution of multi-cellularity; the existence of prokaryotic cells was a pre-condition for the evolution of eukaryotic cells (as the latter are symbiotic unions of the former); and so on. So in this relatively trivial sense, evolution is clearly directional. However this leaves unanswered our first question, which is really a question about whether the laws that govern micro-evolutionary change are time-asymmetric. This question isn't answered by the observation that some evolutionary changes could only have come later than others, since historical contingency, in addition to law, has almost certainly played a major role in actual evolutionary history.

Why does it matter whether evolution is directional, in the sense of being governed by time-asymmetric laws? The question has a certain intrinsic interest, given the importance of time-asymmetry in the philosophy of physics, but there are other

S. Okasha (✉)
Department of Philosophy, University of Bristol, 9 Woodland Road, Bristol BS8 1TB, U.K.
e-mail: Samir.Okasha@bristol.ac.uk

reasons too. A long-standing debate in biology, going back to the nineteenth century, concerns the 'progressiveness' or otherwise of Darwinian evolution. Modern biologists sometimes deride the idea that evolution is progressive, equating it with the anthropomorphic conceit that humans are at the pinnacle of the tree-of-life, but as Ruse (1996) has emphasised, Darwin himself believed in progress of some sort, and a watered-down, non-anthropomorphic version of the idea persists in some evolutionary circles. Progress is a notoriously hard concept to define; but however defined, it presumably entails directionality. Asking whether evolutionary change is directional may thus help clarify the sense, if any, in which it is progressive.

Replacing the notion of progress with that of directionality was first suggested in an article of Gould (1988); see also Maynard Smith (1988). However, Gould was referring to directionality in relation to macroevolutionary, rather than microevolutionary, change. Specifically, he was concerned with whether or not the pattern of species diversity within clades has a temporal arrow. This is certainly an interesting issue, as is the more general issue of long-term evolutionary trends (cf. McNamara 1990). However for the purpose of studying evolutionary directionality, understood as time-asymmetry of the underlying dynamics, I think that microevolution provides a better theatre. For it is widely agreed that many microevolutionary changes (though not all) are explained by basic Darwinian principles; thus we have a unified, consistent and well-understand theory of microevolutionary dynamics. The same is not true of macroevolution. Certainly, there are non-trivial empirical generalizations about macroevolutionary pattern, but it is unclear whether there are underlying dynamical principles on a par with those that govern microevolution; certainly, no one has discovered them to date. Thus microevolution seems the better arena in which to debate the issue of time-asymmetry.

In his famous book *The Genetical Theory of Natural Selection* (1930), Fisher made an argument which *appears* to have a direct bearing on the issue of microevolutionary directionality. Fisher stated a theorem, of which he offered an elliptical and hard-to-follow proof, called 'The Fundamental Theorem of Natural Selection' (FTNS). According to this theorem, the rate of increase of the 'fitness' of a population evolving by natural selection is equal to its 'genetic variance in fitness' at that time. By the 'fitness' of a population, Fisher meant simply the average fitness of its members. By the 'genetic variance in fitness' he meant what is today called the *additive* genetic variance, i.e., the variance in fitness that is attributable to the independent action of all the genes in the population. (If each gene contributes a fixed amount to the fitness of its host organism, independently of which other genes are in the organism, then all the genetic variance in fitness is additive.) Thus the theorem appears to link the rate at which a population's average fitness changes, under the pressure of natural selection, with the amount of additive genetic variance in fitness present in the population.

The relevance of this for the issue of directionality stems from the fact that the additive genetic variance is necessarily non-negative (and in most real cases will be positive). So if the rate of change of average fitness equals the additive genetic variance in fitness, this implies that average fitness can never decrease (and in most real cases will increase.) And this, if it is true, confers a directionality on the evolutionary process. For it apparently implies that, given a sequence of state-descriptions of a population undergoing Darwinian evolution, one *could* deduce the direction of time – it is the direction of increasing average fitness.

From this quick description of the FTNS, it is apparent that average population fitness plays a role superficially similar to the role of entropy in thermodynamics. Just as closed thermal systems evolve in the direction of higher entropy, according to the second law of thermodynamics, so biological populations evolve in the direction of higher fitness, according to the FTNS. Just as entropy is maximised at equilibrium, so too is average fitness. This analogy was noted by Fisher (1930, p. 47) himself, and was presumably what led him to describe the FTNS as holding "the supreme position among the biological sciences". (However, Fisher took care to stress the limitations of the analogy too, noting a number of respects in which it fails.) Whatever about the extent of the analogy, it is clear that Fisher regarded the FTNS as a highly important scientific discovery.

The FTNS is interesting for a variety of reasons, which go beyond the specific issue of directionality (though are related to it.) One is that the theorem appears to provide a mathematical vindication of a widespread folk belief about natural selection – namely that it is a 'force for the good'. (Herbert Spencer was perhaps the most well-known defender of this view.) The average fitness of a population is a natural measure of its well-being, or how well-adapted it is to the environment. So the FTNS appears to imply that natural selection will always tend to increase a population's well-being (though of course external factors, such as human destruction of the environment, may counteract the trend.) Just as certain economists claim that unfettered market forces will always enhance society's material well-being, so Fisher's theorem appears to claim that natural selection will always enhance a population's biological well-being. (See Edwards (1994) for discussion of the possible economic origins of Fisher's theorem.)

A closely related issue concerns the relation between Fisher's theorem and Wright's 'adaptive landscape'. As is well-known, Wright conceptualised evolution as the movement of a population on a 'surface of selective value', or adaptive landscape, which depicts a population's fitness as a function of its genetic composition.[1] He argued that selection would tend to push populations up (local) peaks in the landscape, i.e., in the direction of higher mean fitness, and he appealed to the FTNS as justification for this. Interestingly, Fisher himself strongly objected to Wright's use of his theorem, and indeed regarded the adaptive landscape as a flawed concept; this should have been an early indication that what Fisher intended by his FTNS was not quite what other biologists had understood. Fisher's opposition notwithstanding, the adaptive landscape emerged as a powerful heuristic for thinking about evolution, and is closely tied up with the idea of directionality.[2]

---

[1] This is one version of Wright's adaptive landscape; the other version depicts the fitness of an organism as a function of its genotype. See Provine (1971) or Edwards (1994) for discussion of the difference.

[2] For recent assessments of the adaptive landscape concept, see Plutynski (2008), Kaplan (2008) and Pigliucci (2008).

I said above that the FTNS *appears* to confer directionality on the evolutionary process, not that it actually does. Indeed, some theorists have explicitly denied that it does. For example, Maynard Smith (1988) writes: "at first sight, Fisher's "fundamental theorem of natural selection" might seem to predict an increase in "mean fitness", but it would be a mistake to think that there is any quantity that necessarily increases, as entropy increases in a closed physical system" (p. 220). Maynard Smith goes on to argue that, given a complete description of a closed biological system at two points in time, "there is nothing in Fisher's theorem that would enable a biologist to say which state was earlier" (p. 220). More recently, Rice (2004), in his discussion of the FTNS, describes the idea that the Darwinian process leads to the maximisation of some quantity as "one of the most widely held popular misconceptions about evolution" (p. 37). If these authors are right, the implications of the FTNS for the directionality question are less obvious than first indications suggest.

The task of figuring out whether the FTNS provides an argument for directionality is complicated by a long-standing controversy over what exactly the FTNS says, and whether it is true. In Fisher's original statement, the theorem *appears* to say that the rate of change of a population's mean fitness, at any time, equals the additive genetic variance at that time. This is how Fisher's theorem was understood for many years. But the problem with this interpretation is that it is simply untrue, at least as a general statement. Work by population geneticists such as Moran (1964), Kimura (1958) and others showed that natural selection does *not* always lead mean fitness to increase, and that even if it does, the rate of increase only equals the additive genetic variance under certain highly restrictive conditions. (Briefly, the conditions are: fixity of genotype fitnesses, no dominance and no epistasis.) As a result of this work, a consensus emerged that Fisher was wrong – the FTNS does not state a general truth about evolving populations at all. Somewhat uneasily, biologists arrived at the view that Fisher's "supreme principle of the biological sciences", was in fact a mistake. The grandfather of evolutionary theory, it seemed, had erred.

This puzzling situation persisted until a landmark paper by Price (1972), which was overlooked for many years, and has only recently come to be widely known. Price showed, through a combination of his own mathematical analysis and careful exegesis of Fisher, that the FTNS had been widely misinterpreted. Fisher was not talking about the total rate of change of mean fitness at all, as most commentators had assumed, but rather the *partial* rate of change, according to Price. The partial rate of change is the change that results solely from natural selection acting in a 'fixed environment', where the notion of environmental fixity is understood in a certain very specific way. Fisher's theorem, according to Price, is the claim that this partial rate of change is given by the additive genetic variance in fitness, a claim which turns out to be true. So interpreted Price's way, the FTNS *is* a true, general statement about evolving populations. However, it doesn't imply that selection will always drive mean population fitness up, but rather that it would always do so if the selective environment were fixed (which in general it is not.) There is now widespread acceptance that Price's interpretation supplies the correct account of what Fisher meant, and explains why he thought he had discovered a deep biological principle (cf. Frank 1997; Edwards 1994; Grafen 2003; Lessard 1997; Okasha 2008; Plutynski 2006).

The idea behind the 'partial change' formulation of the FTNS is that the total change in a population's mean fitness, from one generation to another, is made up of two components.[3] One component is what Fisher called 'the direct effect of natural selection', and the other is what he called 'environmental deterioration' (or change), which in effect means everything else. Fisher's interest was in the first component only; the FTNS shows that this component must be non-negative, as it is directly proportional to the additive genetic variance. This is a highly non-trivial claim, and given how Fisher defined environmental change, it is true.

Fisher's underlying idea – that to assess the effect of natural selection on a population's mean fitness, the selective environment must be held fixed – sounds quite reasonable. For in general, to assess the direct influence of one variable on another, we need to condition on, or hold fixed, other factors. (In statistical terms, this means looking at the partial rather than total correlation between the two variables.) However, Fisher understood the notion of environment in a rather unusual way, as Price and others have pointed out. On Fisher's view, if there is any change in the *average effects* of any of the alleles present in the population, then the selective environment has changed. (An allele's average effect is the partial regression of genotype fitness on the number of copies of the allele present in the genotype; thus it measures the effect on an organism's fitness that an extra copy of the allele would bring, against a fixed genetic background.) Since the average effect of an allele is in general a function of its frequency, it follows that in an evolving population, average effects are unlikely to remain constant from generation to generation.[4] So the fact that the partial change in mean fitness is always guaranteed to be non-negative, as the FTNS interpreted a la Price teaches us, implies nothing about whether mean fitness will in fact increase or decrease over time.

In the light of Price's interpretation, we can understand why Fisher wanted to distance himself from Wright's adaptive landscape concept. For if evolution is a matter of ascending peaks in an adaptive landscape, on which the vertical axis measures mean population fitness, this implies that evolutionary change IS in the direction of greater mean fitness. But the FTNS guarantees no such thing; it tells us only that in a constant selective environment, natural selection will lead populations to climb peaks in the landscape. But since the environment, in the relevant sense, changes whenever average effects change, and since average effects depend on gene frequencies, which are *themselves* affected by natural selection, there is no guarantee that an evolving population will be driven up a peak in a landscape, as Wright had assumed.

The difference between the traditional 'total change' and Price's 'partial change' interpretation of the FTNS helps explains the quotations from Maynard Smith and

---

[3] In talking about the change in mean fitness from one generation to another, rather than the instantaneous rate of change, I have switched to a 'discrete time' formulation of the FTNS. This makes no difference, for as Ewens (1989) showed, the 'partial change' FTNS holds in both discrete time and continuous time models.

[4] The only situation in which average effects are not functions of allele frequency is if there is perfect additivity, i.e. no dominance and no epistasis. See Okasha (2008) for further explanation.

Rice above, in which they downplay the link between the FTNS and directionality. Maynard Smith says that 'at first sight' the FTNS appears to supply such a link, but closer examination suggests otherwise. His point, I presume, is that on the 'partial change' formulation, it does not follow that mean fitness must increase under selection, while on the 'total change' interpretation, the FTNS is not true. Similarly, Rice endorses the FTNS as a correct general statement about evolving populations, but notes the fallacy of assuming that evolutionary equilibrium is attained at maximum mean fitness, or that selectively-driven evolutionary change must be in the direction of higher mean fitness. Again, it is the Pricean 'partial change' interpretation that makes sense of this combination of views.

In effect, Maynard Smith's and Rice's arguments amount to the following dilemma for the proponent of a link between the FTNS and directionality. For the link to be sustained, there must be a true evolutionary law featuring some quantity that is analogous to entropy in the second law of thermodynamics. But on the 'total change' interpretation, the FTNS is not true to start with; which on the Pricean 'partial change' interpretation, the analogy between mean fitness and entropy breaks down. For the second law implies that the *actual* entropy cannot decrease in a closed thermal system; while the FTNS, on Price's interpretation, does not imply that actual mean fitness cannot decrease, but rather that one component of the actual change must be non-negative. So either way, the analogy between the FTNS and thermodynamics fails.

This is certainly a cogent objection to the use of Fisher's theorem as an argument for directionality, but I do not think it is the end of the story. For although Price's interpretation of the FTNS is widely accepted as the correct account of what Fisher meant, there is still substantial controversy over the biological significance of the 'partial change' version of the theorem. Price himself expressed disappointment with the theorem, due to the oddity of Fisher's notion of environment, and the fact that nothing can be concluded about whether mean fitness will actually increase or not. Thus although the FTNS appears to show a "constant improving tendency of natural selection", this in fact "does not necessarily get anywhere in terms of increasing 'fitness' as measured by any fixed standard", Price wrote (1972, p. 131).[5] However, other theorists have taken a more favourable view. Thus for example Grafen (2003) says that the FTNS "isolated the adaptive engine in evolution and made an extraordinary link between gene frequencies and adaptive changes. It really did show how Darwinian natural selection worked simply and consistently and persistently amid the maelstrom of complexities of population genetics" (p. 345). As these quotes show, disagreement persists over the true significance of Fisher's theorem; this suggests that the link between the FTNS and directionality cannot be rejected as quickly as Maynard Smith and Rice suggest.

---

[5] Thus Price wrote: "what Fisher's theorem tells us is that natural selection ... at all times acts to increase the fitness of a species to live under the conditions that existed an instant earlier. But since the standard of 'fitness' changes from instant to instant, this constant improving tendency of natural selection does not necessarily get anywhere in terms of increasing 'fitness' as measured by any fixed standard ... (1972, p. 131)."

In recent work, I have argued that the biological significance of FTNS depends ultimately on one's attitude towards the gene's eye view of evolution (Okasha 2008) My reasoning is as follows. Fisher's basic idea, that the environment must be held fixed in order to assess the impact of selection on mean population fitness, is in itself unobjectionable, as noted above. So everything depends on his particular definition of environmental fixity, as constancy of the average effects of all the alleles in the population. Though odd at first sight, this definition makes good sense from a gene's eye viewpoint. For if we consider individual genes as the real units of selection, a la Dawkins, the notion of selective environment must be modified accordingly, as Sterelny and Kitcher (1988) first argued. From the perspective of an individual gene, it is quite reasonable to say that the environment changes whenever gene frequencies change, as Fisher's definition of environmental fixity in effect requires him to say. (This is no different in principle from regarding *genotype* frequencies as part of the selective environment of each *organism*, as is standard in models of social evolution.) Indeed, I suggest that Fisher's defense of his FTNS should be recognised as an early, and strikingly clear, statement of the gene's eye viewpoint (Okasha 2008).

Where does this leave the issue of directionality? If we are prepared to take the gene's eye view of evolution literally, rather than as mere metaphor, then a link between the FTNS and directionality can be salvaged after all. Certainly, the FTNS supplies no guarantee that mean fitness must always rise, in a population subject to natural selection, *but this is because natural selection is not the only force affecting the population's mean fitness*. Environmental change is also playing a role. The FTNS teaches us that in a fixed environment, understood in Fisher's special sense, selection will always drive mean fitness up. So if we adopt the gene's eye view, and thus accept Fisher's notion of environment fixity, we must recognise that environmental change, as well as natural selection, can impact on mean fitness. (The fact that selection *itself* can cause environmental change, via changing gene frequencies, does not affect the logic of this argument.) There is no great mystery about this. If the 'external' environment, e.g., the ambient temperature, changes, it is obvious that a population's mean fitness may be affected. The same is true of environmental changes in Fisher's more inclusive sense. Therefore, the FTNS shows that natural selection *does* always tend to increase mean fitness; it is just that this increase may be countered, or offset entirely, by changes to the environment. So if we confine attention to the portion of the total change in mean fitness that is directly[6] attributable to selection, setting aside environmental change, the FTNS teaches us that it will always be non-negative.

To appreciate the type of directionality that this salvages, consider again the thought experiment we started with. Given a sequence of state-descriptions of a biological population, could we infer the direction of time? The answer to this question is 'no' – for the FTNS provides no guarantee that average population fitness cannot decrease over time, as we have seen. But this does not automatically sound the death knell for directionality. For the thought experiment focuses only on the

---

[6] The qualification 'directly' is necessary in order to exclude the indirect effect of selection on mean fitness that occurs via changes to the environment.

overall or net state of the population at successive points in time, and it is true that nothing general can be said about this. However, the FTNS, in the partial change formulation, tells us that selection will always *tend* to push mean fitness up, though other factors may counteract this. So if we focus on evolutionary tendencies, rather than net outcomes, we get a sort of directionality. Evolutionary change, in so far as it is driven by the direct action of natural selection, must always be in the direction of greater mean fitness.

In terms of the adaptive landscape metaphor, the point is this. It is not true that a population, under the pressure of natural selection, will always be driven higher up an adaptive peak. However, it *is* true that natural selection will always tend to push populations up peaks, since as the FTNS shows, the component of the change due to selection must always be non-negative. So if we focus on net effects, we see no directionality, but if we focus on underlying dynamical tendencies, we do see directionality. That is the moral of the FTNS, as I see it, though it requires accepting Fisher's notion of environmental change, which in turn requires taking the gene's eye view of evolution literally.

Is the type of directionality that we have salvaged really of any interest? It is easy to feel sympathy with Price's disappointment, given that nothing can be said about whether the overall change in mean fitness is positive or negative. But I think this reaction would be wrong. After all, the interesting question is whether the laws that govern evolutionary change are time-asymmetric or not; and Fisher's theorem, I claim, supplies a positive answer to that question (again, modulo a gene's eye viewpoint). The fact that mean fitness can decrease over time simply reflects the fact that mean fitness is affected by contingencies such as environmental change, in addition to the law-governed process of natural selection, of which the FTNS speaks. So if we accept the FTNS as a fundamental evolutionary law – as we surely should, given its great generality – it follows that there is a time-asymmetry at the heart of the Darwinian process.

This may still not convince a sceptic. For the analogy with the second law of thermodynamics, the classic example of a time-asymmetric law, seems rather thin, once we have acknowledged that mean fitness can decrease over time, as we must. After all, the second law does not merely say that there is a tendency for entropy to increase over time, but that it actually does do. On reflection, however, this disanalogy may be less stark than it seems. For the second law applies only to closed thermal systems; if a system is not closed, then all bets are off as to whether entropy will increase or decrease. So in a sense, environmental fixity in the FTNS plays a similar role to that of energetic closure in the second law. The second law tells us that if a thermal system is closed, then the entropy cannot decrease, and will attain a maximum at equilibrium; the FTNS tells us that if the environment stays constant (in Fisher's sense), then the mean fitness cannot decrease under natural selection, and will attain a maximum at equilibrium. In either case, if the antecedent is not satisfied, then the quantity in question – entropy or mean fitness – can perfectly well decrease, compatibly with the truth of the respective law. So the analogy can be partially restored.

It might be objected that this manoeuvre conceals a further important disanalogy. In the thermodynamic case, it is perfectly possible to ensure a system's closure, simply by isolating it from the external environment. In the evolutionary case, it is not similarly possible to ensure environmental constancy (in Fisher's sense), for natural selection itself is constantly changing the environment, by changing gene frequencies. So the very factor that, according to the law, leads mean fitness to increase – natural selection – is also constantly ensuring that the antecedent of the law – constancy of the selective environment – is not satisfied! (This is reminiscent of Price's remark about the 'paradox of mean fitness always tending to increase, but not doing so' (1972, p. 131).) It must be admitted that there is no analogue of this in the thermodynamic case.

Despite the limitations of the analogy with thermodynamics, of which Fisher himself was aware, I maintain that the FTNS still yields an interesting form of directionality. The oddity exposed in the previous paragraph really reflects the deeply counter-intuitive nature of Fisher's notion of environmental change; but as I have suggested, this notion makes quite good sense on a gene's eye viewpoint. For there can be no general objection to the idea of holding fixed the environment, in order to assess the impact of selection on mean population fitness. As noted, this is just the standard procedure for assessing the direct causal influence of one variable on another. The oddity stems entirely from the particular definition of environment fixity that Fisher operates with.

Another way to put the point is this. Clearly, no-one could hope for an evolutionary law that says that mean fitness, or any other parameter of a population, must always increase over time. For an organism's fitness, and thus mean population fitness, is affected by a myriad of different factors – including the state of the external environment. For example, human-induced habitat destruction must have dramatically altered the fitness of countless biological populations in recent years. So clearly, nothing general can be said about the total change in mean fitness over time, which would apply to every biological population. The contingencies of nature are simply too great. But if we hold fixed the environment, we might hope for a general statement about the effect of natural selection on mean fitness – and this is exactly what the FTNS provides. The rub, of course, is that 'holding fixed the environment' must be understood in Fisher's special sense. This sense is admittedly rather counter-intuitive, but it fits well with a gene's eye viewpoint. In a way, the counter-intuitiveness of Fisher's notion of environmental fixity simply reflects just how radical the 'gene's eye view' of evolution actually is.

Acknowledgements This work was supported by the Arts and Humanities Research Council, Grant No. AH/F017502/1, which I gratefully acknowledge.

References

Edwards AWF (1994) The fundamental theorem of natural selection. Biol Rev 69:443–474
Fisher RA (1930) The genetical theory of natural selection. Clarendon, Oxford

Frank S (1997) The price equation, Fisher's fundamental theorem, kin selection, and causal analysis. Evolution 51:1712–1729
Gould SJ (1988) On replacing the idea of progress with an operational notion of directionality. In: Nitecki MH (ed) Evolutionary progress. University of Chicago Press, Chicago, IL, pp 319–338
Grafen A (2003) Fisher the evolutionary biologist. The Statistician 52:319–329
Kaplan J (2008) The end of the adaptive landscape metaphor. Biol Philos 23:625–638
Kimura M (1958) On the change of population fitness by natural selection. Heredity 12:145–167
Lessard S (1997) Fisher's fundamental theorem of natural selection revisited. Theoret Popul Biol 52:119–136
Maynard Smith J (1988) Evolutionary progress and levels of selection. In Nitecki MH (ed) Evolutionary progress. University of Chicago Press, Chicago, IL, pp 219–230
McNamara K (ed) (1990) Evolutionary trends. University of Arizona Press, Tuscon
Moran PAP (1964) On the nonexistence of adaptive topographies. Ann Hum Genet 27:383–393
Okasha S (2008) Fisher's fundamental theorem of natural selection. Br J Philos Sci 59:319–351
Pigliucci M (2008) Sewall Wright's adaptive landscapes: 1932 vs. 1988. Biol Philos 23:591–603
Plutynski A (2006) What was Fisher's fundamental theorem of natural selection and what was it for? Stud Hist Philos Biol Biomed Sci 37:59–82
Plutynski A (2008) The rise and fall of the adaptive landscape. Biol Philos 23:605–623
Price G (1972) Fisher's fundamental theorem made clear. Ann Hum Genet 36:129–140
Provine WB (1971) The origins of theoretical population genetics. University of Chicago Press, Chicago, IL
Rice S (2004) Evolutionary theory: Mathematical and conceptual foundations. Sinauer, Sunderland, MA
Ruse M (1996) From monad to man: the concept of progress in evolutionary biology. Harvard University Press, Harvard
Sterelny K, Kitcher P (1988) The return of the gene. J Philos 85: 339–61

Chapter 19
Substantive General Covariance: Another Decade of Dispute

Oliver Pooley

19.1 Orthodoxy and a Recent Challenge

Whether Einstein's theory of general relativity (GR) satisfies a substantive principle deserving the name "general covariance" is a notoriously controversial matter. John Norton's masterful review of the matter, published in 1993, was aptly subtitled "eight decades of dispute" (Norton 1993). And yet, despite the continuing controversy, there has been broad agreement about a number of core issues. Two closely related theses are part of the orthodox position: (i) that general covariance does not distinguish general relativity from pre-relativistic theories when the latter are appropriately formulated and (ii) that general covariance, by itself, does not have any physical content.

The first of these theses is almost as old as GR itself. Einstein had sought a gravitational theory that was compatible with special relativity (SR). Soon after 1905 he came to believe that what was required was a generalization of SR's restricted relativity principle. According to SR, all inertial frames are on a par from the point of view of the fundamental laws. What Einstein sought was a theory according to which *all* frames are on a par. General covariance was supposed to implement this. A theory is general covariant if *the equations that express its laws are left form-invariant by smooth but otherwise arbitrary coordinate transformations.*[1] Since these coordinate transformations include transformations between coordinate systems adapted to frames in arbitrary relative motion, it would seem that there can be no privileged frames of reference in a general covariant theory.

This impression, however, is misleading. As Kretschmann famously pointed out, "by means of a purely mathematical reformulation of the equations representing the theory, and with, at most, mathematical complications connected with that reformulation" any physical theory can by made generally covariant, and

---

O. Pooley (✉)
Faculty of Philosophy, Oriel College, University of Oxford, Oxford, OX1 4EW, UK
e-mail: oliver.pooley@philosophy.ox.ac.uk

[1] As discussed below, this is but one of a number of closely related properties that go by the name "general covariance".

this "without modifying any of its content that can be tested by observation" (Kretschmann 1917, 575–6).[2] Generally covariant formulations of pre-relativistic theories (i.e., Newtonian and specially relativistic theories) are now utterly familiar in philosophical and foundational discussion. Even if it can be argued that, so formulated, the physical content of such a theory is somehow different from that of the "standard", non-covariant formulation (a claim I reject), it seems that general covariance cannot be what distinguishes GR from *generally covariant versions* of pre-relativistic theories.

The idea that general covariance per se has no physical content is reinforced when one considers the nature of the controversy dissected in Norton's review. In the conclusion to his paper, Norton claims that there are essentially three views on the question whether a 'principle of general covariance' plays a foundational role in GR (852–3). The third of these views straightforwardly rejects the idea that general covariance has any foundational role at all. The first view seeks to *supplement* general covariance with some other requirement. For example, GR might be distinguished from a rival theory T either because T's *simplest formulation* is not its generally covariant formulation, or because, when the generally covariant formulation of T is compared to (generally covariant) GR, it is seen that GR is the simpler, more elegant theory. Such an approach to identifying the 'principles' that distinguish GR faces a host of problems. But what is important for the current discussion is that, according to the approach, a theory's being generally covariant has nothing to do with its special status. Instead the generally covariant formulations of two theories to be compared merely make manifest the truly distinguishing characteristic, viz., some kind of simplicity.

The second point of view Norton mentions is associated with the so-called Anderson–Friedman programme (Anderson 1967, 73–88; Friedman 1983, 46–61). Here one distinguishes between two types of geometric object that can feature in the formulation of a spacetime theory. There are the truly dynamical objects on the one hand and, on the other, the *absolute objects*: very roughly, objects that do not vary from model to model of the theory. The programme also distinguishes between the *covariance group* of the theory (the group of transformations, defined on the theory's space of kinematically possible models, that leaves the space of dynamically possible models invariant) and the theory's *invariance group*. The latter is that subgroup of the covariance group that includes all and only automorphisms of the theory's absolute objects. One can then differentiate GR from pre-relativistic theories by noting that only GR satisfies a principle of general *invariance*: the invariance group of the theory should include the group of all smooth, but otherwise arbitrary coordinate transformations. For a specially relativistic theory, for example, the invariance group will be the Poincaré group, no matter whether the standard formulation of the theory is considered (in which case the covariance group will also be the Poincaré group) or whether a generally covariant formulation is considered.

---

[2] I gloss over the fact that the claim that "all physical observations consist in the determination of purely topological relations" formed part of Kretschmann's argument. See Norton (1993, 818), from where the translation is taken.

Anderson's definitions validate a sense in which the group of general coordinate transformations (or the diffeomorphism group) is a 'symmetry' group of GR but not of, say, SR (however the latter is formulated). In a generally covariant formulation of a specially relativistic theory, the transformations of a proper subgroup of the diffeomorphism group isomorphic to the Poincaré group have a special status: they leave the theory's absolute objects invariant. In GR, *every* diffeomorphism has this special status and so (it seems) diffeomorphisms in GR differ in status to diffeomorphisms in SR.

On closer inspection, things are not so clear-cut. A group gets to be a symmetry group on Anderson's view if it leaves the theory's absolute objects invariant. Arbitrary diffeomorphisms preserve the absolute objects of GR, and are thus symmetries, only because *GR has no absolute objects* and thus, trivially, *any* transformation preserves GR's absolute objects. It therefore seems that what really differentiates GR from pre-relativistic theories is its *lack of absolute objects*.[3] In particular, nothing in Anderson's approach suggests we should treat two diffeomorphically related models of GR differently from how we might treat two diffeomorphically related models of a generally covariant SR theory.

Thus, on any of the three views that Norton highlights, GR's general covariance has little to do with what distinguishes GR from pre-relativistic theories. Pre-relativistic theories can be given generally covariant formulations and the substantive principles just reviewed have little to do with general covariance per se. They might highlight various special features of GR. They might even highlight differences between the status of diffeomorphisms in GR and in SR. But they do not licence the claim that GR's general covariance is somehow more substantive than that of SR. All this, I claim, is orthodoxy. It has recently been challenged.

Amongst philosophers of physics, the challenge has been spearheaded by Earman (2006a; 2006b). He claims to be following physicists in distinguishing two kinds of general covariance: *merely formal* general covariance and *substantive* general covariance. Generally covariant formulations of pre-relativistic theories are supposed to satisfy only the former of the two. Note that Earman's substantive general covariance is quite distinct from Anderson's 'principle of general invariance' or any other of the notions just reviewed. It will clarify matters to introduce Earman's definition as just one of a number of versions of general covariance. It will also be helpful to introduce a number of 'toy' theories, whose satisfaction of the various versions of general covariance can then be assessed.

---

[3] I should note that whether GR does indeed lack absolute objects in the Anderson–Friedman sense is currently a live topic. In fact, it seems that $\sqrt{-g}$ counts as an absolute object (Pitts 2006; Giulini 2007; Sus 2008, Chapter 3).

19.2 Varieties of General Covariance

Our toy theories are all theories of the Klein–Gordon field. Their sole matter field will be a single, real scalar field. They differ, inter alia, in the geometric structure they posit.

The first theory is a specially relativistic theory written in standard form. It is defined by the equation:

$$\frac{\partial^2 \Phi}{\partial x^2} + \frac{\partial^2 \Phi}{\partial y^2} + \frac{\partial^2 \Phi}{\partial z^2} - \frac{\partial^2 \Phi}{\partial t^2} - m^2 \Phi = 0 \tag{SR1}$$

This equation is only satisfied by descriptions of Φ given with respect to inertial coordinate systems. I.e., if $\Phi(x)$ is a coordinate representation of our scalar field that satisfies equation SR1, then (in general) of those coordinate redescriptions obtained from $\Phi(x)$ via coordinate transformations, only those obtained via Poincaré transformations will also satisfy SR1.

Contrast this theory with generally relativistic Klein–Gordon theory, defined via the following equations:

$$\begin{aligned} g^{\mu\nu} \Phi_{;\nu\mu} - m^2 \Phi &= 0 \\ G_{\mu\nu}(g) &= \kappa T_{\mu\nu}(\Phi, g). \end{aligned} \tag{GR1}$$

Here the equations are intended to be read as identifying the values of the coordinate components of the objects involved. If all the components of the pair (g_{ab}, Φ) with respect to some coordinate chart $\{x\}$ satisfy these equations, then their components with respect to any chart smoothly related to $\{x\}$ will also do so in the region where the charts overlap. The theory might equally be specified via a set of equations relating the geometric objects themselves, rather than their coordinate components:

$$\begin{aligned} g^{ab} \nabla_a \nabla_b \Phi - m^2 \Phi &= 0 \\ G_{ab}(g) &= \kappa T_{ab}(\Phi, g) \end{aligned} \tag{GR2}$$

Now for the first two formulations of general covariance. A theory T is generally covariant iff:

GC1   the equations of motion/field equations of T transform in a generally covariant manner under an arbitrary coordinate transformation, or

GC2   the equations of motion/field equations of T relate "intrinsic, coordinate-free"[4] objects; they are true independently of coordinate systems.

GR1 and GR2, our two formulations of generally relativistic Klein–Gordon theory, satisfy GC1 and GC2 respectively. Our specially relativistic theory satisfies neither. But this is easily corrected via a Kretschmann-type move. We simply rewrite

---
[4] The terminology is Earman's (2006a, 446).

equation SR1 so that it holds good in arbitrary coordinates, making the role of the fixed metric of Minkowski spacetime explicit:

$$\eta^{\mu\nu}\Phi_{;\nu\mu} - m^2\Phi = 0. \qquad \text{(SR2)}$$

Alternatively, rather than equating coordinate components, we can write down an equation referring directly to the geometric object fields themselves:

$$\eta^{ab}\nabla_a\nabla_b\Phi - m^2\Phi = 0. \qquad \text{(SR3)}$$

It is clear that, appropriately formulated, our specially relativistic theory now satisfies GC1 and GC2.

So far we have focused on the transformation properties of a theory's equations. Let's consider models of the theories. Models of GR2 are triples of the form (M, g, Φ), where M is some 4-dimensional differentiable manifold, g is a Lorentzian metric on M and Φ is a scalar field on M. g and Φ must satisfy the equations GR2. Models of SR3 are likewise triples of the form (M, η, Φ), where M is some 4-dimensional differentiable manifold, η is now a flat, Minkowski metric on M and Φ is a scalar field on M. η and Φ must satisfy the equation SR3.

Our third formulation of general covariance is stated in terms of models. A theory T, with models of the form $(M, O_1, O_2, \ldots, O_N)$ is generally covariant iff

GC3 If $(M, O_1, O_2, \ldots, O_N)$ is a model of T, then so is $(M, d^*O_1, d^*O_2, \ldots, d^*O_N)$ for any diffeomorphism $d \in \text{Diff}(M)$.[5]

It is uncontroversial that generally relativistic theories, and hence our theory GR2, satisfy GC3.[6] What of our reformulations of SR1?

The orthodox (philosopher's) answer is that the theory specified via SR3 satisfies GC3 just as much as any generally relativistic theory. For suppose that (M, η, Φ) satisfies SR3. It follows from the fact that this is a tensor equation that $(M, d^*\eta, d^*\Phi)$ also satisfies SR3. I.e., if $\eta^{ab}\nabla_a\nabla_b\Phi - m^2\Phi = 0$ then $d^*\eta^{ab}\nabla'_a\nabla'_b d^*\Phi - m^2 d^*\Phi = 0$, where ∇' is the covariant derivative associated with $d^*\eta$. In their agenda-setting paper on the hole argument, Earman and Norton embraced this equivalence with respect to GC3 of appropriately formulated pre-relativistic theories and generally relativistic theories, arguing that the substantivalist was compelled to classify *all* "local spacetime theories" as indeterministic (Earman and Norton 1987, 524).

---

[5] This matches the definition given by Earman (1989, 47). Diff(M) is the group of M's automorphisms; i.e., the group of all invertible maps from M onto itself that preserve its differentiable structure. GC3 is the requirement that Diff(M) be a subgroup of T's *covariance group* in Anderson's sense.

[6] Uncontroversial, that is, amongst those who classify GR as a generally covariant theory. Maudlin's metrical essentialist (Maudlin 1988, 1990) denies that both $(M, O_1, O_2, \ldots, O_N)$ and $(M, d^*O_1, d^*O_2, \ldots, d^*O_N)$ represent genuine possibilities. But even the metrical essentialist can admit that $(M, O_1, O_2, \ldots, O_N)$ and $(M, d^*O_1, d^*O_2, \ldots, d^*O_N)$ are on a par as models of T. They should claim only that, *relative to the choice of* $(M, O_1, O_2, \ldots, O_N)$ as the representation of a genuine possibility, $(M, d^*O_1, d^*O_2, \ldots, d^*O_N)$ does not represent a possibility (compare Bartels 1996).

19.3 In Search of Substantive General Covariance

Let us return to our general relativistic theory, GR2. A key premise in Earman and Norton's argument is their claim that the substantivalist must interpret (M, g, Φ) and $(M, d^*g, d^*\Phi)$ as representations of *distinct* possibilities. Most commentators (relationalists and substantivalists alike) take the moral of the hole argument to be that (M, g, Φ) and $(M, d^*g, d^*\Phi)$ should be interpreted as representing the same physical state of affairs. This gives us our fourth version of general covariance, Earman's "substantive general covariance" (2006a, 447; 2006b, 4–5).

GC4   A theory T is generally covariant iff:

1. If $(M, O_1, \ldots, O_N)$ is a model of T, then so is $(M, d^*O_1, \ldots, d^*O_N)$ for any $d \in \text{Diff}(M)$.
2. $(M, O_1, \ldots, O_N)$ and $(M, d^*O_1, \ldots, d^*O_N)$ represent the same physical possibility.

In other words GC4 supplements GC3 with the requirement that $\text{Diff}(M)$ is a *gauge group* in the non-technical sense: diffeomorphisms relate distinct representations of one and the same situation.

Does the specially relativistic theory expressed by SR3 satisfy GC4? Not according to Earman. GC4 counts as "substantive" because:

> it is *not* automatically satisfied by a theory that is formally generally covariant, i.e., a theory whose equations of motion/field equations are written in generally covariant coordinate notation or, even better, in coordinate-free notation (Earman 2006a, 444).

Thus, for Earman, the substantive principle embodied in GC4 differentiates GR from pre-relativistic theories, even when these are formulated using generally covariant notation, along the lines of SR3. He is committed to denying that $\text{Diff}(M)$ is a gauge group with respect to the theory expressed by SR3. What justification does he offer?

19.4 When (Not) to See Gauge Freedom

According to Earman, the physics literature contains a "generally accepted apparatus that applies to a very broad range of spacetime theories and that serves to identify the gauge freedom of any theory in the class." This apparatus decrees that GR does satisfy GC4 whereas "formally generally covariant forms of special relativistic theories... need not satisfy substantive general covariance" (Earman 2006a, 445).

The "broad range" of spacetime theories Earman refers to are those whose field equations are derivable from an action principle. Suppose T's r equations of motion are derivable from an action $S = \int d^p x L(\mathbf{x}, \mathbf{u}, \mathbf{u}^{(n)})$. $\mathbf{x} = (x^1, \ldots, x^p)$ are the independent variables (the spacetime coordinates x, y, z, t in the cases we're considering) and $\mathbf{u} = (u^1, \ldots, u^r)$ are the dependent variables (e.g., $g^{\mu\nu}$ and Φ). The term $\mathbf{u}^{(n)}$ indicates that L can depend on the derivatives of $\mathbf{u}$ (with respect to the independent variables) up to some finite order n. A group $\mathscr{G}$ of transformations $g : (\mathbf{x}, \mathbf{u}) \mapsto (\mathbf{x}', \mathbf{u}')$ whose generators leave L form-invariant up to a divergence term is a *variational symmetry group* of S (Earman 2006a, 449–50).

Associated with this notion of a symmetry of S are the (generalized) Noether theorems. The one relevant to Earman's proposal is Noether's second theorem: if the parameters of $\mathscr{G}$ are s arbitrary functions of the independent variables, then there are s independent equations relating the r *Euler expressions*, the r variational derivatives $\delta L/\delta u^i$ of the Lagrangian with respect to each dependent variable. Imposing Hamilton's principle with respect to that variable (i.e., requiring that S is stationary with respect to arbitrary infinitesimal variations of that dependent variable that vanish on the boundary of integration) gives the Euler–Lagrange equation $\delta L/\delta u^i = 0$. Thus Noether's second theorem shows that the equations of motion are not independent and we have fewer independent equations of motion than field variables. When time is amongst the independent variables, this underdetermination manifests itself as apparent indeterminism. The physicist's standard move is to restore determinism by identifying solutions related by the variational symmetries. Thus Earman writes that "the applicability of Noether's second theorem is taken to signal the presence of gauge freedom" (Earman 2006b, 7) and proposes that, according to the physicists' apparatus, "variational symmetries containing arbitrary functions of the independent variables connect equivalent descriptions of the same physical situation, i.e., are gauge transformations." (Earman 2006a, 450).

Applied to our generally relativistic Klein–Gordon theory this gives us the expected result. What seems to me more suspect is Earman's application of the machinery to our specially relativistic Klein–Gordon theory. He notes that our formally generally covariant equation SR3 is derivable from the action:

$$S(\Phi, \eta) = \int \frac{1}{2}(\eta^{ab}\nabla_a\Phi\nabla_b\Phi + m^2\Phi^2)\sqrt{-\eta}d^4x \tag{19.1}$$

where Φ *but not* η is subject to Hamilton's principle. Earman concludes that while "the action admits the Poincaré group as a variational symmetries... the apparatus sketched above renders the verdict that there is no non-trivial gauge freedom in the offing" (Earman 2006a, 452). In other words, Earman suggests that applying his apparatus to this theory yields the verdict that diffeomorphisms are *not* gauge transformations.

There are at least three reasons to be sceptical of the method used to reach this conclusion.

1. The criterion Earman claims to find in the physics literature tells us that *if* some group $\mathscr{G}$ is a variational symmetry to which Noether's 2nd theorem applies, then $\mathscr{G}$ is a gauge group. I.e., it tells us when to see gauge freedom. But to draw the conclusion he does concerning SR3, Earman needs the converse criterion: $\mathscr{G}$ is a gauge group *only if* $\mathscr{G}$ is a variational symmetry. I.e., he needs a criterion that tells us when *not* to see gauge freedom. Earman freely admits that the apparatus is silent on non-Lagrangian theories (e.g., Earman 2006a, 454) which, we shall see, is significant.

2. Physicists' identification of (local) variational symmetries as gauge symmetries is not simply read off from the Lagrangian formalism. As Earman himself carefully explains, the identification is motivated by a desire to avoid indeterminism. But then there are equally good grounds for regarding a theory for which Diff(M) is a symmetry group in the sense of GC3, but which is not derivable from an action principle for which Diff(M) is a variational symmetry group, as also satisfying GC4. I.e., there are equally good grounds for regarding Diff(M) as a gauge group with respect to such a theory too. The reason there is not a good precedent in the physics literature for such a move is indicative of the fact that such theories are almost never discussed (in this literature); it does not indicate that in such theories diffeomorphisms should not be regarded as gauge.
3. Finally, what of Earman's claim that the Poincaré group is a variational symmetry of 19.1? Although he does not explicitly say that Diff(M) fails to be a variational symmetry group, this would appear to be implicit in his discussion. But why should one think this? It is true that one only applies Hamilton's principle to Φ, in order to derive the equation SR3. One does not also consider variations in η. But this is irrelevant to which transformations count as variational symmetries in sense of Noether's second theorem. That theorem applies in full force to 19.1, independently of which of the dependent variables one regards as background structure and which one regards as dynamical. One still obtains mathematical identities relating the Euler expressions. The only difference with the general relativistic case, where all dependent variables are subject to Hamilton's principle, is that the vanishing of the Euler expression corresponding to η (effectively the stress energy tensor of the Klein–Gordon field) is not one of the field equations.[7]

It is true that, if we consider drag-alongs only of Φ under the action of diffeomorphisms, while leaving η unaltered, then only a subgroup of diffeomorphisms isomorphic to the Poincaré group will be symmetries. Why should we consider such transformations? In the next section we will see that there is a reason, and that it connects to whether we regard our specially relativistic theory as derived from an action principle.

19.5 An Alternative Distinction Between Theories

The distinctions between formulations of a theory that we have so far considered have focused on the equations that express the theory, even though our characterization of general covariance has taken a model-theoretic turn. Continuing to think in terms of models is key to making some further, crucial distinctions.

---

[7] For an illuminating discussion of various connections between Noether's theorems and general covariance, see Brown and Brading (2002).

Let us suppose that the models of our theories have the following basic structure.[8] They are functions from a given space, V, into a given space of field values, W. The theories will involve a space, $\mathscr{K}$, of *kinematically possible models* (i.e., the space of all suitably well-behaved but otherwise arbitrary such functions), and a proper subspace $\mathscr{S} \subset \mathscr{K}$ of *dynamically possible models*, normally picked out via a set of equations.

In these terms, a "theory" corresponding to equations GR2 will involve a differentiable manifold M as the space V. A kinematically possible model will assign a (pseudo)metric tensor and real number to each point of M in a suitably smooth way, and consistently with any necessary boundary conditions. The subspace $\mathscr{S}$ of dynamically possible models is picked out by the equations GR2. Call this theory T_{GR}.

When it comes to the specially relativistic theory, however, we have a choice as to how to proceed. In the first version of such a theory, V is taken to be M *equipped with* a particular Minkowski metric. Each model of the theory then simply maps each point of this space into the real numbers. $\mathscr{K}'$ is the space of all suitably well-behaved such functions. The subspace of physically possibly models, $\mathscr{S}'$, is picked out by a suitable equation, constraining how Φ is adapted to the fixed metric structure of V. Call this theory T_{SR1}.

In the second version of the specially relativistic theory, V is taken, as in T_{GR}, simply to be the differentiable manifold M. We may suppose that the space of kinematically possible models is also the same as that of T_{GR}: each point of M is to be mapped to a metric tensor and real number in a manner consistent with boundary conditions and smoothness requirements. The theory will differ from T_{GR} in terms of its subspace, $\mathscr{S}''$, of dynamically possible models. This will be picked out (obviously) by a different set of equations to those that pick out $\mathscr{S}$. In addition to the Klein–Gordon equation, there will be an equation requiring the vanishing of the Riemann curvature tensor: $R_{abcd} = 0$. Call this theory T_{SR2}.

Which of the equations SR1, SR2 or SR3 is suitable to T_{SR1} and T_{SR2}? It is clear that any one of these equations can be understood as picking out the space $\mathscr{S}'$ of T_{SR1}. Provided coordinate charts on V that are adapted to its metric structure are chosen, SR1 will be satisfied by all and only those models in $\mathscr{S}'$. If we allow arbitrary coordinates, and interpret $\eta_{\mu\nu}$ as the coordinate components of the metric structure of V, then SR2 will be satisfied by all and only those models in $\mathscr{S}'$. If we interpret η_{ab} as referring directly to the fixed metric structure of V, then SR3 is satisfied by all and only those models in $\mathscr{S}'$.

This is not quite true for T_{SR2}. SR2 or SR3 are the natural equations to combine with the vanishing of the Riemann tensor. Although, for every model in $\mathscr{S}''$, there is a coordinate chart such that SR1 holds, for two arbitrary models in $\mathscr{S}''$ different coordinatizations will be needed. Nonetheless, I think it is clear that the different covariance properties of SR1, SR2 or SR3 do not track in any perspicuous way the difference between T_{SR1} and T_{SR2}. To repeat, all three equations are equally legitimate ways of specifying T_{SR1}. The difference between the theories (or formulations of the theory) *can* be made out in terms of equations (T_{SR2}, but not

---

[8] The following is, very loosely, based on the much more sophisticated material in Belot (2007, §4).

T_{SR1}, involves an equation constraining the geometry), but this does not seem like the most perspicuous way to do so.

19.6 In Search of Substantive General Covariance Again

It is time to assess the general covariance of our new formulations of special and generally relativistic Klein–Gordon theory. Before doing so, I introduce yet another notion of (substantive?) general covariance, advocated by Carlo Rovelli. According to Rovelli:

> A field theory is formulated in manner invariant under passive diffs (or change of co-ordinates), if we can change the co-ordinates of the manifold, re-express all the geometric quantities (dynamical *and non-dynamical*) in the new coordinates, and the form of the equations of motion does not change. A theory is invariant under active diffs, when a smooth displacement of the dynamical fields (*the dynamical fields alone*) over the manifold, sends solutions of the equations of motion into solutions of the equations of motion. (Rovelli 2001, 122, original emphasis)

Rovelli's terminology of "active" versus "passive" *diffeomorphisms* (as opposed to coordinate transformations) is somewhat novel. Let me make a few, hopefully clarifying, remarks. One should not think of diffeomorphisms (as is sometimes unfortunately suggested) as "moving points around". The map $d : M \to M$ simply *associates* each point of M in its domain with another. This map induces maps on fields defined on M, e.g., $d^* : g \mapsto d^*g$. One can think, perhaps, of *these* maps as 'moving g around' (although even this is a bit picturesque; really we use the map to define a new field in terms of an old one). One set of fields on M are the coordinate charts. This suggests the following way of distinguishing 'active' from 'passive' diffeomorphisms:

1. When d is thought of as inducing a change of coordinate chart, but the *physical fields* are left unchanged, d is a 'passive diffeomorphism'.
2. When d is thought of as inducing changes to all the physical fields, d is an 'active diffeomorphism'.[9]

With the notions of active and passive diffeomorphisms so defined, a theory T satisfies GC1 and GC2 if it is invariant under passive diffeomorphisms. It satisfies GC3 if it is invariant under active diffeomorphisms. But the requirement Rovelli in fact labels "active diffeomorphism invariance" in the quotation above is stronger than this. It relies crucially on a distinction between a theory's *dynamical* and *non-dynamical* fields. Let models of T be of the form (M, A_i, D_i), where the A_i are the non-dynamical fields and the D_i are the dynamical fields. A theory T is then generally covariant according Rovelli's version of substantive general covariance iff:

---

[9] I believe this fits with a more recent characterisation that Rovelli has given (2004, 62–5).

GC5 If $(M, A_1, \ldots, D_1, \ldots)$ is a model of T, then so is $(M, A_1, \ldots, d^*D_1, \ldots)$ for any $d \in \text{Diff}(M)$.[10]

How do our theories measure up against this (and the previous) notions of general covariance? Consider first T_{SR1}, whose space of kinematically possible models $\mathscr{K}'$ was constituted by maps from a manifold equipped with metric structure to the reals. We have already seen that its defining equation can be so formulated that it satisfies both GC1 and GC2. Suppose (M, g, Φ) is a dynamically possible model, i.e., $(M, g, \Phi) \in S'$. In general, $(M, d^*g, d^*\Phi)$ will not be *kinematically* possible, let alone dynamically possible. Hence this theory does not satisfy GC3 (and thus also fails to satisfy GC4). If we consider just dragging-along the sole dynamical field, Φ, we obtain a model $(M, g, d^*\Phi)$ that is in $\mathscr{K}'$. (So restricted, Diff(M) does have a well defined action on $\mathscr{K}'$.) However, $(M, g, d^*\Phi) \notin S'$, hence T_{SR1} also fails to satisfy GC5.

Consider next T_{GR} and suppose that $(M, g, \Phi) \in S$. $(M, d^*g, d^*\Phi) \in \mathscr{S}$ and hence T_{GR} satisfies GC3. (Note that GC3 is just the requirement that the action of Diff(M) on the space of kinematically possible models fixes the solution subspace.) The hole argument, therefore, suggests that it should also be classified as satisfying GC4. Finally, what of Rovelli's GC5? Since there are no non-dynamical fields, the requirement is again that $(M, d^*g, d^*\Phi) \in \mathscr{S}$ and GC5 is satisfied.

Finally, we consider T_{SR2} and suppose that (M, g, Φ) is an arbitrary model in S''. Since $(M, d^*g, d^*\Phi) \in S''$ for arbitrary d it follows that T_{SR2}, unlike T_{SR1}, satisfies GC3, and (*pace* Earman) GC4 (recall the hole argument). Does it satisfy GC5? That depends on whether g counts as a dynamical field. The passage quoted from Rovelli above continues:

> Distinguishing a truly dynamical field, namely a field with independent degrees of freedom, from a nondynamical field disguised as dynamical (such as a metric field g with the equations of motion Riemann[g] = 0) might require a detailed analysis (of, for instance, the Hamiltonian) of the theory (Rovelli 2001, 122).

It is certainly the case that the Anderson–Friedman notion of an absolute object can be invoked to classify g as a non-dynamical field. T_{SR2} then fails to be generally covariant in the sense of GC5 (cf. Giulini 2007). But how much illumination does this piece of classification achieve? In some intuitive sense, the metric structure of a specially relativistic theory plays the role of a fixed background against which the real dynamics is defined and unfolds. There is no such background in GR. Not only is the metric a genuine dynamical player; its dynamical evolution is affected by the material content of spacetime.[11] The action–reaction principle is satisfied. I doubt that the right way to make these ideas more precise is to discover a criterion that, e.g., T_{GR} meets but T_{SR2} fails to meet.

---

[10] Compare Earman's definition of a *dynamical symmetry* (Earman 1989, 45).
[11] The former need not entail the latter. For example, consider the, admittedly somewhat contrived, theory whose field equations are $g^{ab}\nabla_a\nabla_b\Phi - m^2\Phi = 0$ and $R_{ab}(g) = 0$. The metric in this theory has a non-trivial dynamics and constrains, but is unaffected by, the evolution of Φ.

19.7 Conclusion

Recall Earman's claim that, if we restrict attention to the class of Lagrangian field theories, generally relativistic theories, but not specially relativistic theories, satisfy GC4. In light of the previous section we can partially endorse this claim, for of the two specially relativistic theories, only T_{SR1} is a Lagrangian theory. All of its equations can be derived from the action described in Section 19.4. T_{SR2}, on the other hand, does not appear to be a Lagrangian theory.[12] The obvious ways to derive $R_{abcd} = 0$ as an Euler–Lagrange equation (in addition to the Klein–Gordon equation) requires an additional field and thus alters the space of kinematically possible models (Sorkin 2002; Earman 2006a, 455–6). However, it should be stressed that the reason diffeomorphisms do not count as *gauge* symmetries of the relevant formulation of the specially relativistic theory is because *they are not symmetries*; the theory does not satisfy GC3. And when we do consider a formulation of the specially relativistic theory for which diffeomorphisms *are* symmetries, viz. T_{SR2}, Earman's Lagrangian apparatus is simply silent. The points of comparison between T_{SR2} and T_{GR} strongly suggest that – re ontology, the nature of what is observable and the gauge status of diffeomorphisms – what goes for one should go for the other.

There are (at least) two ways of conceiving of pre-relativistic theories: as theories that fail both GC3 and GC5 (such as T_{SR1}) and as theories satisfying both GC3 and GC4 (such as T_{SR2}). Generally relativistic theories appear distinguished in that only the second kind of conception is available. If the second kind of conception of pre-relativistic theories is adopted, and they are then compared to GR, it seems doubtful that the interesting differences between GR and such theories is to be made out in terms of a variety of general covariance, or a difference in the status of the diffeomorphism group.

Acknowledgements I have previously given talks related to the topic of this paper in Oxford, Konstanz, Les Treilles, Barcelona and Montreal. I am grateful to numerous members of those audiences for useful discussion. Support for this research from the Arts and Humanities Research Council (UK) Research Leave Scheme (grant ID No: AH/E506216/1) and from the Spanish government via research group project HUM2005-07187-C03-02 and MICINN project FI2008-06418-C03-03 is also gratefully acknowledged.

References

Anderson JL (1967) Principles of relativity physics. Academic Press, New York
Bartels A (1996) Modern essentialism and the problem of individuation of spacetime points. Erkenntnis 45:25–43

---

[12] David Wallace has pointed out to me that, in the vacuum case, there is no difficulty in principle in obtaining the flatness of the metric via the extremization of an action whose only dependent variables are the components of the metric. What needs to be investigated is whether this observation can be extended to theories involving matter fields. (In vacuum GR the extremization of the gravitational action imposes Ricci flatness but spacetime is certainly not Ricci flat when the theory includes a non-trivial matter Lagrangian.)

Belot G (2007) The representation of time and change in mechanics. In: Butterfield J, Earman J (eds) Philosophy of physics. Handbook of the philosophy of science. Elsevier, Amsterdam, pp 133–227
Brown HR, Brading K (2002) General covariance from the perspective of Noether's theorems. Diálogos (Puerto Rico) 79. http://philsci-archive.pitt.edu/archive/00000821/
Earman J (2006a) Two challenges to the requirement of substantive general covariance. Synthese 148:443–68
Earman J (2006b) The implications of general covariance for the ontology and ideology of spacetime. In: Dieks D (ed) The ontology of spacetime. Elsevier, pp 3–24
Earman J (1989) World enough and space-time: Absolute versus relational theories of space and time. MIT Press, Cambridge MA
Earman J, Norton J (1987) What price spacetime substantivalism? The hole story. Br J Philos Sci 38:515–525
Friedman M (1983) Foundations of space-time theories: Relativistic physics and philosophy of science. Princeton University Press, Princeton
Giulini D (2007) Some remarks on the notions of general covariance and background independence. Lect Notes Phys 721:105–120 http://arxiv.org/abs/gr-qc/0603087.
Kretschmann E (1917) Über den physikalischen Sinn der Relativitätspostulate. Ann Phys 53: 575–614
Maudlin T (1988) The essence of space-time. In: Fine A, Leplin J (eds) Proceedings of the 1988 biennial meeting of the philosophy of science association, vol 2. East Lansing, Michigan, Philosophy of Science Association, pp 82–91
Maudlin T (1990) Substances and space-time: What Aristotle would have said to Einstein. Studies Hist Philos Sci 21:531–561
Norton JD (1993) General covariance anf the foundations of general relativity: Eight decades of dispute. Report Progr Phys 56:791–861
Pitts JB (2006) Absolute objects and counterexamples: Jones–Geroch dust, Torretti constant curvature, tetrad-spinor, and scalar density. Stud Hist Philos Modern Phys 37:347–371
Rovelli C (2001) Quantum spacetime: What do we know? In: Callender C, Huggett N (eds) Physics meets philosophy at the Planck scale. Cambridge University Press, Cambridge, pp 101–122, http://arxiv.org/abs/gr-qc/9903045
Rovelli C (2004) Quantum gravity. Cambridge University Press, Cambridge
Sorkin R (2002) An example relevant to the Kretschmann–Einstein debate. Modern Phys Lett A 17:695–700, http://philsci-archive.pitt.edu/archive/00000565/
Sus A (2008) General relativity and the physical content of general covariance. PhD thesis, Universitat Autònoma de Barcelona

Chapter 20
Relativity, Locality and Tense

Steven Savitt

There is a persistent myth that the special theory of relativity somehow proves or entails that the universe is timeless or static. The mathematician and popularizer of science Rudy Rucker is a particularly enthusiastic partisan of this view, writing in regard to the special theory that "The idea of a block universe is... more than an attractive metaphysical theory. It is a well-established scientific fact" (Rucker 1984, 149). Another well-known popularizer of science, the physicist Paul Davies, is equally adamant. "In short," he writes, the time of the physicist does not pass or flow" (Davies 2006, 42).

The prominent philosopher Hilary Putnam seems to deny coming to be and passing away when he writes that according to the special theory "*all* future things are real... and likewise all *past* things are real, even though they do not exist *now*.... I conclude that the problem of the reality and the determinateness of future events is now solved. Moreover, it is solved by physics and not by philosophy.... I do not believe that there are any longer any *philosophical* problems about Time; there is only the physical problem of determining the exact physical geometry of the four-dimensional continuum that we inhabit" (Putnam 1967, 246–247).

And finally the physicist Olivier Costa de Beauregard (1981, 429):

> In Newtonian kinematics the separation between past and future was objective, in the sense that it was determined by a single instant of universal time, the present. This is no longer true in relativistic kinematics: the separation of space-time at each point of space and instant of time is not a dichotomy but a trichotomy (past, future, elsewhere). Therefore there can no longer be any objective and essential (that is, not arbitrary) division of space-time between "events which have already occurred" and "events which have not yet occurred." There is inherent in this fact a small philosophical revolution.

But this myth of stasis misses the real point concerning time that the special theory of relativity teaches. Late in his life, looking back on his scientific achievements and their philosophical importance, Albert Einstein wrote (Einstein 1949, 61):

We shall now inquire into the insights of definite nature which physics owes to the special theory of relativity.

S. Savitt (✉)
Department of Philosophy, University of British Columbia, Vancouver, BC, Canada
e-mail: savitt@interchange.ubc.ca

(1) There is no such thing as simultaneity of distant events…

We deeply and instinctively believe the opposite. Aristotle simply remarked that "there is the same time everywhere at once" (*Physics* 220^{b}5). But were the world described by the special theory of relativity, this deep and instinctive belief would be shown incorrect.[1] There is no unique hyperplane of simultaneity, as will be noted below, that marks *the* same time everywhere – that is, globally. What might cohere with the special theory, as will also be noted below, is a set of events, a present, that is either point-like or spatially localized.

Recently, a paper has appeared with a title that directly contradicts my – or rather, Einstein's – thesis. The paper by Yuval Dolev is called "How to Square a Non-Localized Present with Special Relativity". My aim in this paper is to present the argument of Dolev's paper and to explain how it fails.

Dolev approaches the present through its role in what is known in the trade as *tense*. The term sits a bit uneasily, in my view, between language and metaphysics, but here is one way the Dolev introduces it:

> Events are always experienced, thought of and spoken of as possessing a tensed location. We do not always know whether a given event is past, present or future, but we cannot help thinking and speaking of it as being either past, present or future. (Dolev 2006, 183)

In contrast to this claim, the special theory entails that there are pairs of events whose time order is indeterminate. That is, if one picks a certain point or event O in Minkowski spacetime, the spacetime of the special theory of relativity, then there is a class of events, those that are spacelike separated from O, that each have the following property: in some inertial frames it is future to O, in some it is past to O, and in exactly one (ignoring rotations) it is simultaneous with O. This class of events is what Costa de Beauregard above refers to as the *elsewhere* (of O).

This division of spacetime at any event O into past, elsewhere, and future and the frame-dependence of the time order of events in the elsewhere with respect to O seem to clash with the universality of tense that Dolev finds essential. "[E]very event," he says, "is tensely located." (Dolev 2006, 187) That is, it is either past, present, or future. And he adds:

> [I]f we accept that tense, far from being a naïve and obsolete intuition, is indeed an indispensable element of our language, thought and experience then, unless we are forced to do otherwise, we should seek an interpretation of relativity theory that accords with this element rather than conflicts with it. (Dolev 2006, 187)

In the absence of such a reconciliation, it would seem that one is forced to the conclusion that tense has no place in the relativistic world view. Without tense there can be no passage or becoming or dynamic time as well, as Costa de Beauregard claimed above.

What interpretation of the special theory does Dolev recommend that will, in his view, reconcile that theory with tense as he understands it? It turns on his remark

---

[1] I believe the situation to be no different in the general theory of relativity, but that is a discussion for another occasion.

that the solution to the problem is "accepting the non-transitivity of co-presentness" (Dolev 2006, 188).

Since 'co-present' is an unusual term that permits a certain elasticity in its meaning, Dolev's remark is open to some interpretation. I will suggest two interpretations, in fact, though I think it is clear that Dolev intended the second. Nevertheless, it is useful to look at the first alternative briefly in order to distinguish it clearly from Dolev's actual view.

Dolev's commitment to the universality of tense commits him *prima facie* to a principle like the following:

$$x < y \text{ or } y < x \text{ or } x \, sim \, y, \tag{20.1}$$

where x and y are *any two* events (or spacetime locations), '<' is the relation of earlier than, and *sim* is a two-place equivalence relation, simultaneity. A tense theorist would normally suppose that the relation '<' is irreflexive, anti-symmetric, and transitive. These conditions, plus the universality of '<' so characterized, are then in conflict with the frame dependence of tense noted above.

Might Dolev wish to reconcile SR with tense by denying that *sim* is transitive (and hence by denying that it is an equivalence relation) while retaining the idea that it is a binary relation? While this seems a possible reading of his words, it is nonetheless doubtful. That is, I doubt that one as committed to tense as Dolev would be willing to give up any of the following general principles. For any events x, y, and z:

$$\text{if } x < y, \text{ then } \sim(x \, sim \, y), \tag{20.2}$$

$$\text{if } x < y \text{ and } z \, sim \, x, \text{ then } z < y, \tag{20.3}$$

$$\text{if } x < y \text{ and } z \, sim \, y, \text{ then } x < z. \tag{20.4}$$

It is easy to show that Eqs. 20.1–20.4 require the transitivity of *sim*, if *sim* is a binary relation. Imagine events a, b, and c such that a *sim* b, b *sim* c, but ∼(a *sim* c). Is a < c? No, since then b < c (by Eq. 20.3) and then ∼(b *sim* c) by Eq. 20.2, contradicting the assumption that b *sim* c. Similarly, it is not the case that c < a. But then by 20.1 we have a *sim* c, contradicting the assumption that ∼(a *sim* c).

The alternative is to suppose that *sim* is a three-place rather than a two-place relation. That is, events are simultaneous (or not) *with respect to an inertial frame* F. On this construal of *sim* it is not transitive. That is, from 'a *sim* b in F' and 'b *sim* c in F′' it does not follow either that a *sim* c in F or that a *sim* c in F′. However, within one fixed inertial frame *sim* is transitive and we have the familiar complete ordering of events as expressed in 20.1. That is, *in a given frame* tense has its familiar structure. It is by relativizing tense to frames that Dolev seeks to preserve his basic intuitions regarding tense.

How does this shift help to preserve one's pre-relativistic intuitions about a global present? If one considers the hyperplane orthogonal to the inertial world line that defines a given frame, one has a structure that looks very much like the classical global present. It seems as if one can retain one's deepest intuitions about time if one relativizes them. But what else should one expect in trying to accommodate one's intuitions to the (special) theory of relativity?[2]

There is a standard objection to relativizing temporal notions in this manner. The notions of past, present and future, it is argued, are somehow tied to ontological notions – for instance, the present is real whereas the past and future are not – and these ontological notions cannot be relativized. Dolev clearly and forcefully (and, I think, correctly) rejects the link to ontology. Following Austin (1962, Chapter 7), he argues that 'real' has meaning only when its contrast is specified, when one is told the way in which a thing may fail to be real. When one tries to do this for temporal notions, the resulting ontological claims inevitably turn out to be pedestrian, if true, or clearly false.

There are, nevertheless, two considerations that reduce the appeal of Dolev's view. First is its arbitrariness. That is, while it is true that in any given inertial frame plus orthogonal hyperplane of simultaneity events can be completely ordered into past, present and future just as in classical or Newtonian spacetime, the principle of relativity insists that all inertial frames at a given event (that is, the inertial frames corresponding to all the straight lines in Minkowski spacetime through the given event) are physically on a par. Even if it is true, then, that events can be divided into past, present and future in frame F, the division will be different in each of a non-denumerable infinity of distinct frames, and the relativization of tense implies that there is no sense to the assertion that an event is (say) present *tout court*.

In fact, there is a second level of arbitrariness beyond our inability to single out a particular inertial frame as somehow "privileged" in respect to the non-denumerable infinity of others corresponding to inertial lines through a given event. Why should one prefer the orthogonal hyperplane through the given event (relative to a given inertial line) as opposed to all the other possible hyperplanes through that event? Of course, if one prefers to use the standard Einstein synchronization of distant clocks (as opposed to all other possible values of Reichenbach's epsilon),[3] then the orthogonal hyperplane (in that frame) will be the set of points with the same t-coordinate as the given event. Why should that fact mark those events out as having some particular "tensed" position – say, as present (relative to a given inertial line and choice of clock synchronization parameter)? Even worse, it's not even clear that one need be restricted to flat hypersurfaces. Why can't any arbitrary global achronal surface[4] that includes the given event be considered a "present" for that event in a given inertial frame, adding yet another level of arbitrariness?

The second line of thought emphasizes the uselessness of Dolev's construction for his stated purpose. Dolev indicated that classical tense was required by our

---

[2] This is more a rhetorical than a real question. The special theory of relativity hinges on the existence of a fundamental four-dimensional invariant quantity, the spacetime interval. Various familiar three-dimensional quantities are then seen to be relative to the way this invariant is projected onto $3 + 1$ spaces.
[3] See Reichenbach (1958: §19).
[4] A surface, that is, in which all events are pairwise spacelike separated.

experience or phenomenology. While it is not quite clear what he intended by these assertions beyond the fact that we do, when employing our commonsense or scientifically untutored world view, believe that events are completely ordered as in Eq. 20.1, the fact that the speed of a light ray *in vacuo* is an upper limit for the transmission of information or causal influence ensures that, given some event e, no events occurring on any of the global achronal spacelike hypersurfaces passing through e can affect e in any way. Given a sentient observer at e, it is *precisely* these events that are irrelevant to its experience at e.[5] In particular, whether an event spacelike separated from e is past, present or future (in some inertial frame, relative to some choice of clock synchronization) can have no effect on the experience of an observer at e. Tense, as preserved by relativization, can not be connected to phenomenology in the usual way, the stated motivation for the preservation.

So why bother relativizing classical tense, especially when there is an invariant notion of tense that survives naturally in Minkowski spacetime? Since Minkowski spacetime is temporally orientable, let us suppose that we have (somehow) chosen an orientation and can designate one of the directions as past and the other as future. Then all events that are timelike or lightlike separated from a given event e can be divided invariantly into those that are future to e and those that are past to e. Granted, this is not a complete ordering of all events in spacetime; it is only a partial ordering. That is Eq. 20.1 does not hold, but why is Eq. 20.1 supposed to indicate an *essential* feature of tense? Why is tense without a complete ordering – as opposed to the partial ordering of events in the special theory–not tense at all? No argument is offered to defend such a claim.

Dolev does offer one objection to the view I have just espoused. He writes, "Moreover, distant events that are future with respect to a given point, become past without ever being present!" (188) (In this he echoes Putnam [1967: 246] down to the exclamation point.) This claim is true. Consider a timelike line Γ and some event e on Γ. Then there are events not on Γ that will have this feature. At some times earlier than e along Γ these events will be future (with respect to these events, of course). At some times later than e along Γ these events will be past with respect to them. But since at first blush the present for e is confined to the given event e itself, these other events will never be present for e.

But this is less an objection to the proposed position than a reiteration of it. Indeed, there is in Minkowski spacetime, as stated above, only a partial ordering of events with respect to '<' rather than a complete ordering. It's not the way we ordinarily think about tense and time, but the special theory of relativity was revolutionary precisely because it abandoned some aspects of the ordinary way we think about tense and time.

Let me close by adding that there is a way to mitigate some of the oddity of thinking of tense in a spacetime of partially ordered events. In the view as presented above, the present is essentially identical with the given point e, the event that divides its future from its past. Recently, a slightly less austere present has

---

[5] This point is emphasized in Dieks (2006).

been proposed as consistent with the special theory. The idea is that, first of all, the present has some temporal extent or duration. Since the present of our experience is temporally extended, this much should fit comfortably with Dolev's insistence on the importance of phenomenology.

The second step is to standardize on some small temporal interval (say, 1 s, which is a good approximation to the variable human psychological present) and consider intervals of that duration on a timelike line Γ. We'll indicate this interval as $[e_0, e_1]$. I then call the interior of the intersection of the future light-cone of e_0 and the past light cone of e_1 the *Alexandroff present* of the interval $[e_0, e_1]$ on Γ (Savitt 2009).[6]

If the usual sort of units are used, in which the speed of a light ray is one spatial unit per temporal unit, then a causal present will look in a spacetime diagram like a diamond with a waist of about 300,000 km (1 light second, if its temporal length is 1). This Alexandroff present will contain all events that can causally interact with any pair of events in the initial interval on Γ. Since it is defined in terms of the light cone structure, it is not frame-dependent. Nor is it subjective, even though its temporal length is scaled to match that of the human psychological present. The foot is also anthropomorphic in origin, but it is a perfectly objective unit of length.

If this is an acceptable notion of present in the special theory of relativity, then, as long $[e_0, e_1]$ is of some reasonable duration (like 1 s), most nearby events will become present after they are future and before they are past. Of course, very distant events still will not ever be present in this sense, even though they are at earlier times future and at later times past; but one can understand how we might have come to think that they should be, given the behaviour of nearby events. This, as far as I can see, is as much reconciliation as is possible between our pre-relativistic notions of tense and the revolutionary insights of the special theory.[7]

References

Austin J (1962) Sense and sensibilia. Oxford University Press, Oxford
Costa de Beauregard O (1981) Time in relativity theory: Arguments for a philosophy of being. In: Fraser JT (ed) The voices of time, 2nd edn. The University of Massachusetts Press, Amherst, pp 417–433
Davies P (2006) That mysterious flow. Sci Am Special Edn Frontiers Phys 15:82–88
Dieks D (2006) Becoming, relativity and locality. In: Dieks D (ed) The ontology of spacetime I. Elsevier, Amsterdam, pp 157–176
Dolev Y (2006) How to square a non-localized present with special relativity. In: Dieks D (ed) The ontology of spacetime I. Elsevier, Amsterdam, pp 177–190

---

[6] I chose this name since the Russian mathematician Alexandroff was, as far as I know, the first to investigate this structure. Pooley and Gibson (2006) call this structure *the Stein present (of the interval $[e_0, e_1]$ on the world line Γ)* in honor of Howard Stein, whose papers on time led many us to think about this structure. It might be best to call this structure *the causal present* (for the interval $[e_0, e_1]$ on Γ).

[7] I would like to thank Yuval Dolev for helpful comments on a previous draft of this paper, though he of course disagrees with its central contentions.

Einstein A (1949) Autobiographical notes. In: Schilpp PA (ed) Albert Einstein: Philosopher-scientist. Open Court, La Salle, IL, pp 2–94
Pooley O, Gibson I (2006) Relativistic persistence. Philos Perspect 20:157–198
Putnam H (1967) Time and physical geometry. J Philos 64:240–247
Reichenbach H (1958) The philosophy of space and time. Dover, New York
Rucker R (1984) The fourth dimension: a guided tour of the higher universes. Houghton Mifflin, Boston, MA
Savitt S (2009) The transient nows. In: Myrvold WC, Christian J (eds) Quantum reality, relativistic causality, and closing the epistemic circle. The western Ontario series in philosophy of science 74, Springer, New York, pp 339–352

Chapter 21
A Weylian Approach Towards Theories of Matter: Dynamic Agents and Geometrisation

Norman Sieroka

21.1 Introduction

Around the middle of the 1920s Hermann Weyl distinguished three different types of theories of matter; namely substance, field, and agency (or "agens") theories. In what follows I shall sketch these distinctions and take some of Weyl's own examples as an illustration in particular for what he calls a "pure" field theory and an agens theory. Then I shall briefly indicate how some more recent treatments of matter in physics relate to this distinction; that is, to what extent they could also be understood as being field or agens theories of matter in Weyl's sense.

Afterwards I shall relate this dichotomy to what one might call the "fundamental theme" in Weyl's philosophical writings of the mid-1920s (and of his book *Philosophy of Mathematics and Natural Science* in particular), which is the gap or tension between activity and passivity or "freedom and constraint", as Weyl put it. This tension – not only in the case of theories of matter – manifests itself as a historical seesaw or oscillation. And the resulting and quite specific intertwining of historical and systematic thought against the background of this tension marks a rather typical German Idealist tradition. Hence, I shall end up my presentation by a brief sketch of how Weyl aims at what might be called a philosophy of nature.

21.2 Matter Since Early Modern Times

According to Weyl's section on matter in his 1927 *Philosophy of Mathematics and Natural Science*, early modern times started with assuming that matter is substance (Weyl 1927, pp. 124–127; note that in the following I refer exclusively to the original German edition, not to the 1949 revised English edition). Matter was assumed to be, one might say, self-contained and self-sufficient and it was assumed to be the

---

N. Sieroka (✉)
ETH Zürich, Professur für Philosophie, RAC G16, CH-8092 Zürich, Switzerland
e-mail: sieroka@phil.gess.ethz.ch

bearer of properties like mass and extension. However, the development of physics in particular since the eighteenth century showed that so-called substantial properties could be understood in terms of dynamics and thus without assuming a separate bearer. In particular, this development was fostered by the growth and consolidation of electromagnetism; and with this classical field theory it became tempting to reduce matter completely to fields (Weyl 1927, pp. 129–131). A very influential example of such a "pure field theory", as Weyl calls it, is the electromagnetic programme by the German physicist Gustav (Mie 1912a, b; see also Weyl 1921; Vizgin 1994, pp. 26–38). Roughly speaking, Mie claimed that matter was nothing but knots in the electromagnetic field.

This was in 1912 and 6 years later an important pure field theory was suggested by Weyl himself (Weyl 1918, 1919). Since it was now 1918 Weyl could start from a different classical field theory, namely general relativity. Also intuitively this might be a more promising approach, since one now tries to dissolve matter into time and space instead of electromagnetism. In his philosophical reflection of 1927 Weyl showed a direct link between the old Cartesian idea that matter is pure extension and modern pure field theoretic approaches like his own one from 1918 (Weyl 1927, p. 136).

Tempting as this reduction of matter might seem prima facie, already around 1920 Weyl became sceptic about it. According to Weyl, what goes wrong in pure field physics is that it only describes what he calls a "silent continuous flowing" of fields and that it is no "dynamic view" of the world (Weyl 1927, p. 130). Weyl, I take it, has two things in mind here (cf. also Weyl 1921, 1924). First, after getting rid of all concepts of substance in a pure field theory, there are no agents left, no particular entities acting and suffering in space and time. All is "quiet", in a sense. Second, field theories are invariant under time reversal and so the tension between past, present and future – which we as human agents in the world do experience – plays no role.

It was Weyl's ingenious move to now solve these problems or to fill in the gaps of a purely field theoretic world view by means of new results from atomic physics, that in itself first also looked problematic. Already in 1920 Weyl argued for acknowledging the genuine probabilistic nature of processes in the atomic realm (Weyl 1920). Since for Weyl the use of statistics in physics is based on a notion of cause and effects and since for him the direction of time is constituted by cause and effects, quantum physics brings time as experienced into the description of nature (cf. also Weyl 1927, pp. 142–143). However, since electromagnetism and general relativity also are empirically successful theories and since they describe the "silent happenings" in space and time, the realm of the quanta cannot be space-time (Weyl 1920, p. 122).

As far as physics is concerned, this conception crucially depends on the usage of Gauss's theorem in field theory. As Weyl himself showed in his paper entitled "Field and Matter" (*Feld und Materie*), one can transform all relevant properties of matter from volume to surface integrals (Weyl 1921). According to Weyl this has to be taken seriously, which means that field physics has to get rid of a hypothetical interior of particles and has to content itself with surfaces or spatio-temporal neighbourhoods. The fields themselves then are caused by matter, but that does not mean that matter is really placed within such spatio-temporal neighbourhoods. Matter itself, Weyl tells us, is rather transcendent or "extramundane". So for Weyl space-time is not a simply connected manifold but, at least topologically, rather something like a cheese with holes in it. By the same token, Weyl gains "room" for quantum physics, because quantum physics works in spaces other than proper space-time (Hilbert-space, fiber bundles, etc.), while at the same time he leaves the success of classical field physics in describing space-time untouched (Weyl 1927, pp. 142–143).

According to Weyl's agens theory, matter is something which "acts and suffers", as he puts it (Weyl 1924, pp. 509–510). In this respect matter is somehow similar to an acting subject or to an "ego" (Weyl explicitly uses Fichtean terminology here; for more details and examples cf. Sieroka 2007). For instance, state reduction in quantum mechanics is described by Weyl as decision making, though importantly not in the sense that in the realm of quantum physics matter *decides* (active voice) but that decisions *are made*; that is, Weyl is keen to use passive voice here and to avoid attributing reflexivity to matter (Weyl 1920, p. 122). Let me skip over the details here and just briefly add that this conception allows Weyl to combine the human experience of freedom and the activity or spontaneity of matter and to account for the absolute coincidences in the quantum realm as a kind of perspective effect. Besides, it closely relates to Weyl's idea of an in- and outward developing space-time, which for him is the only satisfying account of any continuum, be it mathematical or physical, and which is opposed to the finished or tenseless universe of field physics (cf. Weyl 1920 and his letter to Pauli from December 1919, ETH-Archive, Hs176: 1).

To put it in a nutshell then: with his agens theory Weyl aims at a unification of field physics and the newly arising quantum physics. He claims to give back causality its proper place in physics, to get causal experience and the experience of a flow of time back into the description of nature, and to provide a satisfying account of the continuum of space and time (Weyl 1920; Weyl 1927, pp. 132–134, 142–148). Besides, he claims to have finished the old Leibnizian project of an agens theory of matter (cf. Leibniz 1982a, b); indeed Weyl's term "agens" is borrowed from Leibniz. By way of combining the active realm of, if you like, "quantum monads" with field physics (which describes only the *transmission* of forces through space and time) Weyl claims to have found out about "the communication of the monads" (Weyl 1924, p. 510; cf. Sieroka 2008).

Having introduced Weyl's dichotomy between field and agens theories of matter, I shall briefly turn to the question whether Weyl's distinction between different types of theories of matter is also applicable to physics after 1927.

21.3 Post-Weylian Applications

Within the second half of the twentieth century, a prominent attempt to develop pure field physics was John Wheeler's geometrodynamics. Like Weyl's unified field theory of 1918 Wheeler's project was particularly motivated by general relativity and also aimed at reducing matter to geometrical features of space-time (Misner and Wheeler 1957, pp. 595–596). According to geometrodynamics, what appears to be a mass is basically a lump of electromagnetic energy sticking together by its own gravitational force; and what appears to be a charge is nothing but what Wheeler called a "wormhole" in space-time through which a bundle of electromagnetic field lines go and which, if observed from some distance, looks like a charged pointlike particle (Wheeler 1961, pp. 65, 78). Wheeler here correctly refers to Weyl as the originator of the view that space-time might have a complex topology (Misner and Wheeler 1957, p. 532; cf. Weyl 1927, p. 64).

However, the problems of Wheeler's approach are numerous: these lumps of radiation are very, very large and heavy; and so it is extremely difficult to relate them to what we usually think of as being physical entities. Most importantly, Wheeler was only able to elaborate *classical* geometrodynamics; that is, he could give a geometrodynamic account of classical mechanics, general relativity and electrodynamics. However, as he himself emphasised, this was only meant as a kind of stage-setting for the real project, namely *quantum* geometrodynamics (Misner and Wheeler 1957, p. 534; see also Wheeler 1968). But this project was only hinted at and could never be worked out by Wheeler.

A more recent approach claiming to stand in the Wheelerian tradition is the fiber bundle formalism in quantum field theory as presented by Mielke (1987; the general approach itself is, of course, older and goes back to the work of, amongst others, Trautman 1973; Wu and Yang 1975; Drechsler and Mayer 1977). Since fiber bundles are connected to the points in space-time but are not themselves part of space-time, this view also has an element in it taken from Weyl's agens theory – as indeed Mielke explicitly acknowledges (Mielke 1987, pp. 131–132). Take, for instance, strong interaction which is now described as the geometrodynamics of colour space. Arguably this is a geometrical description – but of something that happens "beyond space and time". So, contrary to his own claim, Mielke has left (at least the exact) geometrodynamic framework given out by Wheeler. However, as I like to suggest, according to Weyl's distinction Mielke does not belong to the camp of the agens theorists, for his approach neither knows of genuine material agents nor does it allow for Weyl's notion of "decision", since the fiber bundle formalism itself is classical.

Apart from these two (rather exotic) examples, which I mentioned because Wheeler and Mielke refer explicitly to Weyl, also more common approaches in quantum physics can be meaningfully evaluated against Weyl's dichotomy between "silent" (or passive) field theories and agens theories – although, of course, one has to be careful then with the notion of field which Weyl used only in the classical sense. Let me only mention the two main roads of quantising gravitation here: quantum general relativity (e.g., loop quantum gravity) and string theory. Like Weyl's unified field theory and like Wheeler's and Mielke's geometrodynamics, approaches like loop quantum gravity also are primarily motivated by general relativity and start from the allegedly firmly based concept of background independence (Kiefer 2006; see also Wüthrich 2006, pp. 2–9). So in Weyl's sense they build on a rather field theoretic framework. In contrast to this, and rather similar to Weyl's agens theory, string theory starts from a more general speculation about what matter might be (other than just curved space-time or assumed test particles) and that this matter is not necessarily or outright from the beginning placed in our four-dimensional space-time (Kiefer 2006). It would be tempting here to discuss the transcendent acting of matter further and, for example, relate it to the role of what is called a holographic principle; i.e., Gauss's theorem in Weyl's agens theory and the AdS/CFT conjecture in string theory (cf. Kiefer 2004, p. 279). However, instead of speculating about this, I shall come back to Weyl and to how his description of different theories of matter is part of a bigger philosophical picture he wants to draw.

21.4 Wavering Between Freedom and Constraint

Going back to the conceptual frame of Weyl's historiography of theories of matter a certain dialectic structure is apparent (cf. Weyl 1924; Weyl 1927, pp. 124–137). Putting it very roughly: in their pure form both substance and field theories of matter, although being opposed to each other, cannot provide a satisfying dynamical worldview. Substances (in the classical philosophical sense of the term) do not interact with one another, and pure field physics knows of no dynamic agents at all. Both views are then, as one might say, "preserved" (*aufgehoben*) in an agens theory with its dualism of matter and field and Weyl's resulting concept of "communicating monads".

As Weyl is keen to show, the difference between dynamic views which take matter to be an agent and passive (silent) views which try to dissolve matter into geometry and fields shows itself as a historical seesaw or wavering. According to Weyl this wavering can also be described as that between freedom and constraint which fundamentally characterises human nature: for humans are both spontaneously acting, intelligent beings and also bound to a body in space-time (Weyl 1927, pp. 89–90, 134, 147–149; Weyl 1931, pp. 3, 4, 19). Having this in mind it is perhaps easier to understand why Weyl claimed that with his notion of "decision" spontaneity enters the physical realm, how thereby matter becomes more like an ego; and how on the other hand he talks about field physics as "strict law physics" (*reine Gesetzesphysik*) which is only able to account for the "silent flow" in four-dimensional space-time but not for such crucial concepts as "life" (Weyl 1920, pp. 116, 122).

Indeed one should take these remarks seriously – in particular, since Weyl himself says that it is the tension between freedom and constraint that marks the driving force of his whole book *Philosophy of Mathematics and Natural Science*. And one can easily see this also by looking at what Weyl writes, for instance, about the notion of continuity in mathematics which, as already mentioned, is also intimately connected to what he thinks about physical continua (Weyl 1925). As in the case of matter, Weyl here presents a conceptual development as a seesaw process between passive and active views; views that take a mathematical continuum to be something given (as, for instance, in set theory) and views according to which a continuum has to be actively constructed by the mathematician (as, for instance, in the case of Brouwer's free choice sequences . . . though this example is a little tricky).

For Weyl the wavering between freedom and constraint as a historical process can be found in all major areas of mathematics and physics. This and the way he presents this intertwined systematic and historical considerations turns his 1927 book into something like a philosophy of nature; and it shows Weyl's reception of German Idealism, in particular of the writings of Johann Gottlieb Fichte. Apart from the historical and autobiographical evidence we have for this reception (cf. Sieroka 2007), it can be seen, for instance, from the way Weyl attributes activity but not reflexivity to matter. Weyl is very careful here not to turn into naturalism or into a Schelling-type natural philosophy (cf., e.g., Schelling 1985). Weyl rather stays with the Fichtean slogan that one should attribute as much activity to nature as possible without turning nature itself into a self-aware subject (Fichte 1971, p. 362). More specifically, Weyl himself writes that by looking at the historical wavering between freedom and constraint one recognises that "there arises a third realm" (Weyl 1925, p. 540). This realm is that of what Weyl calls "symbolic construction"; and he states that it was the "born-and-bred constructivist" Fichte who "first entered this realm" (Weyl 1954b, p. 641; Weyl 1925, p. 540).

To me Weyl's constructivist and in part pragmatist reading of Fichte seems to be a sensible and much more interesting alternative than the (today unfortunately rather standard) reading of Fichte as a philosophical foundationalist interested in some rather mystic self-relation of the ego (as, e.g., in Henrich 1993, pp. 57–82). In contrast, I suggest that by telling his oscillating history of, for instance, the concept of matter and of the mathematical continuum Weyl showed a historical dimension of what Fichte called "the wavering of the power of imagination" (*das Schweben der Einbildungskraft*; see, e.g., Fichte 1979, p. 29; Fichte 1997, p. 146). By the same token Weyl fulfills (at least in part) the Fichtean programme of philosophy as being the "pragmatist historiography of the human mind" (Fichte 1991, p. 69; 1997, p. 141).

I hope this look at theories of matter could illustrate Weyl's philosophical framework around 1925. With his concepts of wavering between activity and passivity and of an arising third realm Weyl tries to interpose himself between Husserlian phenomenology (which he pretty much adhered to until about 1920; see Ryckman 2005) and Cassirer's philosophy of symbolic forms (cf. Friedman 2005). For Weyl phenomenology with its key concept of viewing essences was now an altogether too passive view (Weyl 1949, p. 334; cf. Sieroka 2007); and Cassirer lacking a strong or proper concept of life was in danger of being a kind of idle running functionalism or "activism" (Weyl 1954a). So using the Fichtean term which is meant to express exactly this two-sidedness of life as activity and passivity one might say that Weyl *posits himself* between Husserl and Cassirer.

References

Drechsler W, Mayer ME (1977) Fiber bundle techniques in gauge theories. Springer, Berlin
Fichte JG (1971) (1799/1800) Sätze zur Erläuterung des Wesens der Thiere. In: Fichte IH (ed) Fichtes Werke, vol XI. de Gruyter, Berlin, pp 362–367

Fichte JG (1979) (1796) Grundlage des Naturrechts nach Prinzipien der Wissenschaftslehre. Meiner, Hamburg
Fichte JG (1991) (1794) Über den Begriff der Wissenschaftslehre oder der sogenannten Philosophie. Reclam, Stuttgart
Fichte JG (1997) (1794) Grundlage der gesamten Wissenschaftslehre. Meiner, Hamburg
Friedman M (2005) Transcendental philosophy and twentieth century physics. Philos Today 49:23–29
Henrich D (1993) Selbstverhältnisse. Reclam, Ditzingen
Kiefer C (2004) Quantum gravity. Clarendon, Oxford
Kiefer C (2006) Quantum gravity – A short overview. In: Fauser B, Tolksdorf J, Zeidler E (eds) Quantum gravity. Mathematical models and experimental bounds. Birkhäuser, Basel, pp 1–13
Leibniz GW (1982a) (1695) Specimen dynamicum. Meiner, Hamburg
Leibniz GW (1982b) (1714). Monadologie. Meiner, Hamburg
Mie G (1912a) Grundlagen einer Theorie der Materie: Erste Mitteilung. Ann Phys 37:511–534
Mie G (1912b) Grundlagen einer Theorie der Materie: Zweite Mitteilung. Ann Phys 39:1–40
Mielke EW (1987) Geometrodynamics of gauge fields: On the geometry of Yang-Mills and gravitational gauge theories. Akademie-Verlag, Berlin
Misner CW, Wheeler JA (1957) Classical physics as geometry: Gravitation, electromagnetism, unquantized charge, and mass as properties of curved empty space. Ann Phys 2:525–603
Ryckman TA (2005) The reign of relativity: Philosophy in physics 1915–1925. Oxford University Press, Oxford
Schelling FWJ (1985) (1799) Einleitung zu dem Entwurf eines Systems der Naturphilosophie. In: Frank M (ed) Schelling – Ausgewählte Schriften, vol I. Suhrkamp, Frankfurt (Main) pp 337–394
Sieroka N (2007) Weyl's "Agens Theory" of Matter and the Zurich Fichte. Stud Hist Philos Sci 38:84–107
Sieroka N (2008) Hermann Weyl (1885–1955). In: Weber M, Desmond W (eds) Handbook of Whiteheadian process thought, vol. II. Ontos, Frankfurt (Main), pp 539–548
Trautman A (1973) Theory of gravitation. In: Mehra J (ed) The physicist's conception of nature. Reidel, Dordrecht, pp 179–198
Vizgin VP (1994) Unified field theories in the first third of the 20th century. Birkhäuser, Basel
Weyl H (1918) Gravitation und Elektrizität. Sitzungsberichte der Königlich Preußischen Akademie der Wissenschaften zu Berlin 1918:465–480. Cited according to Weyl 1968, II, pp 29–42
Weyl H (1919) Eine neue Erweiterung der Relativitätstheorie. Ann Phys 59:101–133. Cited according to Weyl 1968, II, pp 55–87
Weyl H (1920) Das Verhältnis der kausalen zur statistischen Betrachtungsweise in der Physik. Schweizerische Medizinische Wochenzeitschrift 50:737–741. Cited according to Weyl 1968, II, pp 113–122
Weyl H (1921) Feld und Materie. Ann Phys 65:541–563. Cited according to Weyl 1968, II, pp 237–259.
Weyl H (1924) Was ist Materie? Die Naturwissenschaften 12:561–568, 585–593, 604–611. Cited according to Weyl 1968, II, pp 486–510
Weyl H (1925) Die heutige Erkenntnislage in der Mathematik. Symposion 1:1–23. Cited according to Weyl 1968, II, pp 511–542
Weyl H (1927) Philosophie der Mathematik und Naturwissenschaft. Oldenbourg, München
Weyl H (1931) Die Stufen des Unendlichen. Fischer, Jena
Weyl H (1949) Wissenschaft als symbolische Konstruktion des Menschen. Eranos-Jahrbuch 1948:375–431. Cited according to Weyl 1968, IV, pp 289–345
Weyl H (1954a) Address on the unity of knowledge. Columbia University in the city of New York bicentennial celebration. Cited according to Weyl 1968, IV, pp 623–630
Weyl H (1954b) Erkenntnis und Besinnung (Ein Lebensrückblick). Studia Philosophica, Jahrbuch der Schweizerischen Philosophischen Gesellschaft 1954. Cited according to Weyl 1968, IV, pp 631–649
Weyl H (1968) Gesammelte Abhandlungen, 4 vols. Springer, Berlin

Wheeler JA (1961) Geometrodynamics and the problem of motion. Rev Modern Phys 33:63–78
Wheeler JA (1968) Einsteins Vision: Wie steht es heute mit Einsteins Vision, alles als Geometrie aufzufassen? Springer, Berlin
Wu TT, Yang CN (1975) Some remarks about unquantized non-Abelian gauge fields. Phys Rev D 12:3843–3844
Wüthrich C (2006) Approaching the Planck scale from a generally relativistic point of view: A philosophical appraisal of loop quantum gravity. Ph.D. thesis, Faculty of Arts and Sciences, University of Pittsburgh

Chapter 22
Mirroring and Understanding Action

Corrado Sinigaglia

22.1 Introduction

Mirror neurons are a specific class of neurons that respond when an individual performs a given action and when s/he observes a similar action performed by others. There is now a general consensus that there are at least two mirror neuron systems, one located on the lateral convexity on the brain, the other in the insula and in the cingulate cortex. The first translates observed actions devoid of any emotional content into the corresponding motor representations, while the second converts emotional behaviours into the corresponding viscero-motor responses (Rizzolatti and Sinigaglia 2008a).

It was originally held that the primary function of the mirror mechanism is to enable an individual to understand the actions performed by others, by directly matching the sensory with the motor representations of those actions (di Pellegrino et al. 1992; Rizzolatti et al. 1996; Gallese et al. 1996). A similar interpretation has been offered for the understanding of the emotions of others, since the observation of an emotional expression or context determines the activation of the same cortical sites as the direct experiencing of the same emotion (Wicker et al. 2003; Gallese et al. 2004).

Whereas mirroring in the emotional system has been mostly accepted, mirroring for action has recently become a target of criticisms. Of these, the critical account proposed by Csibra (2007) is particularly worthy of mention. Not only does it constitutes the basis of the majority of the objections raised against action mirroring, starting from those formulated by Jacob (2009) and Wood and Hauser (2008), but more than any other it insists on the relation between the mechanism and the function of mirror neurons, with the objective of demonstrating that interpreting the former in terms of a direct matching must inevitably be in conflict with the interpretation of the latter as critical for action understanding.

C. Sinigaglia (✉)
Department of Philosophy, University of Milan, Via Festa del Perdono 7, 20122 Milan, Italy
e-mail: corrado.sinigaglia@unimi.it

The paper aims to refute these criticisms by arguing that they are mostly due to a partial reading of mirror neuron properties and to a biased construal of both action and action understanding. In the next sections I shall give a brief outline of Csibra's argument and then focus on analyzing functional properties of mirror neurons, starting with the motor ones. I will use this analysis to counter Csibra's objections, illustrating how they presuppose a restricted conception of the directedness of mirror matching and how the latter leads to a misapprehension of the exact nature and range of mirror-based action understanding.

22.2 What Are Mirror Neurons for?

According to the direct-matching hypothesis (DMH), the observation of an action performed by others evokes, in the observer's brain, a motor activation that is alike to that which spontaneously occurs during the planning and effective execution of that action. The difference is that while in the latter case the motor activation becomes an *overt* motor act, in the former it remains at the stage of a *potential* motor act, thus enabling the observer to immediately understand the witnessed motor act. DMH does not exclude that other more complex mechanisms, such as those that are supposed to be at the basis of many inferential or meta-representational processes, may be at work and play a role in this function. It simply maintains the primacy of a direct matching between observation and execution of action, pointing out that observer's ability to understand the actions of others primarily capitalizes on the same 'motor knowledge' that underpins her/his own ability to act (Rizzolatti et al. 2001; Rizzolatti and Sinigaglia 2007).

Recently, however, DMH has been challenged by objections and alternative proposals. Of these the most stimulating challenge was undoubtedly that raised by Csibra, who proposed that "action mirroring cannot be direct but must be based on some kind of interpretation of observed action", on an "emulative action reconstruction", arguing that this implies that action understanding cannot be the primary function of action mirroring, because the former "may precede, rather than follow," the latter (Csibra 2007: 436).

Starting point of Csibra's argument is the assumption that the intuition behind DMH is that action interpretation during mirroring occurs at a relatively low-level (i.e., kinematics) and that it contributes to action understanding by means of a bottom-up motor activation allowing the observer to estimate "what higher level sub-goals and goals might have generated the observed action" (Csibra 2007: 441). He counters such intuition with the idea that mirroring can be achieved at a higher level of action interpretation and therefore that observed actions can be interpreted to the highest possible level before they are passed, via a top-down activation, on to the motor system for their kinematical reconstruction. Some basic mirroring phenomena (e.g., automatic imitation of simple transitive and intransitive movements, motor interferences, and so on) can be considered as highly consistent with low-level action interpretation for mirroring. However, Csibra remarks, while these mirror phenomena can be also construed as generated by emulative action reconstruction, there are several findings that seem incompatible with DMH and can be accounted for *only* by an emulative model of action mirroring.

First of all, single cell recordings showed that monkeys' mirror neurons responded to the observation of hand grasping actions even when the final part of these actions, consisting in the effective object-hand interaction, were hidden behind a screen, whereas they did not respond to the observation of the same hand movements when the experimenter mimed to grasp something in absence of any objects (Umiltà et al. 2001). As Csibra writes, "this finding is puzzling if action mirroring is performed by low-level direct matching because the low-level kinematics of a mimicked action is presumably similar to that of an object-directed action, and is available for mirroring" (Csibra 2007: 443).

Similarly "puzzling" to Csibra would be the finding that mirror neurons might respond to motor acts that the observer is unable to perform (Ferrari et al. 2005) or even to biologically impossible actions (Costantini et al. 2005): how does DMH explain such mirroring? There is no matching action in the observer's repertoire, so that the only possible interpretation appears to be that on the basis of which "observed actions are interpreted outside the motor system and then fed into the observer's action control system for reconstruction" (Csibra 2007: 446).

Finally, Csibra quotes the papers by Fogassi et al. (2005) and Iacoboni et al. (2005) that showed how mirror neurons respond differentially to the individual motor acts according to the overall intention with which it is thought they were carried out. Because both experiments were designed to unable the observer to figure out the intention underlying the observed motor acts from mere kinematical cues, it would be difficult to explain the differential motor activations by appealing to low-level mirroring: how and to what extent should such a mirroring allow the observer to understand further goals or intentions of witnessed motor acts? In contrast, as Csibra held, "these results fit perfectly with the emulation model of action mirroring", since intention understanding would be based on no-motor information (object semantic, contextual cues) processing, so that mirror activation would reflect the motor emulation of an observed action whose underlying intention is coded outside the motor system.

To sum up, the above-mentioned data would be incompatible with DMH as they would show that the fact that an observed motor act belongs to the observer's own motor repertoire can be neither sufficient nor necessary for mirror activation, because the latter can be determined by a high-level interpretation of the observed action, based on no motor and contextual information, or even by the sight of non executable movements, whose goal relatedness could be estimated from visual information only. To quote Csibra, these findings would actually undermine DMH insofar as they would reflect "two conflicting claims about action mirroring" implied by DMH: "The claim that action mirroring reflects low-level resonance mechanism, and the claim that it reflects high-level action understanding. The tension arises from the fact that the more it seems that mirroring is nothing else but faithful duplication of observed actions, the less evidence it provides for action understanding; and more mirroring represents high-level interpretation of the observed actions, the less evidence it provides that this interpretation is generated by low-level motor duplication" (Csibra 2007: 446).

22.3 Motor Goals and Action Mirroring

But is really there such a "tension"? Does the directness of matching truly imply that action mirroring occurs at relatively low-level? In other words: is only the kinematics of an observed motor act that can be directly matched in the case of mirror activation? Does goal (and intention) coding really require leaving motor the motor system? In addition: does the direct matching mechanism, by definition, actually involve the same or similar effectors and biological constrains between actor and observer? What is the effective role of observer's motor repertoire?

To answer to these questions, it is useful to begin with the functional properties mirror neurons share with other motor neurons from the ventral premotor cortex (area F5) and infero-parietal lobule (IPL). Single cell recordings showed that most F5 and IPL motor neurons code goal-related motor acts (such as grasping, holding, manipulating, etc.) and not the individual movements that compose these acts. Indeed, many F5 and IPL motor neurons discharge when the monkey performs a motor act such as grasping a piece of food, irrespective of whether it uses its right or left paw or even its mouth. Others motor neurons are more selective, discharging only for a specific effector or grip. However, even when selectivity is at its highest, the motor responses cannot be interpreted in terms of single movements: neurons discharging during certain movements (the flexing of a finger, for example) performed with a specific motor goal, such as grasping an object, discharge weakly or not at all during the execution of similar movements that compose a different motor act such as scratching (Rizzolatti et al. 1988; Rizzolatti et al. 2000).

A very recent study has shown that this is true not only for hand- and mouth-, but also for tool-mediated motor acts (Umiltà et al. 2008). The experiment was carried out with macaque monkeys, which were trained to grasp objects using two different types of pliers, 'normal pliers', which require typical grasping movements of the hand (opening and then closing), and 'reverse pliers', which require hand movements in the opposite sequence (closing and then opening). All recorded F5 neurons discharged in relation to the goal-related action of the pliers, maintaining the same relation to the different phases of grasping in both conditions, regardless of the fact that diametrically opposite hand movements were required to achieve the goal.

This quick review of the motor properties mirror neurons share with most of F5 and IPL neurons is enough to throw the assumption that lies at the basis of Csibra's criticism into discussion. As mentioned earlier, mirror defining functional characteristic is that these neurons become active not only when an agent performs a given motor act, but also when s/he observes it being performed by another. In the light of their motor properties, there is no reason to assume that mirror neurons "have to duplicate every minute detail of the observed act (including, for example, direction and speed of motion, angles between joints, etc.) in order to facilitate its understanding" (Csibra 2007: 437). Just as there is no reason to assume, as Csibra however does, that the directedness of matching has to be restricted to the kinematics of observed motor acts. And in fact DMH does not assume this. According to DMH, that in the observer's brain, the sight of a motor act performed by another recruits the same neurons that would become active if s/he were planning and effectively executing that act, means that mirror neurons code the motor goal-relatedness that identifies that particular motor act, independently of whether it is executed by the agent him/herself or simply observed while being carried out by another. What is directly matched is the motor-goal relatedness that characterizes both the effective observed and the effective executed motor acts (Sinigaglia 2008a; Rizzolatti and Sinigaglia 2008b).

This explains why mirror neuron activation is not strictly bound to the completeness of the sensory information or to only one sensory modality. Indeed, mirror neurons have been shown to respond to only partially seen motor acts (Umiltà et al. 2001) as well as to sound-producing motor acts (e.g., paper tearing), independently of whether they were seen, heard or both seen and heard (Kohler et al. 2002). As we have seen, Csibra regards as "puzzling" the fact that "observing a reaching act for an occluded target object elicits mirror neuron activation whereas the same movement does not trigger mirror neuron response when the monkey knows that there is no food behind the occluder" (Csibra 2007: 443). However, Umiltà et al. 2001' findings are "puzzling" only if one presupposes that sensory information directly mapped on motor neurons uniquely concern the kinematics of observed action. Only on the basis of such a presupposition it makes sense to contend, as Csibra does, that, "what [these findings] really indicate is that mirror neurons *reflect* action understanding rather than contribute to it" (Csibra 2007: 443).

There is absolutely no doubt that in Umiltà et al.'s experiment the fact that miming did not activate the mirror neurons was due to the circumstance that the monkey was fully aware that there was no object behind the screen. This however would in no way entail a top-down mechanism in which the monkey would access (how we do not know) a high level description of the observed action that would then be transmitted to the motor system for reconstruction. On the contrary, the hypothesis that the visual information relative to the presence (or absence) of the target contributes to the activation of the mirror system to the extent to which it is compatible with the animal's motor repertoire is much more economical (and plausible). In other words, the motor goal 'grasp' implies (at least for the macaque), a reference to an object which is physically present; the absence of the target entails that the observed movement cannot be matched with a corresponding motor goal, irrespective of the fact that the kinematics may be very similar in the two cases.

This can also be explained in the light of the different types of congruency that characterizes mirror neurons. In some cases this congruence can be extremely strict, pertaining not only the motor goal (e.g., grasping), but also the ways to achieve it (e.g., the type of grip). For most mirror neurons, however, the congruence is broader: though not identical, observed and executed actions are clearly connected, sharing the same goal-relatedness. For instance, a mirror neuron that is active during hand-grasping action can be activated by the observation of a mouth-grasping action.

According to Csibra, it would be hard to realize "how low-level motor mirroring could produce such a mismatch"; on the contrary, this kind of mirroring would fit perfectly with the emulation model of mirroring: "If the monkey has 'understood' the immediate goal of the action outside the motor system, from which the motor activation reconstructs the observed action, we would expect exactly this kind of correspondence between observation and execution" (Csibra 2007: 444). However, there is no reason to leave the motor system in order to account for the different degrees of mirror neuron congruency. In fact, giving their motor properties, it is no so "puzzling", as Csibra thinks, that mirror neurons visually (or acoustically) code observed motor acts with different degrees of generality.

The directedness of matching does not imply that it has necessarily to be construed in terms of one-to-one mapping, and even when this is the case, as for the strictly congruent mirror neurons, mirror mechanism does not run at the level of mere kinematics, but at a higher level, that captures the motor goal-relatedness of the observed act. Indeed, both strictly and broadly congruent mirror neurons respond to the goal-relatedness of the observed movements, even if they represent it in a different way, the former being more detailed and the latter more general. As a result, what is matched, and how it is matched, depends on the degree of generality that characterizes the motor responses of a given (set of) mirror neuron(s) as well as on its degree of congruence, and there is no need to leave the motor system to explain these differences.

22.4 From Motor Goals to Motor Intentions

Formulated in these terms, the question of the activation of mirror neurons during the observation of act that the observer had never executed previously or even biologically impossible acts, also takes on a difference significance to that attributed by Csibra. With regards the first point, he cites the study carried out by Ferrari et al. (2005) which showed mirror neurons that became active both when the monkey grasped a piece of food with its own paw and it watched the experimenter using a stick to pick up the food, even though the animal had never been taught how to use a stick in this way.

According to Csibra, this would be "a clear example of mirroring activation [...] which is incompatible with the idea of low-level motor mirroring". Nevertheless, as he himself acknowledged, "the mirroring process was not random", since "mirror neurons responded to the sight of a non-executable action with a different action that the monkey could have used to achieve the same goal". This, however, is not sufficient to justify Csibra's conclusion that a similar activation could *only* be explained within the emulation model of action mirroring. Indeed, the hypothesis that the motor system contributes to the effective understanding of observed actions mapping them on the motor representations that underlie the animal's capacity to act is far more simple (and more plausible too). The goal coded in motor terms is (at least) as general as that which should be coded in purely visual terms, as the mirror neurons that discharge at the sight of the stick being grasped by the experimenter also fire during the execution of the hand- and mouth-grasping action. Thus, more than just representing a reconstruction of the action, the activation of the mirror neurons would reflect the way in which it is understood.

In the case in point, grasping with a stick would have the motor meaning of grasping with the hand or the mouth for the animal observing the act. The situation changes when the animal has a certain degree of familiarity with the tool. In the above-mentioned study on the use of normal and reverse pliers, Umiltà et al. (2008) have shown that part of the recorded neurons had mirror properties and that their discharging coded the distal goal of the pliers as the same (i.e., grasping), even when the observed movement of the fingers were diametrically opposite. Very recently, Rochat et al. (submitted) attempted to weigh the rootedness in the motor repertoire of observed acts and its role in the coding of visual information, comparing the mirror responses determined by the sight of hand-grasping actions (reverse) pliers-grasping actions and stick-grasping actions in monkeys who were accustomed to using pliers but not sticks. The data indicated that although the various types of grasping were all mapped on the observer's motor repertoire, they gave rise to different types of mirror activation, and specifically, the more the observed act was rooted in the observer's motor repertoire, the more anticipated the mirror discharge.

This shows that Ferrari et al.' findings cannot be used as an argument against the directedness of matching, but suggests that the mirror mechanism presents different degrees of generality that enable it to code, in motor terms, a wide range of observed goal-directed motor acts. Indeed, the generality of the motor coding of the mirror neurons allows them to map observed motor acts that appear to violate biological parameters as in the experiments conducted by Costantini et al. (2005) in which a number of volunteers were presented with finger movements that were outside of the normal range of such actions. Quite the opposite to what Csibra thought, there is no need to assume that the directedness of mirror matching implies that such a mechanism has to take into account the biomechanical constraints the observed movement would involve if they were actually executed. On the other hand, if we take the alternative explanation proposed by Csibra into account, according to which the motor system would try to approximate (albeit unsuccessfully) the visual action reconstruction, using the available motor programs, we have to ask why the motor system should attempt to reconstruct a similar act, or what meaning would such an emulation have, particularly as the visual system has guaranteed "appropriate description of the end-state of such an action" (Csibra 2007: 446).

The same argument can be adopted when we consider whole motor actions, identified by determined goal hierarchies as opposed to individual motor acts, characterised by a specific goal-directedness. As Csibra himself recalls in his criticism of DMH, recent studies appear to indicate that the cortical motor system codes not only *what* an individual is doing but also *what* s/he is doing it *for*.

In particular, Fogassi et al. (2005) recorded single IPL neurons during eating and placing grasping actions. Most of the tested hand-grasping neurons appeared to be 'action constrained', forming pre-wired motor chains and discharging differentially depending on whether the grasping was a grasping to carry to the mouth or a grasping to move the piece of food from one place to another. But even more interesting is the fact that most of the recorded IPL 'action constrained' neurons showed mirror properties, responding both to eating and placing actions performed by an experimenter and discharging differentially depending on which action the single observed act of grasping was embedded into (e.g., grasping *for* eating or grasping *for* placing).

According to Csibra, this study would demonstrate, better than any other, the alleged tension between the two conflicting claims about action mirroring implied by DMH. On the one hand, indeed, it would be possible to hypothesise that monkeys' IPL mirror neurons were sensitive to kinematical parameters, so that their activation would represent a low-level mirroring phenomenon: however, Csibra remarks, "nothing in this study would then suggest that the monkeys would have understood the 'intention' behind the observed actions" (Csibra 2007: 447). On the other hand, it might be possible to accept Fogassi et al.'s (2005) argument that the mirror activation was independent of the kinematical parameters, reflecting an 'intention' understanding based on contextual cues: however, Csibra adds, "nothing in this study such an understanding is based on low-level mirroring (i.e., motor resonance)" (Csibra 2007: 447).

Now, it has already been seen that the mirror system does not run at a mere kinematical level, but is capable of coding the goal-relatedness of observed movements, thus allowing the observer to understand immediately the actions of others. Fogassi et al. 2005's data suggest that not only is the motor system sensitive to the goal-directedness that characterises an individual motor act, it also reflects the goal architecture in which that specific act may be embedded or the motor intention with which it had been carried out. In the case of first person execution of an action, the organization of the motor system explains the fluidity of action that is typical of intentional behaviour, since the final motor goal is displayed in the motor sub-goals that are suitable for its achievement from the start: from the first launch of hand movements, grasping a piece of food is a grasping *for* bringing to the mouth or a grasping *for* placing. But the most important aspect of all is that such motor organization extends the reach of mirror mechanism, allowing the observer to understand the motor intention underlying the observed act: indeed, when 'action constrained' neurons discharge, the sight of a hang-grasping motor act evokes much more in the observer than merely a single isolated potential motor act, it evokes an entire chain of potential motor acts which actually prefigure the motor intention that underlies the movements that were observed.

There is no doubt that information processing concerning object semantics (the type of object to be grasped) and/or some relevant contextual cues (e.g., the presence or absence of containers) might play a role in the elicitation of a given motor chain (grasping *for* eating instead of grasping *for* placing). But this does not require that mirror activation had to be construed here in terms of a top-down emulative mechanism deputed to reconstruct the kinematics development of the observed action on the basis of a not very clearly defined visual-inferential understanding of the intention. Nor it does imply that the construal of mirror activation as constitutive of observer's understanding of agent's motor intention represents a behaviouristic drift "cognitive science should resist" (Borg 2007: 18; see also Jacob 2009).

I argued elsewhere (Sinigaglia 2008b) that such interpretations end up missing the specificity of mirror mechanism, that is, the fact that the sensory information concerning the observed scene is mapped onto motor neurons forming pre-wired motor chain, and that only in virtue of this motor chain organization the activation of these neurons can be functionally interpreted in term of motor intention understanding. Here I just will mention the EGM experiments carried out by Cattaneo et al. (2007), showing that motor intention understanding does not rely on a processing of mere object or contextual information, but is primarily rooted in the observer's motor knowledge.

They recorded the activation of the mouth-opening mylohyoid muscle (MH) during the execution and observation of eating and placing actions in both traditionally developed (TD) children and children with autistic spectrum disorders (ASD). Both the execution and the observation of the eating action produced a marked increase of MH activity in TD children as early as the reaching phase. On the contrary, children with ASD showed a much later activation of the MH while eating, with the muscle becoming active only during the bringing-to-the-mouth phase, and, most importantly, no MH activity was recorded during their observation of eating action. These findings suggest that TD children were able to represent the action to be executed as an organized motor chain (grasping *for* eating), while children with ASD could represent the intended action just a simple sequence of unrelated single motor acts (reaching, grasping, *and* bringing-to-the-mouth), and this prevented them from disambiguating the sensory information regarding observed actions and, therefore, from immediately grasping the motor intentions underlying those actions, even when they are able to comprehend the goals of the single observed motor acts.

As well as throwing new light on the relationship between ASD and mirror neurons (see Sinigaglia and Sparaci 2008; Gallese et al. 2009 on this point), this data clearly indicates that the level at which an observed act is described during action mirroring, as well as the range of such description, depends on the observer's motor knowledge: the more refined this is and the more detailed, the more the mirroring will able to capture the intentional dynamics of the observed act. Therefore, contrary to what Csibra maintains, this implies that not only there would be no tension between the level of action mirroring and the level of action understanding, but the efficacy of the latter would be lessened if the former were not present.

22.5 Conclusions

I have devoted the previous sections to replying to Csibra's account of mirroring and understanding action, arguing that his objections to DMH are mostly based on the (unwarranted) assumption that directedness of mirror matching would imply that action mirroring occurs at a very low-level, being confined to a mere kinematical description of action. It is only in virtue of such an assumption that Csibra can portray the mirror mechanism as follows: "The popular conception of the causal role of mirroring in understanding the 'meaning' of actions involves a direct, unmediated, automatic, mandatory, resonance-like transfer mechanism, which miraculously generates a copy of the motor command responsible for the observed action, and forms the basis of bottom-up identification of the goals (or intentions) that have guided that action" (Csibra 2007: 454).

However, once one realizes that what characterizes the cortical motor system is its coding of goal-related motor acts and actions rather than single movements, and that this motor goal-relatedness can be represented with different degrees of generality, then it becomes immediately evident that there is no reason to look elsewhere other than the motor system to account for the motor goal-relatedness that identifies a given motor as such, regardless of whether it is performed by an agent or is witnessed by someone else. In the latter case, what is directly matched, i.e., what level of action description is involved in the direct matching mechanism, depends on the motor properties and degree of congruence of the activated (set of) mirror neurons. Even when the congruence is very strict, the direct matching occurs at the level of the motor goal-relatedness that is shared by the effective observed and effective executed motor acts; it therefore follows that even in this case action mirroring cannot be reduced, as Csibra does, to a mere kinematics resonance.

Moreover, the alternative account proposed by Csibra has a number of hitches. For example, it is not clear *where* and *when* that high-level action description would occur, of which mirror activation would be a mere motor emulation. It is true that, as Csibra mentions, Perrett et al. (1989) recorded pure visual neurons in the monkey's superior temporal sulcus (STS) that responded to the sight of goal-related motor acts. Just as it is true that STS neurons project directly to IPL areas that are endowed mirror properties and are strongly connected to the ventral premotor cortex. This suggests that visual information processing contributes to mirror activation. However, this does not justify the interpretation of action mirroring as a two-step process, where the first, purely visual step would be deputed to recognising the goal of the observed movements while the second step, with its visuo-motor characteristics, would be dedicated to emulating them. Such interpretation cannot be justified because this process could in no way give rise to a top-down action reconstruction, given that, at best, the coding of the STS neurons' goal is characterised by the same degree of generality as the mirror system and is not concerned with the goal architecture which the motor system is able to code, because of its chain organization. Finally, what would be the function of such motor emulation? Monitoring of the actions of others, as Csibra seems to be suggesting? The problem here is how a similar function would be compatible with the various degrees of generality with which the mirror system codes the goals of the actions of others? And how would it be interpreted in the case of anticipation of others' intentions?

With these arguments, I am not denying that descriptions of higher level and in any case different from those based on mirror activation play a key role in action understanding. What I have attempted to do is to demonstrate that mirror-based action understanding represents a specific way of understanding the actions and intentions of others, a way which is original and primary in nature, and this in virtue of a direct matching mechanism that maps the sensory information on the observer's motor repertoire, thus facilitating the grasping of the motor goal-relatedness which makes a certain sequence of movements a given motor act, achieved with a given motor intention, regardless of whether the agent is performing the act or whether s/he is watching it being performed.

References

Borg E (2007) If mirror neurons are the answer, what was the question? J Conscious Stud 14(8): 5–19
Cattaneo L, Fabbi-Destro M, Boria S, Pieraccini C, Monti A, Cossu G, Rizzolatti G (2007) Impairment of actions chains in autism and its possible role in intention understanding. Proc Natl Acad Sci USA 104:17825–17830
Csibra G (2007) Action mirroring and action understanding: an alternative account. In: Haggard P, Rosetti Y, Kawato M (eds) Sensorimotor foundations of higher cognition. Attention and performance XII. Oxford University Press, Oxford, pp 453–459
Costantini M, Galati G, Ferretti A, Caulo M, Tartaro A, Romani GL, Aglioti SM (2005) Neural systems underlying observation of humanly impossible movements: An fMRI study. Cerebral Cortex 15:1761–1767
di Pellegrino G, Fadiga L, Fogassi L, Gallese V, Rizzolatti G (1992) Understanding motor events: A neurophysiological study. Exp Brain Res 91:176–180
Ferrari PF, Rozzi S, Fogassi L (2005) Mirror neurons responding to observation of actions made with tools in monkey ventral premotor cortex. J Cogn Neurosci 17:212–226
Fogassi L, Ferrari PF, Gesierich B, Rozzi S, Chersi F, Rizzolatti G (2005) Parietal lobe: From action organization to intention understanding. Science 302:662–667
Gallese V, Fadiga L, Fogassi L, Rizzolatti G (1996) Action recognition in the premotor cortex. Brain 119:593–609
Gallese V, Keysers C, Rizzolatti G (2004) A unifying view of the basis of social cognition. Trend Cogn Sci 8:396–403
Gallese V, Rochat M, Cossu G, Sinigaglia C (2009) Motor cognition and its role in the phylogeny and ontogeny of action understanding. Dev Psychol 45:103–113
Iacoboni M, Molnar-Szakacs I, Gallese V, Buccino G, Mazziotta J, Rizzolatti G (2005) Grasping the intentions of others with one's owns mirror neuron system. PLoS Biol 3:529–535
Jacob P (2009) The tuning-fork model of human social cognition: A critique. Conscious Cogn 18:229–243
Kohler E, Keysers C, Umiltà MA, Fogassi L, Gallese V, Rizzolatti G (2002) Hearing sounds, understanding actions: Action representation in mirror neurons. Science 297:846–848
Perrett DI, Harries MH, Bevan R, Thomas S, Benson PJ, Mistlin AJ, Chitty AJ, Hietanen JK, Ortega JE (1989) Frameworks of analysis for the neural representation of animate objects and actions. J Exp Biol 146:87–113
Rizzolatti G, Sinigaglia C (2007) Mirror neurons and motor intentionality. Funct Neurol 22(4):205–210
Rizzolatti G, Sinigaglia C (2008a) Mirrors in the brain. How our minds share actions and emotions. Oxford University Press, Oxford
Rizzolatti G, Sinigaglia C (2008b) Further reflections on how we interpret the actions of other. Nature 455:589

Rizzolatti G, Camarda R, Fogassi M, Gentilucci M, Luppino G, Matelli M (1988) Functional organization of inferior area 6 in the macaque monkey: II. Area F5 and the control of distal movements. Exp Brain Res 71:491–507

Rizzolatti G, Fadiga L, Gallese V, Fogassi L (1996) Premotor cortex and the recognition of motor actions. Cogn Brain Res 3:131–141

Rizzolatti G, Fogassi L, Gallese V (2000) Cortical mechanisms subserving object grasping and action recognition: A new view on the cortical motor functions. In: Gazzaniga MS (ed) The cognitive neurosciences, 2nd edn. MIT Press, Cambridge, MA, pp 539–552

Rizzolatti G, Fogassi L, Gallese V (2001) Neurophysiological mechanisms underlying the understanding and imitation of action. Nat Rev Neurosci 2:671–670

Rochat M, Caruana F, Jezzini A, Escola L, Intskirveli I, Grammont F, Gallese V, Rizzolatti G, Umiltà MA (submitted) Hand grasping mirror neurons *know* the goal of action.

Sinigaglia C (2008a) Enactive understanding and motor intentionality. In: Morganti F, Carassa A, Riva G. (eds) Enacting intersubjectivity: A cognitive and social perspective to study of interactions. IOS Press, Amsterdam, pp 17–32

Sinigaglia C (2008b) Mirror neurons: This is the question. J Conscious Stud 15(10–11):70–92

Sinigaglia C, Sparaci L (2008) The mirror roots in social cognition. Acta Philos 17(2):307–330

Umiltà MA, Kohler E, Escola L, Fogassi L, Fadiga L, Keysers C, Rizzolatti G (2001) 'I know what you are doing': a neurophysiological study. Neuron 32:91–101

Umiltà MA, Escola L, Intskirveli I, Grammont F, Rochat M, Caruana F, Jezzini A, Gallese V, Rizzolatti G (2008) How pliers become fingers in the monkey motor system. Proc Natl Acad Sci USA 105(6):2209–2213

Wicker B, Keysers C, Plailly J, Royet J-P, Gallese V, Rizzolatti G (2003) Both of us disgusted in *my* insula: The common neural basis of seeing and feeling disgust. Neuron 40:655–664

Wood JN, Hauser MD (2008) Action comprehension in non-human primates: motor simulation or inferential reasoning? Trend Cogn Sci 12: 461–465

Chapter 23
Absolute Objects and General Relativity: Dynamical Considerations

Adán Sus

23.1 Introduction

General covariance and its role in the formulation of GR has proved to be a rather elusive notion. Once assimilated the import of the Kretschmann objection to a merely formal understanding of general covariance, the challenge is to be able to formulate a notion related to this one powerful enough to capture a key difference between GR and the rest of the physical theories that seem to introduce something like a prior geometry or a fixed background. The difference has been expressed at the intuitive level using several metaphors – prior geometry theories introduce a stage where the dynamics takes place or they introduce objects that act on the others while not being acted upon.

One of the most prominent attempts to answer this question is what can be called the Anderson–Friedman program. It pivots around a formal definition of absolute object, capturing those geometrical objects that are the same across the models of the theory in a certain technical way. Going back to the metaphorical language, one can see these objects as fixed in the theory and, in this sense, as not being affected by the changes of the other objects. Then, one can go on to think that whatever a background is, it is going to be encoded by one or several of these objects and that, certainly, absence of them is going to be a clear indication of background independence. But this cannot be the whole story, not unless one has very compelling reasons to think that every object meeting the definition of absoluteness is a good candidate for background. Because, as the metaphor goes, a background should be an object that, besides not being acted upon, acts on the other objects or, one can say, has a certain kind of relevance for the dynamical behaviour of the other objects. The problem will be then to specify which objects, if not all, from those that are absolute can be suitable for the role of background, which goes in hand with giving a clear cut account of what absolute objects are expected to do, namely, in what

A. Sus (✉)
Department of Philosophy, Autonomous University of Barcelona, 08193 Bellaterra (Cerdanyola del Vallès), Barcelona, Spain
e-mail: adansus@gmail.com

sense a background "acts". Perhaps, at this stage one should agree with Pitts[1] in his reservations about the present metaphor; in typical examples acting is not a good image for what absolute objects do, although this will depend on how much one is ready to stretch the meaning of a word like "acting".

That the definition of absolute object is not the whole story for an elucidation of the notion of background independence is already suggested by Anderson's (and also Friedman's) formulation of the program.[2] Together with the mentioned definition, they provide a distinction between covariance and invariance, and a characterisation of the symmetries of a theory as those transformations that leave the absolute objects invariant. This is followed by an assertion about the effect of absolute objects: they select a subgroup of the covariance group, the largest that leaves the absolute objects invariant, as the symmetry group. It has never been made clear, at least to me, what the exact status of this additional ingredient of the Anderson–Friedman program is. In principle, the definition of absolute object is independent of the symmetries of the theory. But, what about its suitability to be a background? I think that the answer implicit in the usual presentations of this program is that any absolute object makes a good background, at least good enough to spoil background independence. And the official story to defend this goes like this: any spacetime theory can be formulated in a coordinate independent manner, so as having the full diffeomorphism group as its covariance group. If the theory has absolute objects, it is going to have necessarily a symmetry group that is smaller than the covariance group (due to the fact that the only diffeomorphism invariant geometrical objects are constant scalars). Therefore, presence of absolute objects and reduction of the symmetry group, or absence of them and equality of the symmetry and covariance groups, are equivalent. So, even if one could have thought that the reference to symmetries suggested a new element on top of the definition of absolute objects, at the end all the weight seems to be on such definition.

One of the aims of this paper is to discuss the logic of this argument by defining in a slightly different way the symmetry group of a theory. I intend to argue that the invariance group, when properly defined, expresses the extent to which absolute objects, when dynamically relevant, "act". This goes hand in hand with stating that absolute objects do not spoil background independence when the symmetry group is equivalent to the whole diffeomorphism group and that their relevance increases as the subgroup of diffeomorphism group that conforms the invariance group gets smaller.

The second objective of the paper will be to use this conceptual scheme in the discussion of the most recently proposed counterexample to the Anderson–Friedman program. It has been pointed out by Geroch and Giulini that a local definition of absolute object together with the consideration of tensor densities as legitimate candidates for absoluteness, produces the result of GR having an absolute object and therefore, according to the standard use, the theory would not be background

---

[1] Pitts (2006).
[2] See Anderson (1964, 1967, 1971) and Friedman (1973, 1983).

independent. By using the alternative definition of invariance, and helped by a comparison to Unimodular Relativity (UR), I will argue that this result must be tempered; at least GR with no sources and no cosmological constant should be declared background independent according to the program.

The Anderson–Friedman program captures a meaningful difference when read in the right way, which is not to say that the novelty of GR consists in its lacking of absolute objects but in a substantive statement about the symmetries of the theory. This is also, I think, closer to Anderson's spirit when originally formulating the program.

23.2 Anderson–Friedman Program: Standard Use

Anderson's main motivation to introduce the absolute objects approach was to provide a physical notion of symmetry that would allow one to match unambiguously physical theories and symmetry groups. Part of the confusion about the status of symmetry principles in physical theories comes from its association with the notion of covariance. To define covariance, Anderson firstly introduces the notion of a kinematically possible trajectory (k.p.t.) as the set of the values that the mathematical quantities of the theory can take. A dynamically possible trajectory (d.p.t.) is a k.p.t. that actually satisfies the equations of motion of the theory. Then the covariance group G of a theory will be one for which the k.p.t. constitute the basis of a faithful realization of G and associates d.p.t. to d.p.t.[3] The covariance group divides the k.p.t. in equivalence classes in which all the elements in it are related by a transformation of the group. Anderson identifies each equivalence class with an internal state of the physical system, while the element of the classes would represent external states of it (possible measures of a given internal state).

It is clear that covariance is a theory dependent notion in the following sense: two theories describing the same physical system can have, in general, different associated covariance groups, even if they have the same number of equivalence classes. This means that covariance is a notion dependant on the means for representing a physical system postulated by a theory and not unambiguously determined by the internal states that the theory attributes to the physical system. If one thinks of two theories as equivalent when they assign the same internal states to a physical system, two equivalent theories can have, in principle, different covariance groups. Nonetheless, it must be noticed that Anderson introduces a condition that makes covariance not devoid of all physical content, namely, that the quantities used to describe the k.p.t. be in principle measurable.[4]

The physical content of Anderson's covariance is not enough to escape an objection of the Kretschmann's type. To overcome this difficulty, Anderson introduces

---

[3] Anderson (1967).
[4] Anderson (1967), p. 81.

first the notion of absolute object. We find two, in principle, different definitions of absolute object in Anderson's writings but both of them amount to the fact that an absolute objects is one that appears with all its transforms under the covariance group in every equivalence class of d.p.t.

With this in hand, Anderson defines the symmetry group or invariance group of a theory as the group of transformations that leaves the absolute objects unchanged. This is the concept of symmetry that Anderson considers associated to the structure of internal states of the physical system. For classical examples of theories with an absolute spacetime structure, Anderson's definition renders the expected symmetry group associated with them, even if one starts with formulations that differ in their covariance. For theories with no absolute objects, the covariance and invariance groups coincide. If this is the case for a theory that is already generally covariant, lack of absolute objects implies an invariance group as big as the diffeomorphism group; this is the closest that one finds in Anderson's definitions to a notion of background independence.

Friedman reformulated Anderson's program and introduced some changes in the definitions, but the core of the proposal is the same. It has the two ingredients: First, a definition of absolute object with the aim of capturing the way in which the field equations of theories with fixed background determine the objects that constitute these structures (although, in this case, and object is absolute if is the same, up to diffeomorphisms, in the local neighbourhoods around every point of the manifold). Second, a way of classifying spacetime theories by assigning to them a symmetry group (the group formed by the transformations that are symmetries of their absolute objects).

23.3 The Meaning of the Invariance Group

Sameness across models provides a way of characterising the intuition of absolute objects not being acted upon: they are fixed structures present in every model of the theory. But so far we have not provided a justification for the notion of invariance. For an object to be absolute, given the definitions of Anderson and Friedman, it is completely irrelevant the symmetries of the object in question; it is not difficult to envisage absolute objects with different (even null) degrees of symmetries. From the point of view of the original motivations of the program, the notion of invariance seems to have the explanatory weight by providing a physically meaningful concept of symmetry, in opposition to the merely formal given by the covariance group.

So it can be argued that, at least in its original formulation, the program connects the question of background independence with the invariance properties of the theories. Nevertheless, in the usual discussions of the Anderson–Friedman program, the notion of background independence is linked directly to the lack of absolute objects. If a generally covariant theory has no absolute objects, then it is going to be automatically background independent. But, is it true that presence of absolute objects always spoils background independence? In Friedman we find the simple argument that supports this claim; as no geometrical object, besides constant scalars, is invariant under Diff(M), the only way of achieving diffeomorphism invariance, given the definition of the invariance group, is by the lack of absolute objects.[5]

This equivalence between lack of absolute objects and background independence obviously has its roots in the definition of the invariance group. It seems strange to me that the justification for such a fundamental part of the Anderson–Friedman program has received so little attention, starting with the same creators of the scheme. I suppose that the reason for this silence is that Anderson's definition of invariance group, as the symmetry group of the absolute objects, connects well with that part of our intuitions that attribute the symmetry properties of pre-GR spacetime theories to the symmetries of the absolute spacetime structure. Nevertheless, after a closer look, Anderson's definition of invariance reveals what I think are some undesirable features.

To see this, let us think of two different theories with the same absolute object, the Minkowski metric for instance. Only with this information and with Anderson's definition of invariance we can assign to both theories the same symmetry group (the Poincarè group) and be sure that the theories will not be background independent. But there is something strange in the fact that we can give all this information without knowing how the Minkowski metric enters into the theory. One could have theories as different, from the point of view of the role of the metric, as a special relativistic Klein-Gordon field theory and a Rosen's bimetric type relativistic theory and still say that the invariance group for both is the same. Even more extremely, one can think of a falsely bimetric GR with an idle second metric and the verdict is going to be the same. At this point I can be accused, quite rightly, of having performed an illicit move; Pitts[6] notes that Anderson's scheme contains a clause to eliminate this kind of irrelevant objects from the theories; objects which do not interact with any of the other variables of the theory. Using this strategy removes the danger of having extreme cases with completely idle objects, but still does not have enough sensitivity to distinguish between different degrees of relevance. The problem can be stated in the following way: having made the invariance group dependant only on the symmetries of the absolute objects, all the substantive information that the program can give you about differences between theories rests on what absolute objects each theory has. Then one can eliminate aberrant cases of irrelevancy by including a third category or by forbidding objects that from the point of view of the equations of motion are not doing any work, but this manoeuvre is not very subtle: either the object is completely irrelevant and therefore eliminated or its relevance is completely determined by its transformation properties. It seems desirable to have something in the middle.

---

[5] It must be said that this is strictly true only if one understands, as Friedman does, that under the category of geometrical object only tensors and connections are allowed.
[6] Pitts (2006), p. 12.

23.4 Definitions of the Invariance Group

John Earman[7] distinguishes between spacetime symmetries and dynamical symmetries. His definition of spacetime symmetry is basically Anderson's definition of invariance group, with the condition that the absolute objects under consideration must be part of the characterisation of the spacetime structure. As a dynamical symmetry Earman understands a diffeomorphism f such that $< M, A_i, O_j >$ is a dpt of the theory iff $< M, A_i, f^*O_j >$ is a dpt (where A_i are the absolute objects).

After introducing his definitions Earman discusses two symmetry principles that in conjunction would assert the equivalence of spacetime and dynamical symmetries. As he argues, in principle, these two notions of symmetry are different and the symmetry principles point to the ideal adequacy of laws of motion and spacetime structure or, in other words, the situation in which the formulation of the theory employs no more spacetime structure than that that is necessary to support the laws. In such cases, the symmetries of the absolute objects are a faithful reflection of the dynamical symmetries of the theory.

For the purposes of giving a definition of background independence, I think that the difference between these two ways of defining symmetries is going to be essential. We must remember that the objective is to find an unambiguous way of matching theories and symmetry groups that is informative about its background dependence and we cannot presuppose that the theory is going to be free of surplus structure; what we need, precisely, is some criteria to identify irrelevancies coming from the formalism. Using something like the spacetime notion of symmetry presupposes then that one has an independent mechanism of eliminating irrelevant structure, perhaps Anderson's idea. What I want to defend is that such mechanism comes from the same definition of invariance group. But first, let us reflect on the differences between the two ways of defining the invariance group that match the two definitions of symmetries given by Earman.

Giulini[8] uses the notion of dynamical symmetry to define the invariance group of a theory. Later in his paper he gives an equivalent way of defining G-invariance (invariance under the group of transformations G), given the case that the theory is already G-covariant:

$< M, A_i, O_j >$ is a dpt of the theory iff $< M, f^*A_i, O_j >$ is a dpt, for every f that belongs to G (IG^*)

Then he notes that in "generic" cases, the group of transformations that meets this last definition is formed by the transformations that leave the absolute objects invariant. One can take Giulini's reference to generic situations as similar to Earman's use of the two symmetry principles in order to assert the equivalence between spacetime and dynamical symmetries. Now, from a conceptual point of view, the differences between a definition based on the symmetries of absolute objects and definition (IG^*) are worth noting. The latter depends not only on which absolute objects the theory has, but also in what the other geometrical objects in the theory are and what

---

[7] Earman (1989), pp. 45–48.
[8] Giulini (2006).

the interactions between these and the absolute objects are. Therefore, it is free, in principle, of one of the defects that I attributed to the definition of the invariance group used in the Anderson–Friedman program; given different theories with the same absolute objects, definition (IG^*) needs not give the same resulting invariance group and the difference in the verdict is going to depend on the dynamics encapsulated in the equations of motion.

A further reason to prefer a definition of type (IG^*) comes from its possible connection to the metaphor guiding the Anderson–Friedman program. One could say that the way in which a geometrical object acts on others is by imposing changes on them when it changes. Obviously certain changes on the geometrical objects are going to be irrelevant for the others and one could introduce the idea that the degree of relevance of a certain geometrical object is going to be related to the amount of changes of that object that, by virtue of the relations expressed in the field equations, imposes changes on the others. In the extreme case of an object for which no changes whatsoever induce changes in any of the other objects, we can say that it is dynamically irrelevant. One can apply this idea to express the way in which absolute objects act: when under a transformation the field equations dictate that the change induced on the absolute objects imposes changes on the other objects, then this possible change on the absolute object is showing a dimension of the acting of such absolute object. Furthermore, if one is allowed to implement changes on the absolute object without changing the other objects, these changes express degrees of irrelevance of the absolute object in question. The more degrees of irrelevance that an absolute object in a given theory has, the more inefficient or dynamically irrelevant it will be. Again, if no transformation on the absolute object imposes changes on the other objects, the absolute object can be said to be dynamically irrelevant.

To the extent that this definition of invariance group is equivalent to the one usually associated to the Anderson–Friedman program, the latter captures also the active role of absolute objects. This is so in the cases in which the spacetime symmetries are well tuned to the dynamical ones. Nevertheless there are other theories where there is not such equivalence and the difference between the two definitions will be essential for the program to be informative about background (in)dependence in those cases.

23.5 GR and Absolute Objects: The Scalar Density Counterexample

According to the standard use of the A-F program, background independence is the lack of absolute objects. This is, at least, the usual reading of it. The whole discussion of this paper suggests that it would be more convenient to say the following: a theory is background independent if its symmetry group (invariance group) is equivalent to the whole $Diff(M)$; the former formulation implies the latter but, as I have argued, there could be absolute objects that do not reduce the symmetry group, if one is ready to modify the definition of invariance group. It was taken that lack of absolute objects is enough to characterise background independence in GR until the recent awareness of the fact that, according to Friedman's local definition of absolute object, the scalar density $\sqrt{(-g)}$ is an absolute object in GR and the invariance group is the volume preserving diffeomorphisms group. This fact is presented in Pitts'[9] paper under suggestion of Robert Geroch and, independently, in Giulini's[10] one.

It is useful to compare this with what happens in a, in principle, different theory: Unimodular Relativity. Following Anderson and Finkelstein[11] this is a theory with a fixed volume element where the only variable is a metric density. One can write a generally covariant action for this theory and the field equations obtained are Einstein field equations for pure gravity with a cosmological constant (the cosmological constant here is a constant of integration rather than given a priori) and a global coordinate condition.

This theory, according to Anderson and Finkelstein, has an absolute object, the volume element, and the invariance group is the subgroup of the diffeomorphism group that leaves this object invariant; the group of volume preserving diffeomorphisms. In fact, this volume element is not variational and is fixed completely by virtue of the field equations (for every two models of the theory it is the same up to diffeomorphism). Of course, this object also meets Friedman's definition of absoluteness.

The Anderson–Friedman program using Friedman's definition of absolute object and Anderson's definition of invariance group (what I have called the standard use of the program) produces the same result for GR and UR. Both theories contain the same absolute object and the invariance group is also the same for both; this is a consequence of having a definition of invariance that depends only on intrinsic properties of the absolute objects. But this result is strange, beyond the undesirable, and supposedly counter-intuitive, fact of GR not being background independent: the weirdness comes from the fact of Anderson–Friedman scheme not being able to distinguish between two theories that are in principle different, at least in the suppositions that they make about the status of the object representing a local volume element. In UR, it is supposed to be fixed and given a priori, while nothing like this is assumed in GR. This puzzlement could be overcome simply by saying that in GR this fact is hidden and the strength of the program is bringing it up to the surface. If this were true, there would be a second question to be answered; if one is ready to take the equivalence of UR and GR, one must explain why this equivalence is only valid in presence of a cosmological constant. Use of the alternative definition of invariance helps to clarify these issues.

Before regarding the scalar density counterexample using the alternative definition of invariance, it is worth spelling out what this is exactly a counterexample of. We have identified the standard use of the Anderson–Friedman program with the claim that background independence can be equated to lack of absolute objects. If

---

[9] Pitts (2006).
[10] Giulini (2006).
[11] Anderson and Finkelstein (1971).

one assumes that GR must be a background independent theory, then it constitutes a counterexample to this claim, because it has in fact an absolute object. But one must notice that this is true under Friedman's definition of absolute object; Anderson's definitions would recognise the scalar density as absolute in Unimodular Relativity but not in GR. Now, there seems to be good reasons to prefer a definition of absoluteness that introduces the requirement of locality incorporated by Friedman (to avoid dependence on a fixed topology), but the price to pay is that one is also going to admit as absolute objects that, at least intuitively, do not break background independence. So, one must keep in mind that a possible way out of the undesirable effects of the counterexample would be to go back to a global definition of absoluteness, but that this would exclude other theories with legitimate local absolute objects.

Another way of being immune to the counterexample is not to allow objects like tensor densities to be candidates for absoluteness, but this again would have the consequence of excluding many other cases where the failure of background independence is due to tensor densities. If one incorporates locality in the definition of absolute objects and admits tensor densities, keeping Anderson's definition of the invariance group, then the program produces the counter-intuitive consequence of GR not being background independent.

Of course, one could try to think of further refinements on the definition of absolute object that would avoid this counterexample. It seems to me that a simpler option comes from allowing the use of the modified definition of invariance; my intuition is that any good effect achieved by an eventually better definition of absolute object is also achievable through modifying the definition of invariance, with the extra benefit of having a way to explain the idea of absolute objects acting in a certain way on other objects. So let us see what this modification says about the counterexample.

Let us concentrate now in a theory that is GR with no sources and no cosmological constant. The theory is scale invariant, meaning that if a given metric is a solution of Einstein field equations, any global rescaling is going to be also a solution. If we take a formulation of this theory in terms of the scalar and metric densities, a scale transformation will change the scalar density by some constant factor, while leaving the metric density unchanged; given a model of sourceless GR $\langle \mu, \tilde{g}_{\mu\nu} \rangle$, $\langle S\mu, \tilde{g}_{\mu\nu} \rangle$ will also be a solution of the field equations (with S indicating a scale transformation). This means that according to definition (IG^*) the invariance group will now be $VDiff(M) \otimes V$. One can prove that this group is isomorphic to $Diff(M)$. So, using the alternative definition of invariance (IG^*), the result is an invariance group as big as the whole diffeomorphism group; GR without sources and with no cosmological constant would be then fully background independent.

Another way of expressing this is by noticing that the field equations fix the local value of $\sqrt{(-g)}$ completely, but that this fixes the value of the volume element only up to a constant. So once fixed the scalar density, one has the freedom of changing the value of the volume element by changes of coordinates that do not preserve the volume. Unless the dynamics has a way of fixing this value too, the symmetries of the theory should be taken to be the diffeomorphism group.

The contrast with UR is clear now; the field equations of this theory include a gauge condition that fixes the value of the volume element; transformations that change this value are not to be taken as symmetries of the theory. Due to the fact that a cosmological constant breaks scale invariance, the same can be said for the theory given by Einstein field equations with a cosmological term. In both cases the two definitions of the invariance group coincide in determining the volume preserving diffeomorphisms as the symmetries of the theories.

23.6 Concluding Remarks

The Anderson–Friedman program provides a strategy to explicate a notion of substantive general covariance or background independence based on two components: a definition of absolute object and a concept of symmetries of a theory. Although the definition of absolute object is independent of the concept of symmetries, both ingredients are necessary to give a satisfactory account of background independence. Usually, reference to the second ingredient is omitted, and background independence, according to this account, seems dependent only on the presence or absence of absolute objects. According to this reading, GR would be background independent insofar the theory lacks absolute objects. Once one has been found, and this is what the scalar density counterexample does, the theory would not be, at least fully, background independent.

My proposal is that one must take into account the reference to the symmetries of the theory, as it was originally thought by Anderson and Friedman, and that this is equivalent to adding certain dynamical considerations on top of the formal definition of absolute object. The formal definition can identify as absolute objects that are locally fixed in a certain way but that are not relevant to capture a notion of background independence. I have defended that the definition of invariance group, at least when using Friedman's definition of absolute object, must be different to the one implicit in the standard use of the program; Anderson's definition of the symmetry group of a theory as the largest subgroup of the covariance group that leaves its absolute objects unchanged. With this in mind one gets the conclusion that the scalar density counterexample is only a counterexample to the claim of GR having no absolute objects but not to one stating that the symmetry group of GR (with no sources and no cosmological constant) is the whole diffeomorphism group.

Besides its relevance for the scalar density counterexample, the discussion about the definition of the invariance group has a conceptual import in understanding the meaning of background independence. The Anderson–Friedman program is a good attempt at explaining how certain structures do impede that a theory be fully background independent. But when taken with Friedman's definition of absolute object one might be identifying as absolute structures that meet the definition just because of the gauge freedom of the theory.[12] Here, instead of trying to refine the definition

---

[12] See Giulini (2007).

in order to avoid this unwanted result, I propose to vary the characterisation of invariance group. The reasons are the following: First, one finds a way to give content to the meaning of the invariance group as expressing the way in which absolute objects act on others. Second, one sees how absolute objects that are just there due to gauge freedom for some theories become active for other theories. Third, one gets rid of the rigidity of the standard use of the program by relating the degree of background dependence to the size of the invariance group instead of saying that background independence is the lack of absolute objects.

Acknowledgements This paper was completed with financial support from a grant (FPU) provided by the Spanish Ministry of Education and Science, and research project FFI2008-06418-C03-03.

References

Anderson JL (1964) Relativity principles and the role of coordinates in physics. In: Chiu H-Y, Hoffmann WF (eds) Gravitation and relativity. W. A. Benjamin, New York
Anderson JL (1967) Principles of relativity physics. Academic, New York/London
Anderson JL (1971) Covariance, invariance and equivalence: a view point. Gen Relat Gravit 2:161
Anderson JL, Finkelstein D (1971) Cosmological constant and fundamental length. Am J Phys 39:901
Earman J (1989) World enough and space-time: absolute versus relational theories of space and time. Massachusetts Institute of Technology, Cambridge, MA
Friedman M (1973) Relativity principles, absolute objects and symmetry groups. In: Suppes P (ed) Space, time and geometry. Reidelm, Dordrecht
Friedman M (1983) Foundations of space-time theories. Princeton University Press, Princeton, NJ
Giulini D (2006) Some remarks on the notions of general covariance and background independence, arXiv:gr-qg/0603087.
Giulini D (2007) Remarks on general covariance and/or background independence. Talk given in Causal and classical concepts in science. 3rd International workshop: Causality and relativity, UAB, Barcelona
Pitts JB (2006) Absolute objects and counterexamples: Jones-Geroch dust, Torreti constant curvature, tetrad-spinor and scalar density. Stud Hist Philos Modern Phys 37:347

Chapter 24
Empirical Foundation of Space and Time

László E. Szabó

24.1 Introduction

The central issue of special relativity is the comparison of space and time tags of physical events, defined in *different* inertial frames of reference. However, the question of how these space and time tags are defined *in one single* frame of reference is considered as unproblematic and is usually neglected. In this paper, I will focus on this second question.

When I say "definition", I mean *empirical* definition, somewhat similar to Reichenbach's "coordinative definitions", Carnap's "rules of correspondence", or Bridgman's "operational definitions"; which give an empirical interpretation of the theory.

Einstein, at least in his early writings, strongly emphasizes that all spatio-temporal terms he uses are based on operations applying measuring rods, clocks and light signals. In his 1905 paper, he describes the measurement of the length of a rod in an arbitrary (moving) inertial frame of reference as follows:

> The observer moves together with the given measuring-rod and the rod to be measured, and measures the length of the rod directly by superposing the measuring-rod, in just the same way as if all three were at rest.

And this is a typical description of the empirical meaning of length or distance. However, these usual operational definitions so often suggested in the textbook literature are untenable; they are full of obvious circularities. It is not my aim here to address the problems in question, because the upshot of these considerations is also quite common in the more sophisticated part of the literature of space–time physics: In order to avoid these obvious circularities and to minimize the conventional elements in the empirical foundation of our physical theory of space and time, we must avoid using standard measuring rod in the definition of distance and using

---

L.E. Szabó (✉)
Department of Logic, Institute of Philosophy, Eötvös Loránd University,
Múzeum krt. 4/i, 1088 Budapest, Hungary
e-mail: leszabo@phil.elte.hu

slow transportation of the standard clock in the definition of time tags, and the likes. We must also abstain from relying on the concept of rigid body, reference frame, and inertial motion. Instead, we have to use *one standard clock and light signals*.

Of course, using one standard clock and light signals for coordination of space–time is an old idea; as old as the widespread belief that the task is as trivial as it seems from the two-dimensional textbook examples, and that the resulted spatio-temporal structure is, at least locally, necessarily identical with the standard space–time geometry of special relativity. What will be new in our analysis is the consequent performance of this task without operational circularities. As we will see, the task is not trivial; and the analysis of the spatio-temporal conceptions so obtained will raise some still open – although experimentally testable – questions.

24.2 Empirical Definition of Space and Time Tags

First we chose an *etalon* clock. That is to say, we chose a system (a sequence of phenomena) floating somewhere in the universe. Without loss of generality we may stipulate that this is an equipment having a pointer and the readings are real numbers. There is no assumption that this is a clock measuring "proper time". There is no assumption that it "runs uniformly". And there is no assumption that it is "at rest" relative to anything, or that it is of "inertial motion". The reason is that none of these concepts is defined yet.

We will call "marker" an equipment which can be triggered by a physical event and can transmit and receive modulated radio waves containing some information. Assume we have as many markers as we need, with the following functions:

1. There is a distinguished marker floating together with the standard clock and continuously transmitting the actual reading of the standard clock.
2. The others continuously receive the regular time signals from the standard clock.
3. They can transmit radio signals containing the following information: (a) an ID code of the device and information about the standard clock reading, so from the signal they send it always can be known which device was the transmitter and what was the standard clock reading received by the transmitter at the moment of the emission of the signal, (b) information about the event on the occasion of which the signal was transmitted.
4. They can receive the signals transmitted by the others.

By the emission of a radio signal the marker marks an *event*. It is far from obvious, however, what must be regarded as an event in general – prior to the concepts of time and distance. (See Brown 2005, pp. 11–14.) We do not dwell on this problem here. The reader can easily imagine various operational solutions of how to use a marker for marking various physical events/phenomena.

Fig. 24.1 Operational definition of time tags (this is just a symbolic sketch, not a real "two dimensional space–time diagram" or the like)

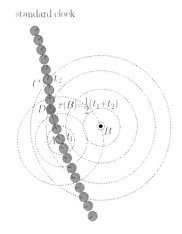

24.2.1 Time

Consider the experimental arrangement in Fig. 24.1. The marker at the standard clock emits a radio signal at clock-reading t_1 (event A). The signal is received by another marker which immediately emits another signal (event B). This "reflected" signal is detected by the marker at the standard clock at t_2 (event C). We assume, as an empirical fact, that the clock we have chosen is such that a given reflected signal is received by the standard clock only once, at reading t_2, and

$$t_2 \geq t_1 \tag{24.1}$$

by which we have chosen, conventionally, an "arrow of time" (not the arrow of physical processes in time; see Price 1996, pp. 16, 58). (In fact, we made two choices here. One is the choice of the direction of the parametrization of the clock's pointer positions (24.1). There is however a more important one: by applying the terms "sending" and "receiving" a signal, we previously determined the causal order of events A and C. To what extent this causal order is purely conventional? How can we – without prior spatio-temporal conceptions – distinguish whether an event is a "sending" or a "receiving" of a signal? How is this choice of causal order related to the change of information content of the signal? To what extent this choice is determined by our free will and free action experience at the modulation of the radio waves? Is this freedom an objective openness of future or merely a subjective experience? These are delicate metaphysical questions into the discussion of which it is not our present purpose to enter.)

Definition (A1) The *absolute time* tag of event B is the following:

$$\tau(B) := t_1 + \varepsilon(t_2 - t_1) \tag{24.2}$$

where $\varepsilon = \frac{1}{2}$ by convention. (Of course, it could be a contingent fact of nature that $t_2 = t_1$, in which case the choice of the value of ε would not matter.)

It is important to emphasize that the choice of using radio signals in definition (A1) is purely conventional. This choice is by no means justified by the "constancy and isotropy of the (round-trip) velocity of light"; simply because we are prior to any spatio-temporal concepts that would make any statement about the "velocity" of light meaningful.

24.2.2 Distance and the Problem of "Rest"

Denote S_τ the set of simultaneous events with time tag τ. One might think that we are ready to define the spatial distance between two points of space, that is distance between two simultaneous events. Surely, we can define the distance between the simultaneous events D and B in Fig. 24.1 as $\frac{1}{2}(t_2 - t_1)c$, where the value of c is taken as a convention. In this way however, as a little reflection shows, we can define the distance only *from the standard clock*, but there is no way to extend this definition for arbitrary pair of simultaneous events. In order to define the distance between two arbitrary simultaneous events we need further preparations.

We would like to base the definition of distance to the definition of time: the distance between two points in a given S_τ will be defined through the period of time in which a radio signal runs "from the one point to the other". Therefore, instead of signals sent and received by the marker at the standard clock, we will use radio signals "sent from the one point and received at the other". However, we encounter the following difficulty. We would like to define distance between *simultaneous* events; but the travel of the signal takes some time; the emission of the signal and the receiving of the signal are not simultaneous events. Whose distance is the one measured by the time of travel of the signal – and when? The distance obtained by means of the time of travel of the signal depends on the concept of "rest"; the concept of "being at the same place at different times" (Fig. 24.2). So, in order to

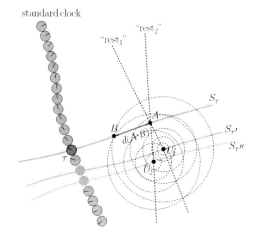

Fig. 24.2 The distance defined by means of the time of travel of the radio signal depends on the concept of "rest"; the concept of "being at the same place at different times". In general,
$\tau(B) - \tau(U_1) \neq \tau(B) - \tau(U_2)$

define the distance of simultaneous events we need a previous concept of "rest"; and, moreover, we have to define this concept by the only means of the standard clock and radio signals.

It is necessary to be careful of a possible misunderstanding. Although they are close to each other, the problem we are addressing here is different from the problem of persistence of physical objects (Butterfield 2005). What we would like to define is the identity of two locuses of space at two different times, and not the genidentity of the physical objects occupying these locuses. One might think that some definition of genidentity of physical objects must be prior to our operational definition of space and time tags, at least in the case of the standard clock. This is, however, not necessarily the case. The standard clock is just an ordered (ordered by the clock readings) sequence of physical events, but without the further metaphysical assumption that these events belong to the same physical object. (We definitely do not make such assumption in the case of a "clock-like" sequence of events that we will call a time sequence below.)

Definition (A2) A one-parameter family of events $\gamma(\tau)$ is called *time sequence* if $\gamma(\tau) \in S_\tau$ for all τ.

One has to recognize that a time sequence is a "clock-like" sequence of events. For every event, one can define a time-like tag in the same way as (A1): Event A (Fig. 24.3) is marked with the emission of a radio signal at time $\tau(A)$. The signal is reflected at event B. Event C is the first detection of the reflected signal at time $\tau(C)$. We define the following time-like tag for event B:

$$\tau^\gamma(B) := \tau(A) + \varepsilon\left(\tau(C) - \tau(A)\right)$$

(If there is no detection of the reflected signal at all, then, say, $\tau^\gamma(B) := \infty$.)

It is an empirical fact that $\tau^\gamma(B) \neq \tau(B)$ in general. It is another empirical observation however that for some particular cases $\tau^\gamma(B) = \tau(B)$.

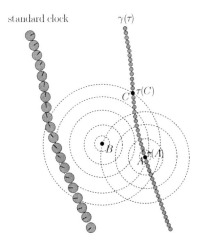

Fig. 24.3 Clock-like time sequence

Definition (A3) A time sequence $\gamma(\tau)$ is a *rest time sequence* if for every event B $\tau^\gamma(B) = \tau(B)$.

Whether or not there exist rest time sequences is an empirical question. We stipulate the following:

Emprical fact (E1) For any event A there exists a unique rest time sequence $\gamma(\tau)$ such that $A = \gamma(\tau(A))$.

Rest time sequence is a concept defined only by means of the standard clock and radio signals. It singles out a "world line" through every event, that will play the role of the "world line of a particle being at rest relative to the standard clock".

Now we are ready to define the distance between simultaneous events.

Definition (A4) The *absolute distance* between two simultaneous evens $A, B \in S_\tau$ is operationally defined in the following way. Take a rest time sequence γ such that $A = \gamma(\tau)$ (Fig. 24.4). Let $U = \gamma(\tau(U))$ be an event marked with the emission of a radio signal at absolute time $\tau(U)$, such that the signal is received and reflected at event B. The detection of the reflected signal marks the event $V = \gamma(\tau(V))$ of time tag $\tau(V)$. The absolute distance is

$$d_\tau(A, B) := \frac{1}{2}(\tau(V) - \tau(U))c \tag{24.3}$$

where $c = 299792458\frac{m}{s}$ by convention.

We know from (24.1) that for all $A, B \in S_\tau$

$$d_\tau(A, B) \geq 0 \tag{24.4}$$
$$d_\tau(A, A) = 0 \tag{24.5}$$

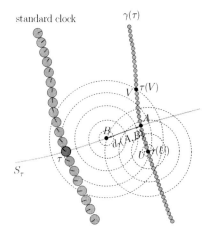

Fig. 24.4 The distance between two simultaneous events

However, the following facts cannot be known without further empirical observations:

Emprical fact (E2) For all $A, B, C \in S_\tau$

$$d_\tau(A, B) = 0 \text{ only if } A = B \qquad (24.6)$$
$$d_\tau(A, B) + d_\tau(B, C) \geq d_\tau(A, C) \qquad (24.7)$$
$$d_\tau(A, B) = d_\tau(B, A) \qquad (24.8)$$

The following proposition is however derivable:

Lemma 24.1 *Let γ_1 and γ_2 be arbitrary two rest time sequences. For any two moments of absolute time τ and τ'*

$$d_\tau(\gamma_1(\tau), \gamma_2(\tau)) = d_{\tau'}(\gamma_1(\tau'), \gamma_2(\tau')) \qquad (24.9)$$

Having distance defined on a given S_τ, we introduce the following abbreviations:

$$Cong_\tau(A, B, C, D) \iff d_\tau(A, B) = d_\tau(C, D)$$
$$Bet_\tau(A, B, C) \iff d_\tau(A, C) = d_\tau(A, B) + d_\tau(B, C)$$

In terms of these abbreviations we formulate the following – not necessarily new – empirical facts:

(E3) $\forall A \forall B \ Cong_\tau(A, B, B, A)$

(E4) $\forall A \forall B \forall C \ Cong_\tau(A, B, C, C) \to A = B$

(E5) $\forall A \forall B \forall C \forall D \forall E \forall F \ Cong_\tau(A, B, C, D)$
$\land Cong_\tau(C, D, E, F) \to Cong_\tau(A, B, E, F)$

(E6) $\forall A \forall B \ Bet_\tau(A, B, A) \to A = B$

(E7) $\forall A \forall B \forall C \forall D \forall E \ Bet_\tau(A, D, C) \land Bet_\tau(B, E, C))$
$\to \exists F \ (Bet_\tau(D, F, B) \land Bet_\tau(E, F, A))$

(E8) $\exists E \forall A \forall B \ A \in \alpha \land B \in \beta \to Bet_\tau(E, A, B)$
$\to \exists F \forall A \forall B \ A \in \alpha \land B \in \beta \to Bet_\tau(A, F, B)$
where α and β are two sets of events in S_τ.

(E9) $\exists A \exists B \exists C \exists D \exists E \ \neg D = E \land Cong_\tau(A, D, A, E)$
$\land Cong_\tau(B, D, B, E) \land Cong_\tau(C, D, C, E)$
$\land \neg Bet_\tau(A, B, C) \land \neg Bet_\tau(B, C, A) \land \neg Bet_\tau(C, A, B)$

(E10) $\forall A \forall B \forall C \forall D \forall E \forall F \ \neg D = E \land \neg D = F \land \neg E = F$
$\land Cong_\tau(A, D, A, E) \land Cong_\tau(A, D, A, F)$
$\land Cong_\tau(B, D, B, E) \land Cong_\tau(B, D, B, F)$
$\land Cong_\tau(C, D, C, E) \land Cong_\tau(C, D, C, F)$
$\to Bet_\tau(A, B, C) \lor Bet_\tau(B, C, A) \lor Bet_\tau(C, A, B)$

(E11) $\forall A \forall B \forall C \forall D \forall E \forall F \ Bet_\tau(A, B, F) \wedge Cong_\tau(A, B, B, F)$
$\wedge Bet_\tau(A, D, E) \wedge Cong_\tau(A, D, D, E)$
$\wedge Bet_\tau(B, D, C) \wedge Cong_\tau(B, D, D, C)$
$\rightarrow Cong_\tau(B, C, F, E)$

(E12) $\forall A \forall B \forall C \forall D \forall E \forall F \forall G \forall H \ \neg A = B \wedge Bet_\tau(A, B, C)$
$\wedge Bet_\tau(E, F, G) \wedge Cong_\tau(A, B, E, F)$
$\wedge Cong_\tau(B, C, F, G) \wedge Cong_\tau(A, D, E, H)$
$\wedge Cong_\tau(B, D, F, H) \rightarrow Cong_\tau(C, D, G, H)$

(E13) $\forall A \forall B \forall C \forall D \exists E \ Bet_\tau(D, A, E) \wedge Cong_\tau(A, E, B, C)$

The quantification runs over S_τ. In brief, we stipulate, as an empirical fact, that the two relations $Cong_\tau$ and Bet_τ, determined by the distances of simultaneous events, satisfy the axioms of three-dimensional Euclidean geometry, namely Tarski's axioms of three-dimensional Euclidean geometry (Tarski 1999). It must be emphasized that all the statements (E3)–(E13) are stipulated, via inductive generalization, merely on the basis of observations about distances of simultaneous events.

24.2.3 Spatial Coordination

Within this axiomatic framework, one can define the basic geometrical concepts in the usual way; and one can derive a body of theorems, well known from the textbooks on Euclidean geometry. Below are a few of the typical definitions and theorems we will use in the construction of space tags.

Definition A subset $\sigma \subset S_\tau$ is called (straight) *line* if satisfies the following conditions (Fig. 24.5):

1. For any $A, B, C \in \sigma$ exactly one of the following three relations hold:

$$d_\tau(A, C) + d_\tau(C, B) = d_\tau(A, B)$$
$$d_\tau(A, B) + d_\tau(B, C) = d_\tau(A, C)$$
$$d_\tau(B, A) + d_\tau(A, C) = d_\tau(B, C)$$

2. σ is maximal for property 1.

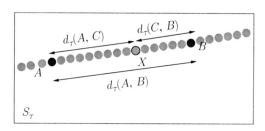

Fig. 24.5 Straight line

Fig. 24.6 Orthogonal lines

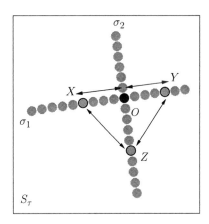

Definition Let σ_1 and σ_2 be two lines in S_τ such that $\sigma_1 \cap \sigma_2 = \{O\}$ (Fig. 24.6). σ_2 is *orthogonal* to σ_1 if for every $Z \in \sigma_2$ and for every $X, Y \in \sigma_1$

$$d_\tau(X, O) = d_\tau(O, Y) \Leftrightarrow d_\tau(X, Z) = d_\tau(Y, Z)$$

Theorem *For every $A, B \in S_\tau$ there exists a unique line containing A and B.*

Theorem *Let $A \in S_\tau$ be an arbitrary event and let $\sigma_1 \subset S_\tau$ be an arbitrary line. There always exists a line σ_2 orthogonal to σ_1, such that $A \in \sigma_2$.*

Definition Using the notations of the above theorem, let $\sigma_1 \cap \sigma_2 = \{O\}$. Event O is called *the orthogonal projection of A to σ_1*. Distance $d_\tau(A, O)$ is called *the distance of A from σ_1*.

Definition Let $\sigma_1 \subset S_\tau$ be a line. A line σ_2 is *parallel* to σ_1 if for all $X \in \sigma_2$ the distance of X from σ_1 is the same.

Theorem *Let $\sigma_1 \subset S_\tau$ be a line and let $C \in S_\tau$ be an arbitrary event. There exists exactly one line σ_2 such that $C \in \sigma_2$ and σ_2 is parallel to σ_1.*

Definition Let $A, B \in \sigma$ be two events on line σ. *Line segment* between events $A, B \in S_\tau$ is the following subset of σ:

$$\sigma(A, B) := \{ X \in \sigma \mid d_\tau(A, X) + d_\tau(X, B) = d_\tau(A, B) \} \tag{24.10}$$

These are however only examples. In what follows, the whole usual system of definitions and theorems of Euclidean geometry are supposed to be known.

Now we are going to define the standard Cartesian coordinates in S_τ. First we need a 3-frame.

Definition (A6) A *3-frame* in S_τ consists of three pairwise orthogonal lines σ_x, σ_y, σ_z in S_τ, such that $\sigma_x \cap \sigma_y \cap \sigma_z = \{O\}$ and three events $X, Y, Z \neq O$ such that

Fig. 24.7 Cartesian coordinates in S_τ

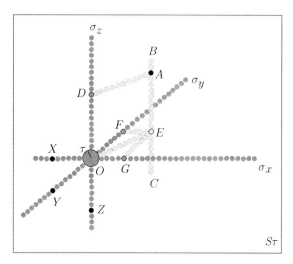

$X \in \sigma_x$, $Y \in \sigma_y$ and $Z \in \sigma_z$. O is called the origin of the frame (Fig. 24.7). Let us introduce the following notations:

$$\sigma_x^+ := \{P \in \sigma_x | Bet_\tau(X, O, P)\}$$
$$\sigma_x^- := (\sigma_x \setminus \sigma_x^+) \cup \{O\}$$
$$\sigma_y^+ := \{P \in \sigma_y | Bet_\tau(Y, O, P)\}$$
$$\sigma_y^- := (\sigma_y \setminus \sigma_y^+) \cup \{O\}$$
$$\sigma_z^+ := \{P \in \sigma_z | Bet_\tau(Z, O, P)\}$$
$$\sigma_z^- := (\sigma_z \setminus \sigma_z^+) \cup \{O\}$$

The origin of the 3-frame is arbitrary, although it is a natural choice to take the "τ-event" of the standard clock as origin.

In the following definition we give the operational definition of space tags in one given S_τ. Let us call them τ-space tags.

Definition (A7) Let A be an arbitrary event in S_τ. Take a line segment $\sigma(B, C) \ni A$ parallel to σ_z (Fig. 24.7). Take another line segment $\sigma(A, D)$ orthogonal to σ_z such that $D \in \sigma_z$. Let $\sigma(O, E)$ be a line segment parallel to $\sigma(A, D)$ such that $E \in \sigma(B, C)$. Finally, take the line segments $\sigma(E, F)$ and $\sigma(E, G)$ such that $\sigma(E, F)$ is parallel to σ_x and $F \in \sigma_y$, and $\sigma(E, G)$ is parallel to σ_y and $G \in \sigma_x$. Now, the τ-space tags are defined as follows:

$$x_\tau(A) := \begin{cases} d_\tau(G, O) & \text{if } G \in \sigma_x^+ \\ -d_\tau(G, O) & \text{if } G \in \sigma_x^- \end{cases}$$

$$y_\tau(A) := \begin{cases} d_\tau(F, O) & \text{if } F \in \sigma_y^+ \\ -d_\tau(F, O) & \text{if } F \in \sigma_y^- \end{cases}$$

24 Empirical Foundation of Space and Time

$$z_\tau(A) := \begin{cases} d_\tau(D, O) & \text{if } D \in \sigma_z^+ \\ -d_\tau(D, O) & \text{if } D \in \sigma_z^- \end{cases}$$

It must be emphasized that with the above definitions we only defined the space tags in a given set of simultaneous events S_τ. Yet, we have no connection whatsoever between two S_τ and $S_{\tau'}$ if $\tau \neq \tau'$. In principle, there exist "infinitely" many possible bijections between the different S_τ's. This is true, even if we prescribe that the bijection must be an isomorphism preserving distances.

Intuitively, a time sequence $\gamma(\tau)$ satisfying that

$$x_\tau(\gamma(\tau)) = \text{const.} \qquad (24.11)$$
$$y_\tau(\gamma(\tau)) = \text{const.} \qquad (24.12)$$
$$z_\tau(\gamma(\tau)) = \text{const.} \qquad (24.13)$$

corresponds to a localized physical object being at rest. "At rest" – relative to what? The actual behavior described by these equations depends on how the different 3-frames are chosen in the different S_τ's. One might think that an object is at rest if Eqs. 24.11–24.13 hold in one and the same 3-frame in all S_τ. But, what does it mean that "one and the same 3-frame in all S_τ"? When can we say that a line segment σ_x' in $S_{\tau'}$ is *the same* 3-frame axis as σ_x in S_τ? When can we say that an event A' is in the same place in $S_{\tau'}$ as event A in S_τ?

When we are seeking for a correspondence between S_τ and $S_{\tau'}$, our aim is not simply to find a mathematically "canonical" bijection – whatever it means. What we wish is a one-to-one map

$$\mathbb{T}_\tau^{\tau'} : S_\tau \to S_{\tau'}$$

of natural physical meaning:

(a) It must be defined by means of physical operations.
(b) For all $A, B \in S_\tau$, we require that $d_{\tau'}\left(\mathbb{T}_\tau^{\tau'}(A), \mathbb{T}_\tau^{\tau'}(B)\right) = d_\tau(A, B)$.
(c) It must reflect our intuition about being "at rest". (For example, in our traditional language, if the standard clock moves along a time-like straight line of the Minkowski space-time, $\mathbb{T}_\tau^{\tau'}$ must be equal to the map $(\tau, x, y, z) \mapsto (\tau', x, y, z)$, in the frame of reference of the standard clock. Of course, this example should be understood only intuitively.)

We have already defined a concept of the unique rest time sequence through every event. So, condition (c) basically means that for any rest time sequence γ we require that $\mathbb{T}_\tau^{\tau'}(\gamma(\tau)) = \gamma(\tau')$. In fact, we will base the connection between different time slices on the rest time sequences:

Definition (A8)

$$\mathbb{T}_\tau^{\tau'} : S_\tau \to S_{\tau'}$$
$$A \mapsto \mathbb{T}_\tau^{\tau'}(A) = \gamma(\tau')$$

where γ is a rest time sequence such that $A = \gamma(\tau)$. Let us call $\mathbb{T}_\tau^{\tau'}$ the *time shift* between S_τ and $S_{\tau'}$. It follows from (E1) and Lemma 24.1 that this definition is sound and $\mathbb{T}_\tau^{\tau'}$ is a distance preserving bijection. Now we have everything at hand to define the space tags of events:

Definition (A9) Let A be an arbitrary event. The *absolute space tags* of A are defined as follows:

$$\xi_1(A) := x_0\left(\mathbb{T}_{\tau(A)}^0(A)\right)$$

$$\xi_2(A) := y_0\left(\mathbb{T}_{\tau(A)}^0(A)\right)$$

$$\xi_3(A) := z_0\left(\mathbb{T}_{\tau(A)}^0(A)\right)$$

Thus, we are given *the absolute space and time tags* for every event: $\xi_1(A)$, $\xi_2(A)$, $\xi_3(A)$, $\tau(A)$.

24.3 Inertial Motion

A remark is in order on the empirical facts (E1)–(E13) to which we refer in constructing the space and time tags. When I call them empirical facts I mean that they ought to be true according to our ordinary physical theories. The ordinary physical theories are however based on the ordinary, problematic, space and time conceptions, relaying on "reference frames realized by rigid bodies" and the likes, without proper, non-circular, empirical definitions. Thus, especially in the context of defining the two most fundamental physical quantities, distance and time, we must not regard our ordinary physical theories as empirically meaningful and empirically confirmed claims about the world. Whether these statements are true or not is, therefore, an *empirical* question, and it is far from obvious whether they would be completely confirmed if the corresponding experiments were performed with higher precision, similar to the recent GPS measurements, especially for larger distances. Strangely enough, according to my knowledge, these very fundamental facts have never been tested experimentally; no textbook or monograph on space-time physics refers to such experimental results.

So, the best we can do is to *believe* that our physical theories based on the usual sloppy formulation of spatio-temporal concepts are true (in some sense) and to consider the predictions of these theories as empirical facts. However, as the following analysis reveals, it is far from obvious whether the predictions of the believed theories really imply (E1)–(E13).

Throughout the definition of space and time tags, we avoided the term "inertial", and because of a good reason. First of all, if "inertial" is regarded as a kinematical notion based on the concept of straight line and constancy of velocity, then it cannot be antecedent to the concept of space-time tags. If, on the other hand, it is understood as a manner of existence of a physical object in the universe, when the object is undergoing a free floating, in other words, when it is "free from forces", then the concept is even more problematic. The reason is that "force" is a concept defined through the deviation from the trajectory of inertial motion (first circularity), and neither the inertial trajectory nor the measure of deviation from it can be expressed without spatio-temporal concepts; consequently, they cannot be antecedent to the definition of space and time tags (second circularity). So there is *no* precise, non-circular definition of inertial motion. It is to be emphasized that this operational/logical circularity is a problem even in a special relativistic/flat/local space-time; and, therefore, it has nothing to do with the problem of conventionality of demarcation between "inertial" or "geodetic" motion versus gravitation as universal force (cf. Märzke and Wheeler 1964).

According to our believed special relativistic physical theory, space-time is a four-dimensional Minkowski space and inertial trajectory is a time-like straight line in the Minkowski space. Since we are prior to the empirical definitions of the basic spatio-temporal quantities, we cannot regard this claim as an empirically confirmed physical theory. Nevertheless, let us assume for a moment that our special relativistic theory is the true description of the world "from God's point of view". It is straightforward to check that all the facts (E1)–(E13) are true if (1) the standard clock moves along an inertial world line in the Minkowski space-time and (2) it reads the proper time, that is, it measures the length of its own word line, according to the Minkowskian metric. However, we human beings can know neither whether the standard clock (chosen by us) is of inertial motion in God's Minkowskian space-time nor whether it reads the proper time. What if these conditions fail? What does special relativistic kinematics say about (E1)–(E13) if the standard clock is accelerated and/or it does not read the proper time?

In order to answer this question, we have to follow up the operational definitions (A1), (A2),... and *calculate* whether statements (E1), (E2),... are true or not if the standard clock moves along a given world line γ and the "time" it reads is, say, a given function of the Minkowskian coordinate time, $\chi(t)$. Although the task is straightforward, the calculation is too complex to give a general answer in details. Fortunately, we do not need all the details: the essential fact is that if we really can go through the whole operational proccdurc, and (E1)–(E13) arc true, then, at the end, we obtain a coordination of events such that the equation describing the trajectory of a signal in the space of the four coordinates is

$$(\xi_1(\tau) - \xi_{10})^2 + (\xi_2(\tau) - \xi_{20})^2 + (\xi_3(\tau) - \xi_{30})^2 = c^2 \tau^2$$

Now, due to the Alexandrov–Zeeman theorem (Alexandrov 1950; Zeeman 1964), one can derive the following results.

Theorem *Facts (E1)–(E13) are true if and only if the standard clock moves along an inertial world line and reads a time $\chi(t)$ which is a linear function of the Minkowskian coordinate time.*

Due to this theorem, in accord with our intuition based on the believed physical theories, we can give an objective meaning to "inertial motion" by means of correct – neither logically nor operationally circular – experiments: *the standard clock is of inertial motion if statements (E1)–(E13) are true*. Assuming that the standard clock is inertial, one can extend the concept for an arbitrary time sequence $\gamma(\tau)$ of events: $\gamma(\tau)$ corresponds to an inertial motion if the absolute space tags $\xi_1(\gamma(\tau)), \xi_2(\gamma(\tau)), \xi_3(\gamma(\tau))$ are linear functions of the absolute time tag τ.

There is a trivial but very important corollary of the above theorem: Imagine that we successfully perform two different coordinations of events by means of two different standard clocks. The theorem implies that the two coordinations are identical up to an almost Lorentz transformation.

Of course, the Alexandrov–Zeeman theorem applies if all of (E1)–(E13) are satisfied. It is perhaps interesting, that the essential condition is (E1). From an analysis by computer one finds the following result:

Result 24.1 There are no unique rest time sequences if the standard clock moves non-inertially in a Minkowski space.

Still, one must emphasize, whether (E1)–(E13) are true or false is an open *empirical* question. Imagine that the standard clock is not inertial; for example (E1) is not satisfied. It would also mean that the clock chosen by us would be inappropriate for the definition of space-time tags. More exactly, we should have to stop at definition (A1). We could define the time tags but could not define the spatial notions, in particular the distances between simultaneous evens. Consequently, it is meaningless to talk about "non-inertial reference frame", "space-time coordinates (tags) defined/measured by an accelerated observer", and the likes. In the light of these consequences, it is an intriguing question whether the standard clock contemporary physical laboratories use for the coordination of physical events satisfies conditions (E1)–(E13), in particular (E1).

24.4 Absolute, Relative, Conventional

I call $\tau(A)$ "absolute time" not in the sense of what Newton called "absolute, true and mathematical time", that is independent of any empirical definition, but in the sense of what the twentieth century physics calls absolute time; it is "independent of the position and the condition of motion of the system of co-ordinates" (Einstein 1920, p. 51). The space and time tags $\xi_1(A), \xi_2(A), \xi_3(A), \tau(A)$ are *absolute in the sense that they are not relative to a reference frame* but prior to any reference frame. (The concept of "reference frame" is still not defined, and actually we do not need it.)

Absolute space and time tags are, of course, "relative" to the trivial semantical *convention* by which we define the meaning of the terms. They are "relative" to the *etalon* clock-like process we have chosen in the universe; and to the particular way in which the space and time tags are defined, including the usage of radio signals, the choice of "$\varepsilon = \frac{1}{2}$", etc. This kind of "relativism" is however common to all physical quantities having empirical meaning.

But there are two things that do not follow from this kind of conventionality. On the one hand, it does not follow that these physical quantities cannot describe *objective* features of physical reality; in spite of the obvious fact that these conventions play a constitutive role in the conceptual representation of the world. On the other hand, it does not follow either that there are no *objective* constraints on the semantical conventions themselves. In this last passage, I would like to give an example of how these objective constraints can restrict the semantical conventions defining absolute space and time tags.

There has been a long discussion in the literature about the conventionality of simultaneity. As it is obvious from (24.2), we chose the standard "$\varepsilon = \frac{1}{2}$-synchronization". This choice was *a part of the trivial semantical convention* defining the term "absolute time tag". It is, therefore, prior to any claim about the one-way or even round-trip speed of electromagnetic signals, because there is no such a concept as "speed" prior to the definition of time and space tags; it is, of course, prior to "the metric of Minkowski space–time", in particular to the "light-cone structure of the Minkowski space–time", because we have no words to tell this structure prior to the space and time tags; and it is prior to the causal order of physical events, because – even if we could know this causal order prior to temporality – we cannot know in advance how causal order is related with temporal order (which we have defined here). It is actually prior to any discourse about two locuses in space, because there is no "space" (S_τ) prior to definition (A1) and there is no concept of a "persistent space locus" prior to definitions (A3) and (A8).

So far, it seems, we are entirely free in the choice of the value of ε, that is in the choice of *which* objective feature of the physical reality – time$_\varepsilon$ – we want to deal with. One might think that starting with some ε, that is with some time$_\varepsilon$ tags $\tau_\varepsilon(A)$ and the corresponding ε-simultaneity slices S_τ^ε, one finally obtains some space$_\varepsilon$ tags $\xi_1^\varepsilon(A)$, $\xi_2^\varepsilon(A)$, and $\xi_3^\varepsilon(A)$, corresponding to the given value of ε. This is true only if we can go through all the operational definitions (A1)–(A13), and all the empirical facts (E1)–(E13) are true for the given $\varepsilon \neq \frac{1}{2}$.

This is, however, not necessarily the case. For, imagine we repeat the operations described in (A1), (A2) and (A3) with some $\varepsilon \neq \frac{1}{2}$, and obtain the concept of a (rest time sequence)$_\varepsilon$. Then, we encounter the question of whether the crucial empirical fact (E1) is true or not. Normally, in case of $\varepsilon = \frac{1}{2}$, we assumed that there exists a unique rest time sequence through every event. This assumption was confirmed by Result 24.1 derived from our believed physical theories. But, a similar computer calculation in case of $\varepsilon \neq \frac{1}{2}$ leads to the following result:

Result 24.2 Fact (E1) is never true if $\varepsilon \neq \frac{1}{2}$, no matter if the standard clock moves along an inertial world line, and no matter if the clock reads the proper time along its world line.

Again, whether or not (E1) is true is an open *empirical* question in both the $\varepsilon = \frac{1}{2}$ and the $\varepsilon \neq \frac{1}{2}$ cases. Nevertheless, assuming that the future empirical findings will confirm what our present physical theories tell about (E1), there seems no way to build up the *spatial* concepts (rest$_\varepsilon$, distance$_\varepsilon$, space$_\varepsilon$ tags, etc.) operationally, if

$\varepsilon \neq \frac{1}{2}$. And, given that our aim is to define not only the temporal but also the spatial concepts, this is a strong experimentally testable argument against the $\varepsilon \neq \frac{1}{2}$-synchronization.

Acknowledgment The research was partly supported by the OTKA Foundation, No. K 68043.

References

Alexandrov AD (1950) On Lorentz transformations. Uspekhi Math Nauk 5:187 (in Russian)
Brown HR (2005) Physical relativity. Space–time structure from a dynamical perspective. Clarendon Press, Oxford
Butterfield J (2005) On the persistence of particles. Found Phys 35:233
Einstein A (1920) Relativity: The special and general theory. H. Holt and Company, New York
Märzke RF, Wheeler JA (1964) Gravitation as geometry I: The geometry of spacetime and the geometrodynamical standard meter. In: Chiu HY, Hoffmann WF (eds) Gravitation and relativity. W. A. Benjamin Company, New York
Price H (1996) Time's arrow and Archimedes' point: New directions for the physics of time. Oxford University Press, New York
Tarski A, Givant S (1999) Tarski's system of geometry. Bull Symbol Logic 5:175
Zeeman EC (1964) Causality implies the Lorentz group. J Math Phys 5:490

Chapter 25
Making Contact with Observations

Ioannis Votsis

25.1 Introduction

Jim Bogen and James Woodward's 'Saving the Phenomena', published only 20 years ago, has become a modern classic. Their centrepiece idea is a distinction between data and phenomena. Data are typically the kind of things that are publicly observable or measurable like "bubble chamber photographs, patterns of discharge in electronic particle detectors and records of reaction times and error rates in various psychological experiments" (p. 306). Phenomena are "relatively stable and general features of the world which are potential objects of explanation and prediction by general theory" and are typically unobservable (Woodward 1989, p. 393). Examples of the latter category include "weak neutral currents, the decay of the proton, and chunking and recency effects in human memory" (Bogen and Woodward 1988, p. 306). Theories, in Bogen and Woodward's view, are utilised to systematically explain and predict phenomena, not data (pp. 305–306). The relationship between theories and data is rather indirect. Data count as evidence for phenomena and the latter in turn count as evidence for theories. This view has been further elaborated in subsequent papers (Bogen and Woodward 1992, 2005; Woodward 1989) and is becoming increasingly influential (e.g., Basu 2003; Psillos 2004; Mauricio Suárez 2005).

In this paper I argue contrary to Bogen and Woodward that data serve as evidence for theories, not only for phenomena. Bogen and Woodward seem to forget the old Duhemian dictum that 'theories cannot be tested in isolation'. That is, they seem to forget that theories require the help of auxiliary hypotheses to make contact with data. When augmented with suitable auxiliaries, theories do entail, predict and potentially explain the data. I say 'potentially explain the data' because my focus in this paper is only on the inferential and predictive relations between theories, phenomena and data. To demonstrate my claim I examine four cases from physics,

I. Votsis (✉)
Philosophisches Institut, Heinrich-Heine-Universität Düsseldorf, Universitätsstraße 1, Geb. 23.21/04.86, 40225, Düsseldorf, Germany
e-mail: votsis@phil-fak.uni-duesseldorf.de

chemistry and astronomy: (i) a controversy between Lavoisier and Priestley, (ii) the calculation of lead's melting point, (iii) the prediction of the Poisson spot, and (iv) the discovery of Neptune. The first of these is discussed in Basu (op. cit.) and the second in Bogen and Woodward (1988). The last two have not yet been discussed in the context of Bogen and Woodward's work but they are widely discussed in confirmation theory as paradigmatic examples of novel predictions. The choice of cases reflects my desire to assess Bogen and Woodward's view (1) under the best light by considering one of their principal examples as well as a meticulously discussed example from one of their devotees and (2) under the most stringent confirmation criteria by considering two exemplary cases of novel prediction.

25.2 The Lavoisier–Priestley Controversy

Basu (op. cit.) argues that for observations to be of use in theory testing, they first need to be transformed into evidence. Since the transformation, according to him, involves the introduction of theoretical vocabulary, the end-product is theory-laden. Basu motivates his claims using a distinction between raw (observational) data and evidence that is explicitly modelled on Bogen and Woodward's distinction.[1] Following Bogen and Woodward, he claims that theories do not entail, predict or explain observation statements or data, not even with the help of suitable theoretical auxiliaries. This prevents any direct observational assessment of theories (plus auxiliaries).[2] To support his claims, Basu considers in detail a rather well-known controversy between Antoine Lavoisier and Joseph Priestley.

The controversy concerns two conflicting results emanating from what appear to be the same experiments (i.e. heating iron in oxygen) independently carried out by the two scientists. Both scientists were in agreement that the observable result of the experiments was the production of a black powder with certain properties.[3] Since their respective theories of oxygen and of phlogiston do not speak of (or indeed entail) the presence of black powder, the observable result cannot immediately be used for theory adjudication. The raw observational data first has to be theoretically treated. This is where the disagreement arose. For Priestley, who advocated the phlogiston theory, when iron is heated in dephlogisticated air it leads to the production of iron calx. For Lavoisier, an advocate of the oxygen theory, the heating of

---

[1] Although Basu agrees with much of what Bogen and Woodward have to say, he thinks that their distinction "is inadequate in handling cases of 'revolutions' in science" (p. 354).

[2] Observations, Basu claims, need not be theory-laden but they cannot play a direct role in confirmation: "...although one could legitimately hold that there are observations that are not theory infected, such observations *cannot* be employed for theory resolution" (p. 356) [my emphasis].

[3] Priestley and Lavoisier agreed on various other observable results such as balance readings. They disagreed on whether the reaction only led to the production of black powder. Priestley thought that carbon dioxide was also produced. This disagreement is not important for our current discussion – Basu similarly sidelines it – as we are only interested in the inferential links between evidence and (commonly shared) observation statements.

iron in oxygen leads to the production of iron oxide. Yet, the presence of iron calx is only entailed by the phlogiston theory and the presence of iron oxide is only entailed by the oxygen theory. In other words, the same observation (i.e., the presence of a particular kind of black powder) is theoretically interpreted – out of necessity, for on its own, Basu claims, it is not evidentially potent – as two different evidential statements, each only confirming its respective theory.

Although Basu takes theoretical auxiliaries as necessary for the transformation of observations into evidence, he insists that they cannot help infer the relevant observation statements from the given theory. In the case at hand, this means that the presence of that particular kind of black powder cannot be inferred from either of the two theories. To see this point, let's formalise the aforementioned statements. Let O_1: Iron is heated in oxygen, O_2: Iron is heated in dephlogisticated air, E_1: Iron oxide is produced, E_2: Iron calx is produced, B: Black powder with certain observable properties is present, L: $O_1 \rightarrow E_1$, P : $O_2 \rightarrow E_2$, A_1 : B $\rightarrow E_1$ and A_2 : B $\rightarrow E_2$. L is a central theoretical claim in Lavoisier's theory and P the one in Priestley's theory. A_1 and A_2 are theoretical auxiliaries that respectively allow each scientist to go from observation to evidence.[4] Consider Lavoisier's theory. From O_1 and L, we can infer E_1 but not B. To confirm Lavoisier's theory we must assume A_1 which together with B entail E_1. Thus, to confirm Lavoisier's theory (or at least one of its parts, i.e., L), we must first transform B into an evidentially relevant statement (i.e., E_1) using theoretical auxiliary A_1. Notice that if we add A_1 to the set of statements $\{O_1, L\}$ we still cannot infer B. This seems to vindicate Basu's point that even with the help of theoretical auxiliaries we cannot infer the observational statement from the given theory. In his own words, "...the construction of E_1 in (1) [i.e., the proposition that B and A_1 imply E_1] is asymmetrical. The fact that iron oxide is produced does not entail (along with [A_1]) that a black powder is produced" (p. 361). The same asymmetry afflicts Priestley's evidential inferences. Note that we cannot judge Priestley's theory on E_1 and Lavoisier's theory on E_2. Each evidential statement is at best irrelevant to the other theory, at worst it disconfirms it.

Basu does ponder at one point "whether it is possible to predict the (raw) data from the hypothesis by employing suitable auxiliary assumptions" (p. 362). He dismisses this possibility two pages later, roundly asserting that "(raw) data *never* have any evidential bearing" (p. 364) [my emphasis]. In what follows, I contest this assertion by finding the requisite auxiliary assumptions that let us derive, predict and potentially explain observational report B. I do so by presenting a general strategy for constructing suitable auxiliaries that has applicability to a broad range of cases. This, as we shall shortly see, takes us through a detour via set-theory. If my strategy is compelling it undermines not only Basu's particular project but more generally Bogen and Woodward's which the former is firmly grounded in.

Sets can be partitioned into various disjoint parts. More formally we say that a set P is a partition of a set S if and only if (1) all of P's members are non-empty subsets

---

[4] A_1 and A_2 have a more complicated structure that for the sake of simplicity I leave out. This should not affect the conclusion of my argument since both auxiliaries appeal to the same Stahlian hypotheses to determine the purity of samples.

of S, (2) the union of P's members is co-extensional to S, and (3) the intersection of any two members of P is empty.[5] A peculiar aspect of this standard definition is that any set S (that can be partitioned) will have {S} among its partitions. For those interested in splitting the original set into two or more disjoint parts, a partition containing the original set as a member will of course be unwanted. To overcome this problem, let's define another notion that prohibits such partitions, call it 'partition*'. A set P is a partition* of a set S if and only if P fulfils the above three conditions (i.e., it is a partition of S) and P does not contain S as a member. Let's denote such a set as Part*(S). Sets with less than two members cannot be partitioned*. For a set S with n members, the number of partitions* is given by the bell number of that set minus one.

Predicates denote properties. Extensionally understood, properties are sets. That means that for any set there is one and only corresponding (natural or artificial) property, and vice-versa. This allows us to partition* properties by partitioning* their corresponding sets. Thus a partition* of a set S will have as members non-empty non-intersecting sets, each of which can be assigned a different property. Indeed, any property applicable to more than one object can be partitioned* into two or more properties each of which is distinct from one another and applicable to at least one object. Take the property of being a mammal. It can be partitioned* into a great number of properties, some of them corresponding to natural, others to artificial properties. Examples of (presumably) natural properties are the properties of primate, rodent, bat and dolphin. Examples of artificial properties are the properties of being a mammal half a meter long, being a mammal named 'Alexa' and weighing more than 500 kg.[6]

To remove any lingering unclarity, let us take a closer look at an example of a set being partitioned*. Suppose $S = \{1, 2, 3\}$. We know that this set has four partitions*, i.e., $Part_1^*(S) = \{\{1\}, \{2\}, \{3\}\}$, $Part_2^*(S) = \{\{1, 2\}, \{3\}\}$, $Part_3^*(S) = \{\{1, 3\}, \{2\}\}$, $Part_4^*(S) = \{\{2, 3\}, \{1\}\}$. Observe that each partition* contains as members sets that are mutually disjoint and whose union is set S. Qua sets, each member of a partition* of S can be assigned a property. Take for example $Part^*_1(S)$. It contains three members, namely sets {1}, {2}, {3}. Each of these can be assigned a different property; we can use the predicates R_1, R_2 and R_3 to denote these properties. Now if R is the predicate denoting the property corresponding to set S, then (x) (Rx ≡ ($R_1 x \oplus R_2 x \oplus R_3 x$)) where $\oplus$ stands for exclusive disjunction. All the partitions* of S can be given the same treatment. What is more, since partitioning* decomposes properties into mutually exclusive and exhaustive parts, inclusive disjunction formulations of such biconditionals – in the case at hand (x) (Rx ≡ ($R_1 x \vee R_2 x \vee R_3 x$)) – are logically equivalent to their exclusive disjunction counterparts.

---

[5] An alternative first condition does not exclude non-empty subsets of S, thereby allowing for partitions such as {S, ∅}.
[6] Overlapping properties such as being a mammal half a meter long and being a mammal named 'Alexa' do not of course belong to the same partitions* of the property mammals.

With these tools and results in mind, let us turn to the problem at hand. Given our move to predicate logic, atomic propositions O_1, O_2, E_1, E_2 and B are now taken to be predicates while complex propositions L, P, A_1 and A_2 are now quantified propositions. For example, predicate O_1 now reads 'is iron heated in oxygen' and theoretical auxiliary A_1 now reads: (x) (Bx $\rightarrow$ E_1x). Crucially, this universal generalisation implies that either E_1 is co-extensional to B or B is a non-empty proper subset of E_1.[7] In the former case, this amounts to the bi-conditional statement A_3 : (x) (Bx $\equiv$ E_1x). If we add A_3 as an auxiliary to our original set of propositions $\{O_1 a, \text{ L} : (x) (O_1 x \rightarrow E_1 x)\}$ we can derive the desired sentence Ba, where a denotes the particular object that bears these properties. In the latter case, we can turn to the concept of partition* to derive an equally suitable statement. We know that B, qua a non-empty proper subset of E_1, belongs to at least one partition* of E_1.[8] Take such a partition*, let's call it 'C'. C is co-extensional to E_1. It contains B as a member but also one or more other sets that are disjoint from B. We can assign a property and hence a predicate to each of them. Let us call these 'C_1', ..., 'C_m', where m is determined by the number of disjoint sets in C other than B. The following auxiliary can now be formulated A_4 : (x) ($E_1 x \equiv$ (Bx $\oplus C_1 x \oplus \cdots \oplus C_m x$)). The properties on the right side of the biconditional are jointly co-extensional to the property on the left side. If we add A_4 to our original set of propositions we can derive the following statement B$a \oplus C_1 a \cdots \oplus C_m a$.[9] Since B$a$ is one of the exclusive disjuncts, the observation that a has property B can confirm the theory and auxiliaries used in the derivation.[10] Contra Basu, Ba need not first be transformed into theory-laden evidence.

Technicalities aside, the conclusion is supported by a very simple logical point. Suppose we are faced with the sort of asymmetry Basu talks about, i.e., we have a statement of the form 'All F's are G's' but we really want a statement of the form 'All G's are F's' or at least some statement that allows us to go from G's to F's. If we know that all objects with property F have property G, we can infer that either some objects with property G have property F or all of them do. The latter case plays straight into our hands. The former needs a little spelling out. That's where the partition* notion comes in, as it facilitates the spelling out by letting us decompose properties like G into F and non-F parts. Doing so allows us to conclude that

---

[7] For simplicity, I use the same letters to denote predicates and their corresponding properties and sets. Context will determine which one I have in mind.

[8] Although some partitions* of E_1 might not have B as a member, their members' union will contain all the objects that are contained in B. From these we can reconstruct B, e.g., by further partitioning* the members of a given partition* and then taking the relevant union of the resulting partitions*. That means that the partition* choice does not really matter for the purposes of inferring something about B from E_1. Choosing a partition* that includes B as a member just makes the point easier to communicate.

[9] The complex proposition B$a \oplus C_1 a \ldots \oplus C_m a$ need not be thoroughly observational, but at least one of its atomic components, i.e., Ba, will be.

[10] I say 'can confirm' instead of 'confirms' to avoid a controversial issue in confirmation theory, i.e. whether or not derived observational statements *always* have confirmational power. The received view has been that they do always have such power but Laudan and Leplin (1991), amongst others, have challenged this view.

an object with property G will also possess a property from a finite selection of mutually disjoint properties (partitioned* from G) that includes F. Thus finding an object with property F can confirm a theory which predicts the existence of objects with property G. To put things in perspective, suppose 'G' is an unobservable property and 'F' an observable one. Theories supplemented with the auxiliary 'All F's are G's' can be confirmed by observational reports of objects possessing property F.

In a sense what I have argued for is unsurprising. An auxiliary of the form 'evidence or phenomenon x implies observation y' or something weaker like 'evidence or phenomenon x implies (or raises the probability of) an exclusive disjunction one of whose disjuncts is an observation y' is implicit in the scientists' thoughts when they employ an inverse conditional, i.e., when they infer from their observations some evidential report. Indeed, on pain of inconsistency, the scientists must have a biconditional or even an identity relation in mind. They take it that one of the manifestations of iron oxide (or iron calx) is black powder, hence they are in effect accepting a statement like 'An object is iron oxide (or iron calx) iff it is black powder with certain observable reactions to other substances xor it is a red-brownish solid with certain observable reactions to other substances xor ...'. The availability of such auxiliaries and the inferential relations they engender undermines Bogen and Woodward's view that theories do not entail, predict or even potentially explain observation statements.

Theories can and do make direct contact with observation reports. It should be obvious that by 'direct contact' I do not mean anything that violates Duhem's thesis that theories can never be tested in isolation. Rather, I mean that theories plus suitable theoretical auxiliaries can and do entail, predict and potentially explain observation statements or data. In short, the view developed in this section is perfectly compatible with various forms of holism.[11]

It is worth noting that auxiliaries A_3 and A_4 are not merely stipulated but derived from the existing auxiliary A_1. We can similarly derive auxiliaries A_5, (x) (Bx $\equiv$ E_2x), and A_6, (x) (E_2x $\equiv$ (Bx $\oplus$ D_1x $\oplus \cdots \oplus D_k$x)), from A_2 to allow Priestley's theory to be tested by observations. Indeed, with the help of A_5 and A_6, Priestley's theory can be confirmed by Ba. Since Ba can confirm both theories it cannot be used to discriminate between them. This problem is of no concern to us here since we are frying an altogether different fish. The aim was to show that theories plus suitable auxiliaries can be tested by observations, i.e., it was not to show that the presence of black powder discriminates between Lavoisier's and Priestley's theories. At any rate, in terms of theory testing we are not worse off than when we started since E_1 and E_2 are also unable to discriminate between the two theories. Moreover, the fact that one observation report cannot adjudicate between two theories (plus associated auxiliaries) does not entail that (1) it cannot adjudicate between those theories and others and (2) all observation reports are similarly impotent.

Alas, things are even more complicated than I have let on so far. Auxiliaries A_1 and A_2 seem to have been ad-hoc stipulations since no independent reasons were

---

[11] In my view, some form of partial holism is highly plausible.

given to support the claims that one of iron oxide's or iron calx's manifestations is black powder with certain observable properties. It is always preferable to have independent confirmation for a hypothesis prior to its utilisation but it is not absolutely necessary. Nowadays we can independently confirm Lavoisier's auxiliary since we have distinct methods of analysing the chemical structure of the black powder residue. If no independent confirmation existed, the relevant auxiliary would be confirmationally impotent. Put differently, either the original auxiliaries that go from data to phenomena enjoy independent support and then so do the derived auxiliaries that go from phenomena to data or the original auxiliaries are ad-hoc postulations that play no genuine confirmational role but then they are of no interest to any party in the debate.

25.3 Calculating the Melting Point of Lead

We turn now to one of Bogen and Woodward's most prominent examples. The sentence 'Lead melts at 327.5 °C' can presumably be explained, derived and predicted from theories of molecular structure. In Bogen and Woodward's view the sentence is not an observation report but rather a report about the phenomenon of the melting point of lead. The relevant observations or data come in the form of scatter points of temperature readings generated by a series of measurements. Provided various experimental conditions hold, e.g., that there is no systematic error, that small uncontrolled causes of variation "operate independently, are roughly equal in magnitude, are as likely to be positive as negative, and have a cumulative effect which is additive" the mean of the data can be considered to be a good estimate of lead's true melting point (1988, p. 308). The data thus serve as evidence for the phenomenon but they cannot be explained by, derived or predicted from the relevant theories of molecular structure because the mean of a given distribution "does not represent a property of any particular data point" and "it will not, unless we are lucky, coincide exactly with that value [i.e., the true value of the melting point]" (1988, pp. 308–309). On the basis of these two reasons, Bogen and Woodward conclude that the data in this and similar cases cannot serve as evidence for the corresponding theories.

Let us consider more closely the two reasons Bogen and Woodward cite to prop up their conclusion. As I understand it the first holds that we cannot explain, derive or predict a datum from a mean because the latter represents a property of a set of data but not of any one of its members. The second reason holds that we cannot explain, derive or predict a given mean from the theoretically predicted value of the melting point – which in the example above Bogen and Woodward suppose to also be its true value – since the mean and theoretically predicted values need not be identical. As before, derivations can be pulled off with the help of suitable auxiliaries. To wit, we can derive exclusive disjunctions whose disjuncts include the desired mean and datum.

Take the second claim first. Suppose we want to derive a particular mean m_1, which we assume for the sake of simplicity to satisfy the aforementioned experimental conditions, from a particular theoretically predicted value p_1. We know that every mean m with some standard error ε corresponds to a different range r of theoretically predicted values of the melting point of lead such that $r = \{x : m - \varepsilon \leq x \leq m + \varepsilon\}$. Now take only those pairs of m and ε that fulfil the aforementioned experimental conditions, i.e., the pairs that are typically good estimates of lead's true melting point. Let us call these 'the selected pairs' and their corresponding ranges 'the selected ranges'. Since Bogen and Woodward assume that the theories of molecular structure determine lead's true melting point – this is not an essential assumption but it simplifies the derivation – we can infer that the selected pairs are typically good estimates of the theoretically predicted value of lead's melting point p_1. This means that the majority of selected ranges contain p_1 as one of their members. Let us denote that set of selected ranges by R and the corresponding set of selected pairs by M. We can obviously derive M from p_1. Provided m_1 and standard error ε_1 are good estimates of lead's true melting point and p_1 is lead's true melting point, as we have assumed above, $(m_1, \varepsilon_1) \in M$. To express this in a more familiar format, the pair (m_1, ε_1) will be one of several disjuncts in an exclusive disjunction that, contrary to Bogen and Woodward, we can derive from the theories of molecular structure plus the foregoing auxiliaries.

The first claim can be handled similarly. Suppose we want to derive a datum d_1 from a mean m_1 which is determined by a particular data set one of whose members is d_1. Like before suppose for simplicity's sake that m_1 satisfies the stated experimental conditions. We know that every mean m with some standard error ε corresponds to a unique range q of data sets of temperature readings of lead's melting point. Obviously different data sets can have the same mean. That's why the relevant auxiliary assigns to each mean a range of data sets, i.e., a set of data sets. Take those pairs of m and ε in M. Each of these has a corresponding range of data sets. Let us denote the set of all such ranges by D. We can obviously derive D from p_1 and the other auxiliaries. We know already that the pair (m_1, ε_1) has a corresponding range of data sets, at least one of which contains d_1. Since $(m_1, \varepsilon_1) \in M$ we can infer that d_1 is contained in at least one of the data sets contained in D. In other words, d_1 will be one of several disjuncts in an exclusive disjunction that, against Bogen and Woodward's view, can be derived from the theories of molecular structure plus some suitable auxiliaries.

25.4 Novel Predictions

A significant gap exists in the writings of Bogen and Woodward. Nowhere do they systematically and explicitly discuss the role of novel predictions, considered by many as the Holy Grail in confirmation, in the relationship between data, phenomena and theories.[12] In this section, I will argue that novel predictions are particularly

---

[12] Woodward (1989) makes some cursory remarks about novel predictions.

damaging to Bogen and Woodward's claim that data cannot serve as evidence for theories. To make this point I will look into two paradigmatic cases of novel prediction.

The notion of novel prediction can be understood in a handful of competing ways. These can roughly be classified under two broad categories: temporal and use. *Temporal novelty* requires that what is predicted be in some sense unknown prior to a theory's prediction of it.[13] The sense of unknown depends on the particular temporal restrictions advocated. For instance, one may require that what is predicted must not be widely known or that it must be unknown to the theoretician who makes the prediction. Examples that satisfy both stringent and liberal criteria of temporal novelty include the two cases that I will shortly be examining, namely the prediction of the Poisson spot and the prediction of the existence and properties of Neptune. These two cases can also be accounted for by the notion of *use novelty* which requires that what is predicted is not in some sense used in the construction of the theory that makes the prediction.[14] As before, the sense of used depends on the particular restrictions advocated. For instance, some require that what is predicted must not be the explanatory target of the individual who designed the theory, while others that it must merely not be used to fix the value of one or more of the theory's parameters.[15] An example that perhaps satisfies both stringent and liberal criteria of use novelty is Newton's prediction of the rate of precession of the equinoxes. This example does not qualify under any temporal novelty account since the rate of precession of the equinoxes was not only widely known to scientists at the time but also known to Newton himself.

Some scholars have questioned the idea that theories can be confirmed at all. Of those who accept that theories can be confirmed, however, none denies that at least some of the examples cited as cases of novel prediction have sharp confirmational power.[16] The Poisson spot and Neptune cases were chosen precisely because they are generally acknowledged to have acute confirmational power. As I already alluded, both cases satisfy stringent and liberal criteria of temporal and use novelty. For this reason, they present a first-rate test of Bogen and Woodward's view in the arena of novel predictions.

Let us first consider the Poisson spot case. In 1819 Augustin Fresnel entered his wave theory of light in the French Academy of Science competition on the diffraction of light. The panellists consisted mostly of supporters of the particle theory of light, which was dominant at the time. One such panellist, Siméon-Denis Poisson, attempted to disprove Fresnel's theory by deriving from it what he and others con-

---

[13] Duhem ([1914]1991, p. 28) can be interpreted as being an advocate of temporal novelty.

[14] Mayo (1991, p. 525) states the relationship between the two notions clearly when she says "most scientific cases are equally accommodated by (and hence fail to discriminate between) temporal and use-novelty, unsurprising since temporal novelty is sufficient, though not necessary, for use-novelty".

[15] The first suggestion can be found in Zahar (1973) while the second in Worrall (2002).

[16] Mayo (op. cit.), for example, criticises the notion of novel prediction but does not deny that many of the cases that qualify as novel predictions have sharp confirmational power.

sidered to be an absurd consequence. If Fresnel's theory was right, a bright spot should appear in the middle of a disk's circular shadow when illuminated by a narrow beam of light. François Arago, one of the other panellists, performed the experiment and to everyone's disbelief observed the bright spot. As a result Fresnel's wave theory received a hard-earned confirmational boost. To make sense of this confirmational boost it is necessary that theories and data are more proximal than what Bogen and Woodward would have us believe. After all, without some guidance from suitable auxiliaries Poisson and Arago would not have known what to look for in order to judge whether Fresnel's theory was right. This guidance came in the form of an auxiliary hypothesis that connects the theoretical prediction of constructive interference in the centre of the disk's circular shadow to the observation of a bright spot. In other words, it was acceptable to both parties in the debate that constructive interference implies brighter regions. Without this assumption, which incidentally still stands today, Poisson would not have been able to predict the bright spot that he thought would undo Fresnel's theory.

The same point can be raised in the context of the discovery of Neptune. Urbain Jean Joseph Le Verrier and John Couch Adams worked independently on explaining Uranus' irregular orbit. Both men hypothesised the existence of a planet with enough mass to gravitationally perturb Uranus's orbit and employed Newtonian calculations to identify its properties and whereabouts. Le Verrier sent his predictions to J.G. Galle at the Berlin Observatory, who detected the planet on September 23 1846 at approximately the exact location forecasted by Le Verrier – the predicted true longitude was at $326°0'$ whereas the observed one was at $326°57'$ (see Brookes 1970). Soon after the discovery, but not before some wrangling, the planet was named 'Neptune'. Once again to make sense of the prediction it is necessary that theories and data enjoy a close relationship. Without some guidance from suitable auxiliaries Le Verrier, Adams and Galle would not have known what to look for in the telescopic observations that led to Neptune's discovery. The requisite auxiliary connects the theoretical prediction of a massive object with a specific orbit to the telescopic observation of a bright dot that appears in a particular part of the sky at a particular time at night. Without this assumption, which also stands today, Galle and others would not have been able to detect the planet via telescopic observations.

It ought to be painfully obvious that almost without exception suitable auxiliaries connecting theories, phenomena and data need to be at hand in the cases of novel prediction. Such auxiliaries play the crucial role of informing scientists about the observable manifestations of physical phenomena. To put the point about novel predictions in the realist's vocabulary: The view that suitable auxiliaries are required in the case of novel predictions is the only (or at least the best) view that does not make our knowledge of what observations to make in order to confirm or disconfirm a theory a miracle. Even well entrenched theories can be undone when the right data comes along. Within a few years of Poisson's prediction and Arago's observation the wave theory became the dominant theory of light.

25.5 Conclusion

It has not been argued here that theories always make contact with the observational ground. Instead, it has been argued that in those cases where the phenomena are inferred from the theories and the data do indeed serve as evidence for the phenomena, the data also serve as evidence for the theories. Four cases were analysed in support of this claim. In each of these cases suitable auxiliaries were available to effect the derivation and prediction of data from the theories. In the novel prediction cases, the suitable auxiliaries were not merely available to the scientists but rather formed an integral part of their reasoning in making the predictions. This lends more credence to the view that observations and theories enjoy much more direct contact than Bogen and Woodward are willing to admit.

Acknowledgements I am grateful to Ludwig Fahrbach and Gerhard Schurz for helpful feedback. I am also grateful for funding from the German Research Foundation (Deutsche Forschungsgemeinschaft).

References

Basu PK (2003) Theory-ladenness of evidence: a case study from history of chemistry. Stud Hist Philos Sci Part A 34:351–368
Bogen J, Woodward J (1988) Saving the phenomena. Philos Rev 97(3):303–352
Bogen J, Woodward J (1992) Observations, theories and the evolutions of the human spirit. Philos Sci 59(4):590–611
Bogen J, Woodward J (2005) Evading the IRS. In: Jones MR, Cartwright N (eds) Poznan studies in the philosophy of the sciences and the humanities, idealization XII: Correcting the model. Rodopi, Amsterdam, pp 233–268
Brookes CJ (1970) On the prediction of neptune. Celest Mech 3:67–80
Duhem P [1914] (1991) The aim and structure of physical theory. Princeton University Press, Princeton, NJ
Laudan L, Leplin J (1991) Empirical equivalence and underdetermination. J Philos 88:449–472
Mayo DG (1991) Novel evidence and severe tests. Philos Sci 58(4):523–552
Psillos S (2004) Tracking the real: Through thick and thin. Br J Philos Sci 55:393–409
Suarez M (2005) The semantic view, empirical adequacy, and application. Crítica Revista Hispanoamericana de Filosofía 37(109):29–63
Woodward J (1989) Data and phenomena. Synthese 79(3):393–472
Worrall J (2002) New evidence for old. In: Gärdenfors P et al (ed) In the scope of logic, methodology and philosophy of science. Vol. 1, Kluwer Academic Publishers, Dordrecht, pp 191–209
Zahar E (1973) Why did Einstein's programme supersede Lorentz's? (I&II). Br J Philos Sci 24:95–123, 223–262

Chapter 26
The Formulation and Justification of Mathematical Definitions Illustrated By Deterministic Chaos

Charlotte Werndl

26.1 Introduction

The definitions mathematicians work with are not arbitrary, i.e., usually there are good reasons why a definition is regarded as worth considering. This thought motivates the following general question, which will be at the centre of this article: *in what ways are definitions justified in mathematical practice, and are these ways of justifying definitions reasonable?* By a justification of a definition I mean the reasons which are given for the definition. In this article we will only consider explicit definitions which introduce a new expression by stipulating that it is equivalent to an already known expression.

Asking about the way a definition is justified means to ask about the kinds of reasons given for this definition. And reasoning is a core philosophical theme. Consequently, the possible ways of justifying mathematical definitions is an important philosophical theme.

We speak of the formulation of a definition when a mathematician generates a definition she or he has not known before. The initial reason why a mathematician accepts a definition usually makes clear how the formulation of this definition was guided. Hence each kind of justification of definitions gives us a corresponding kind of formulation of definitions. Since the guidance of the formulation of definitions and the justification of definitions are connected in this way, it suffices in what follows to focus only on the justification of definitions.

Apart from Imre Lakatos's work on the justification of definitions, there is very little philosophical discussion on the justification of definitions in the light of actual mathematical practice. As I intend to show in this article, there is much more to say on this issue, and also Lakatos's account of justifying definitions is limited.

What I will have to say on the above general question will be based on a case study of topological definitions of chaos. In Section 26.2 I will introduce this case study. In Section 26.3 I will then identify the three kinds of justification which

---

C. Werndl (✉)
The Queen's College, Oxford University, High Street, Oxford OX1 4AW, UK
e-mail: charlotte.werndl@queens.ox.ac.uk

are important for topological definitions of chaos. To my knowledge, two of them have not been identified before. After that, in Section 26.4 I will first explain the main theory about the justification of definitions in the light of mathematical practice, namely Lakatos's account of proof-generated definitions. I will then go on to criticise Lakatos's account: as for topological definitions of chaos, in nearly all mathematical fields various kinds of justification are important.

Some of the theoretical ideas and arguments of this article are developed in more detail in my paper Werndl (2009), where I investigate how notions of randomness in ergodic theory shed light on the justification of definitions.

26.2 Case Study: Topological Definitions of Chaos

Chaotic systems are deterministic systems which are nevertheless highly unstable and show irregular, or even random, behaviour. Due to their instability, chaotic systems exhibit sensitivity to initial conditions (SIC). By SIC we mean the property that small errors in initial conditions lead to considerably different outcomes.

It was not until the late 1960s that the phenomenon of deterministic yet highly unstable, irregular and sometimes even random behaviour was systematically investigated: catalysed by the development of electronic computers an area of research called 'chaos research' developed (cf. Aubin and Dahan-Dalmedico 2002; Dahan-Dalmedico 2004). At the end of the twentieth century chaos research boomed. Since then it has been hailed as having led to extraordinarily interesting scientific results and insights, especially in mathematics and physics but also in other scientific disciplines (Ruelle 1991). The Lorenz system, the Hènon system and the logistic map are some of the paradigm chaotic systems.

The theoretical insights of this article on the formulation and justification of mathematical definitions will be based on a case study of *topological definitions of chaos*, which are discussed in the mathematical field of topological dynamical systems theory. It is widely agreed that the main topological definitions of chaos are:

> Devaney chaos (three versions), Devaney chaos without periodicity, and the definition of chaos based on the topological entropy, of which there are three versions. (cf. Berger 2001, 40; Ott 2003, 151; Robinson 1995, 82–83; Smith 1998, Chapter 10)[1]

I investigated how the formulation of these definitions was guided, how they are justified and whether these kinds of justification are reasonable. I chose this case study because the kinds of justification which play a role for topological definitions of chaos seemed to me widespread in mathematics but different to the ones usually discussed in the philosophy literature.

Before we can turn our attention to the way definitions are justified, I have to introduce the mathematical framework necessary to formulate topological definitions

---

[1] There are two main branches of dynamical systems theory which discuss chaotic behaviour, namely topological dynamical systems theory and measure-theoretic dynamical systems theory (ergodic theory). For more on notions of chaos in ergodic theory see Werndl (2009).

of chaos, namely the concept of a topological dynamical system. To understand the mathematics that follows, it will be enough to have basic knowledge of the theory of metric spaces.

A *dynamical system* is a mathematical model consisting of a *phase space*, the set of all possible states of the system, and an *evolution equation* which describes how solutions evolve in phase space. Dynamical systems often model natural systems.

Definition 26.1. A topological dynamical system is a triple (X, d, T), where X is a set (the phase space), d is a metric on X and T: $X \to X$ is a continuous map (the evolution equation).[2]

The dynamics of the system is given by $x_{n+1} = T(x_n)$, $x_0 \in X$, $n \in \mathbf{N}_0$, and the solution through x is the sequence $(T^n(x))$, where $n \in \mathbf{N}_0$.[3]

With this background we are now ready to look at the kinds of justification which play a role for topological definitions of chaos.

26.3 Kinds of Justification

26.3.1 *Natural-World-Justification*

First, I assert that topological definitions of chaos are often justified because they capture a preformal idea regarded as valuable for describing or understanding the natural world. I will call such definitions natural-world-justified definitions.

Natural-world-justified definitions are a subgroup of preformal-justified definitions – definitions which are justified because they capture a preformal idea regarded as valuable. Several philosophers endorse the general idea that mathematical definitions should capture a valuable preformal idea (cf. Brown 1999, 109).

We of course assume that it is important to describe and understand the natural world. *Thus, if the preformal idea is valuable for describing and understanding the natural world, natural-world justification is a reasonable kind of justification.* Clearly, there can be debates about what makes a preformal idea valuable in this sense.

If a definition is natural-world-justified, or generally preformal-justified, this does not imply that it is a 'best' definition of a vague idea. This may, or may not be the case. For instance, as we will see, several definitions of chaos are natural-world-justified, but there is no 'best' definition among them (cf. Smith 1998, 175).

---

[2] To characterise chaotic behaviour, we want to be able to measure the distance between points in phase space. Hence we not only need a topology but also a metric on the phase space.
[3] To be precise, we are discussing here topological dynamical systems where time increases in discrete steps. There are also topological dynamical systems where time varies continuously, and they usually derive from differential equations. Definitions of chaos are essentially the same for dynamical systems with discrete and continuous time; hence it suffices to treat the discrete case.

Many topological definitions of chaos are natural-world-justified, namely a version of Devaney chaos, Devaney chaos without periodicity, and the definition of chaos based on the topological entropy, of which there are three versions (Bowen 1978, 17; Ott 2003, 145–151; Petersen 1983, 266–267, Robinson 1995, 83–84). Let me now discuss a version of Devaney chaos to illustrate the idea of natural-world-justified definitions.

26.3.1.1 Devaney Chaos

Definition 26.2. (X, d, T) is Devaney chaotic if and only if

(i) it exhibits sensitive dependence on initial conditions:

$$\exists \delta > 0 \, \forall x \in X \, \forall \varepsilon > 0 \, \exists y \in X \, \exists n \in \mathbf{N}_0 (d(x, y) < \varepsilon \text{ and } d(T^n(x), T^n(y)) > \delta);$$

(ii) it is transitive:

$$\forall U \subseteq X, U \neq \emptyset \text{ and open}, \forall V \subseteq X, V \neq \emptyset \text{ and open}, \exists n \geq 0 \, T^n(U) \cap V \neq \emptyset;$$

(iii) the set of periodic points of X is dense in X.

The standard justification for condition (i) is that it captures SIC: for every initial condition there is another arbitrary close initial condition such that the solutions originating from these initial conditions eventually separate considerably (more than δ) (Devaney 1986, 49; Robinson 1995, 82; Smith 1998, 167). As to (ii), Banks et al. (1992, 332) and Berger (2001, 34) follow Devaney in justifying transitivity as corresponding to indecomposability, i.e., that the 'system cannot be decomposed into two disjoint open sets which are invariant under the map' (Devaney 1986, 49–50). Transitivity is also justified as capturing the idea of irregularity that any bundle of initial conditions wanders all over the phase space (Smith 1998, 169). The usual justification of condition (iii) is that it captures the idea that in any arbitrary small region of phase space there are periodic points (e.g., Devaney 1986, 50).

Devaney (1986, 50) combined these three conditions so as to propose a definition of chaos. In the literature the conjunction of these three conditions is regarded as important because it captures a preformal idea which is valuable for describing or understanding the natural world. Thus Definition 26.2 is natural-world-justified (Devaney 1986, 49–50; Robinson 1995, 82–84; Smith 1998, 167–170).

As explained above, condition (ii) is sometimes justified as indecomposability. Yet I think that many systems are not transitive but, beyond doubt, indecomposable (i.e., also indecomposable according to Devaney's characterisation given above). For instance, the system defined by iteration of $T(x) = x^2$ on $(0, 1)$ is clearly indecomposable but not transitive.[4] Hence this justification and the corresponding

---

[4] It is not transitive because no point in $(0, 1/2)$ ever enters $(1/2, 1)$.

26 Definitions By Deterministic Chaos

justification of Devaney chaos is problematic. Generally, if the definition does not capture the idea it is said to capture, the justification is problematic because it is unclear why exactly this definition is chosen.

Let me now turn to the second kind of justification I have identified.

26.3.2 Condition-Justification

I claim that another kind of justification plays a role for topological notions of chaos, namely: a definition is justified by the fact that it corresponds to a mathematically valuable condition, i.e., it is equivalent in an allegedly natural way to a previously specified condition which is regarded as mathematically valuable. I will refer to these definitions as condition-justified definitions.

If the previously specified condition is mathematically valuable and the equivalence is mathematically natural, condition-justification is a reasonable kind of justification. To illustrate the idea of condition-justification, let me now discuss a version of Devaney chaos – the only topological definition which is condition-justified.

26.3.2.1 Devaney Chaos

Recall Definition 26.2 of Devaney chaos. This definition is conjunctive and for reasons of simplicity one might search for a single condition that is equivalent to this conjunctive definition. Guided by this, Touhey (1997) arrives at the following definition; he aims to define chaos for an infinite X, and for this domain his definition is equivalent to Definition 26.2 of Devaney chaos.

Definition 26.3. The system (X, d, T), where X is infinite, is Devaney chaotic if and only if $\forall U \subseteq X$, $U \neq \emptyset$ and open, $\forall V \subseteq X$, $V \neq \emptyset$ and open, $\exists$ periodic point $p \in U$, $\exists n \in \mathbf{N}_0$ such that $T^n(p) \in V$.

Touhey (1997, 411) expresses concerns that this definition does not capture a valuable preformal idea of chaos. Smith (1998, 176) remarks that the emphasis of this definition on periodicity speaks against Devaney chaos as a definition of chaos.

Yet I think that Touhey and Smith are misguided here. True, Definition 26.3 does not capture a preformal idea of chaos, but the question is whether it has to. Let us assume that Devaney's original definition (Definition 26.2) is accepted as a definition of chaos and that our aim is to define chaos for an infinite phase space. Then Definition 26.3 can be justified by the fact that it is a single condition being equivalent to Definition 26.2, regardless of whether Definition 26.3 expresses a valuable preformal idea. This makes Definition 26.3 condition-justified. For other examples of condition-justified definitions, see my paper Werndl (2009).

Generally, condition-justified definitions may in other contexts also capture a valuable preformal idea. However, as for Definition 26.3, often this won't be the case. Then there is the danger of not appreciating that a definition is condition-justified and asserting instead that this definition captures a meaningful preformal idea, when in fact this is not the case.

Let us now turn to the third kind of justification which plays a role for topological notions of chaos.

26.3.3 Redundancy-Justification

I claim that another kind of justification is important for topological notions of chaos, namely that a definition is justified because it eliminates at least one redundant condition in an already accepted definition. Eliminating redundant conditions is often reasonable. Then if the already accepted definition is mathematically valuable, this kind of justification is reasonable. To illustrate the idea of redundancy-justification, let us look at the only topological definition of chaos which is redundancy-justified, namely again a version of Devaney chaos.

26.3.3.1 Devaney Chaos

Recall Definition 26.2 of Devaney chaos. For an infinite phase space X it can be proven that a transitive system with dense periodic points exhibits sensitive dependence on initial conditions, i.e., that the conditions (ii) and (iii) imply (i) (Banks et al. 1992). Consequently, many who want to define Devaney chaos for an infinite phase space choose the following definition with the justification that it eliminates a redundant condition (e.g., Banks et al. 1992):

Definition 26.4. The system (X, d, T), where X is infinite, is Devaney chaotic if and only if it is transitive and has dense periodic points.

Hence Definition 26.4 is redundancy-justified.

As in the case of condition-justified definitions, redundancy-justified definitions may or may not capture a specific valuable preformal idea. If not, there is the danger of not understanding that a definition is redundancy-justified and asserting that it captures a valuable preformal idea, when this is not the case.

26.3.4 The Role of These Kinds of Justification

To conclude, I have identified three kinds of justification which play a role for topological notions of chaos, namely natural-world-justification, condition-justification and redundancy-justification. To the best of my knowledge, condition-justification and redundancy-justification have never been identified before.

Already our above discussion makes clear that these three kinds of justification are different in the sense that there are definitions which are only justified in one way but not in any other way. For instance, as a definition of chaos Definition 26.2 is only natural-world-justified, Definition 26.3 is only condition-justified and Definition 26.4 is only redundancy-justified.

Furthermore, *I claim that each of these three kinds of justification are widespread in mathematics, and thus are among the most important ones in mathematics*.

Let us now see how these insights contribute to the philosophical debate on the justification of mathematical definitions.

26.4 Lakatos and the Importance of Proof-Generated Definitions

Generally, there is little philosophical reflection on the actual practice of how definitions are justified and formulated in mathematics. In the relatively recent literature Larvor (2001, 218) and Feferman (1978, 321) acknowledge the importance of researching the formulation and justification of definitions. Furthermore, as already mentioned, several philosophers argue that definitions in mathematics should be an adequate explication of a preformal idea (cf. Brown 1999, 109).

But the main philosopher to have written on the mathematical practice of justifying definitions is still Lakatos (1976 and 1978). Lakatos's contribution here is the concept of a *proof-generated definition*. His main example for proof-generated definitions are definitions of polyhedron, which are justified because they are needed to make the proof of the Eulerian conjecture go trough, which says that for every polyhedron the number of vertices minus the number of edges plus the number of faces equals two:

> PI: *Proof-generated concepts* are neither 'specifications' nor 'generalisations' of naive concepts. The impact of proofs and refutations on naive concepts is much more revolutionary than that: they *erase* the crucial naive concepts completely and *replace* them by proof-generated concepts.
>
> The naive term 'polyhedron', even after being stretched by refutationists, denoted something that was crystal-like, a solid with 'plane' faces, straight edges. The proof-ideas swallowed this naive concept and fully digested it. In the different proof-generated theorems we have nothing of the naive concepts. That disappeared without trace. Instead each proof yields its characteristic proof-generated concepts, which refer to stretchability, pumpability, photographability, projectability and the like. (Lakatos 1976, 89–90, original emphasis)[5]

Unfortunately, Lakatos does not state exactly what proof-generated definitions are (cf. Lakatos 1976, 89–92, 127–154; Lakatos 1978, 95–97). Obviously, a mathematical definition justified in any way is eventually involved in some proofs. If this were not the case, the definition would not be of interest. Therefore, Lakatos cannot have meant that a proof-generated definition is simply a definition which is eventually involved in proofs.

---

[5] It is these properties of stretchability, pumpability, photographability, projectability etc., which eventually define 'polyhedron' and which are needed to make the different versions of the proof of the Eulerian conjecture work.

The following characterisation of a proof-generated definition most plausibly captures what Lakatos means and applies to all his examples: a *proof-generated definition is a definition that is needed in order to prove a specific conjecture regarded as valuable* (Lakatos 1976, 88–92, 127–133, 144–154).[6]

Lakatos thinks that in the case of his examples proof-generation is a *reasonable* way of justifying definitions: with proof-generation the mathematical aim to prove interesting theorems has been reached (Lakatos 1976, 90–92, 128, 148–149, 153). Generally, proof-justification is a reasonable way of justifying definitions if the conjecture that should be established is mathematically valuable.

Lakatos also never states clearly how widely he thinks that his account of proof-generated definitions applies. As Leng (2002, 11) remarks, his way of writing in his book 'Proofs and Refutations' (1976) suggests that definitions in mathematics should be generally proof-generated and after the discovery of proof-generation are also proof-generated. However, Lakatos's (1976) book derives from his Ph.D. thesis (Lakatos 1961); and as Larvor (1998) has pointed out, in this thesis Lakatos emphasises that he does not think to have discovered a unique logic of discovery. But given the remarks in his book (Lakatos 1976, 91–92, 144), he might have thought that his ideas are representative for several mathematical fields. Therefore, *Lakatos might have held that mathematical fields where proof-generation is, and also should be, the sole important way of how definitions are justified are not exceptional.*

Yet, as our case study of topological notions of chaos suggests, *this seems wrong*. More specifically, *I think that, as in our case study, for nearly all mathematical fields after the discovery of proof-generation many different ways of justifying definitions are found. And I also think that for nearly all mathematical fields various different ways of justifying definitions are reasonable. Actually, I would wonder if a mathematical field could be found where all definitions are, or should be, proof-generated.* As I have argued in my paper Werndl (2009), even for the mathematical fields Lakatos (1976) discusses, proof-generation is not the only important form of justification. Thus, to summarise, while Lakatos's ideas are profound, his account is limited because he focused only on proof-generated definitions.

26.5 Conclusion

The general theme of this article has been the actual practice of how definitions are justified in mathematics. The theoretical insights of this article were based on a case study of topological definitions of chaos, and this case study was introduced

---

[6] Lakatos discusses other ways of justifying definitions too: *monster-barring, exception-barring, monster-adjustment* and *monster-including*. Like proof-generation these kinds of justification are ways of dealing with counterexamples to conjectures, and Lakatos regards them as inferior to proof-generation (Lakatos 1976, 14–33, 83–87). Moreover, Lakatos (1976, 14–42, 136–140) claims that they were only employed when the "better" kind of justifying definitions, namely proof-generation, was not yet known. For these reasons and since these other kinds of justification do not play a role for topological definitions of chaos, we will not discuss them further in this article.

in Section 26.2. In Section 26.3 I have identified the three kinds of justification which are important for topological definitions of chaos: natural-world-justification, condition-justification and redundancy-justification. To the best of my knowledge, the latter two have not been identified before. I have argued that these three kinds of justification are reasonable and that they are among the most important ones in mathematics. Finally, in Section 26.4 I have discussed the main philosophical theory about the justification of definitions in the light of actual mathematical practice, namely Lakatos account of proof-generated definitions. I have criticised Lakatos's account as being limited and also misguided: as for topological definitions of chaos, in nearly all mathematical fields various kinds of justification are found and are also reasonable.

Acknowledgments Most thanks should go to Jeremy Butterfield and Peter Smith for their valuable feedback on previous versions of this manuscript. Many thanks also to Roman Frigg, Franz Huber, Brendan Larvor, Mary Leng, Paul Weingartner and the audiences at the 1st London-Paris-Tilburg Workshop in Logic and Philosophy of Science and the 1st Conference of the European Philosophy of Science Association for helpful and interesting discussions and comments. I am grateful to St John's College, Cambridge, and to The Queen's College, Oxford, for financial support.

References

Aubin D, Dahan-Dalmedico A (2002) Writing the history of dynamical systems and chaos: longue durèe and revolution, disciplines and cultures. Hist Math 29:273–339
Banks J, Brooks J Cairns G, Davis G, Stacey P (1992) On Devaney's definition of chaos. Am Math Month 99:332–334
Berger A (2001) Chaos and chance: an introduction to stochastic aspects of dynamics. De Gruyter, New York
Bowen R (1978) Topological entropy. In: Markley NG, Martin JC, Perrizo W (eds) The structure of attractors in dynamical systems Proceedings, North Dakota State University June 20–24, 1977 Springer Berlin pp 17–24
Brown JR (1999) Philosophy of mathematics: an introduction to the world of proofs and pictures Routledge, London
Dahan-Dalmedico A (2004) Chaos, disorder, and mixing: a new fin-de-siècle image of science? In: Wise MN (ed) Growing explanations, historical perspective on the sciences of complexity Duke University Press, Durham, pp 67–94
Devaney R (1986) An introduction to chaotic dynamical systems. Addison-Wesley, New York
Feferman S (1978) The logic of mathematical discovery vs. the logical structure of mathematics. PSA Proc Biennial Meet Philos Sci Assoc 2:309–327
Lakatos I (1961) Essays in the logic of mathematical discovery. PhD thesis, Cambridge
Lakatos I (1976) Proofs and refutations. In: Worrall J, Zahar E (eds) The logic of mathematical discovery. Cambridge University Press, Cambridge
Lakatos I (1978) Mathematics, science and epistemology. In: Worrall J, Currie G (eds) Philosophical papers vol 2. Cambridge University Press, Cambridge
Larvor B (1998) Lakatos: An introduction. Routledge, London/New York
Larvor B (2001) What is dialectical philosophy of mathematics? Philos Math 9:212–229
Leng M (2002) Phenomenology and mathematical practice. Philos Math 10:3–25
Ott E (2003) Chaos in dynamical systems Cambridge University Press, Cambridge
Petersen K (1983) Ergodic theory. Cambridge University Press, Cambridge

Robinson C (1995) Dynamical systems: stability, symbol dynamics and chaos. CRC Press, Tokyo
Ruelle D (1991) Chance and chaos. Princeton University Press, Princeton, NJ
Smith P (1998) Explaining chaos. Cambridge University Press, Cambridge
Touhey P (1997) Yet another definition of chaos. Am Math Month 104:411–414
Werndl C (2009) Justifying definitions in mathematics – going beyond Lakatos. Philos Math 17:313–340

Chapter 27
Do We Need Some Large, Simple Randomized Trials in Medicine?

John Worrall

27.1 Introduction: Why Randomize?

In a randomized clinical trial (RCT), a group of patients, initially assembled through a mixture of deliberation (involving explicit inclusion and exclusion criteria) and serendipity (which patients happen to walk into which doctor's clinic while the trial is in progress), are divided by some random process into an experimental group (members of which will receive the therapy under test) and a control group (members of which will receive some other treatment – perhaps placebo, perhaps the currently standard treatment for the condition at issue). In a 'double blind' trial neither the patient nor the clinician knows to which of the groups a particular patient belongs. The results of double blind randomized controlled trials are almost universally regarded as providing the 'gold standard' for evidence in medicine. Fairly extreme claims to this effect can be found in the literature. For example the statistician Tukey wrote (1977, p. 679) "almost the *only* source of reliable evidence [in medicine] ... is that obtained from ... carefully conducted randomised trials". And the clinician Victor Herbert claimed (1977, p. 690) "...the only source of reliable evidence rising to the level of proof about the usefulness of any new therapy is that obtained from well-planned and carefully conducted randomized, and, where possible, coded (double blind) clinical trials. [Other] studies may point in a direction, but cannot be *evidence* as lawyers use the term evidence to mean something probative ... [that is] tending to prove or actually proving". Finally, the still very influential movement in favour of 'Evidence Based Medicine' (EBM) that began at McMaster University in the 1980s was initially often regarded as endorsing the claim that only RCTs provide real scientifically telling evidence.

EBM now explicitly endorses a more guarded view involving a hierarchy of evidence of different weights. But although these hierarchies[1] explicitly allow that

---

J. Worrall (✉)
London School of Economics
www.ahrq.gov

[1] A 2002 study identified no less than 40 such systems of grading evidence (Agency for Healthcare Research and Quality. 2002. *Systems to rate the strength of scientific evidence*. Rockville MD:AHRQ,); while a 2006 survey found 20 more (Schünemann, Holger J., Atle Fretheim and

other forms of evidence can legitimately play a probative role, they still (all) unambiguously place evidence from RCTs at the top (sometimes along with systematic reviews or meta-analyses of RCTs).[2]

So although the extreme view that the only truly scientific evidence for the effectiveness of some treatment is that from an RCT seems to have been largely abandoned, nonetheless RCTs continue to be regarded as carrying special epistemic weight. Why? In previous work,[3] I identified five different types of answer:

1. Fisher's argument that randomization is necessary to underwrite the logic of the classical statistical significance test.[4]
2. Randomization controls for *all* possible confounders – known and unknown. This is sociologically speaking the argument that has carried most weight. Clearly a central issue in evaluating the weight of evidence supplied by any clinical trial in which the experimental group does better on average is whether there might be some other overall difference between those in that group and those in the control group – a difference that played a role (possibly the major role) in the 'positive' outcome. In principle the groups could be deliberately matched for 'known confounders' – factors, like age, sex, absence or presence of comorbidities, etc. that background knowledge makes it plausible might play a role in the outcome. But clearly this leaves open the possibility that there are 'unknown confounders' – factors that also play a role in outcome but that background knowledge gives us no reason to think do so – and which may be (of course, by definition, unknown to the clinicians) unbalanced between the two groups. Randomization's many admirers believe that it (and only it) solves this problem.[5]
3. Randomization controls for the *particular* possible confounder: 'selection bias'. Since selection bias is sometimes used in a number of (often very wide) senses, it is important to emphasise that by this term I mean specifically any bias that is, or may be, introduced as a direct result of the clinicians' having control over which group a particular patient goes into.
4. It is just an empirical fact that non-randomized trial designs exaggerate positive treatment effects.
5. Only an RCT can distinguish a real causal connection (between intervention and outcome) from a 'mere correlation' between the two. This argument arises from the burgeoning literature on probabilistic causality, but on analysis is quite quickly revealed to be simply argument 2 under a rather different guise.[6]

Andrew D. Oxman. 2006. Improving the use of research evidence in guideline development: 9. Grading evidence and recommendations. *Health Research Policy and Systems.* 4:21).
[2] Meta-analyses and systematic reviews are attempts to amalgamate different studies on the 'same' intervention into one overall result. They face many interesting methodological problems.
[3] Worrall (2002, 2007a, b, 2008).
[4] For references and an especially clear account of this argument of Fisher's – together with an especially clear demonstration that the argument fails even on its own terms, see Howson (2000).
[5] So for example the Director of the UK Cochrane Centre, Mike Clarke, states on the Centre's Web-site that "[i]n a randomised trial, the *only* difference between the two groups being compared is that of most interest: the intervention under investigation." http://209.211.250.105/docs/whycc.htm. Accessed 18 December 2008.
[6] See in particular Worrall (2007a) and references therein.

Only argument 3 clearly survives critical scrutiny – or so I argued in the previous work alluded to. In this paper I want to look in more depth at argument 3 and its impact.

27.2 Selection and 'Treatment' Bias

The natural home of selection bias in the sense that I am understanding it is the non-randomized clinical trial, where the clinicians have direct control over which patients go into the experimental and which into the control group. Classic illustrations of how selection bias might operate to make the weight of a trial result highly questionable are provided by various early comparisons of patients treated surgically for some condition C and patients treated medically for the same condition. Of course, being considered operable forms a potentially very powerful selection bias – to be considered operable the patient will need to be in comparatively good condition, exhibit particular anatomical conditions and suffer from no major co-morbidity. Hence any 'evidence' from such a trial that surgery is the better treatment for condition C would be clearly suspect.

Leaving it to clinicians to decide the group a patient goes into is clearly fraught with epistemological danger – especially, though not exclusively if they have a vested interest in a positive outcome. (Indeed as Hill [1937/71] pointed out there is even the 'opposite' danger that the clinician may 'bend over backwards' to be fair and produce a control group that is better overall in terms of positive prognostic factors and thus provide an overly severe test of the treatment, one which might well *underestimate* that treatment's virtues!) Moreover if the clinicians select then they will also know to which group a particular patient belongs, and may lavish particular attention on those they know to be in the experimental group (especially if the comparison is placebo). Hence what might be called '*treatment* bias' might be added to any baseline imbalances in the two groups ahead of treatment. (It is useful to have treatment bias as a separate category as we shall see.)

In historically controlled trials (sometimes also rather dismissively categorised as one kind of 'observational study'), the control group is provided by previous patients with the condition under investigation who were treated (preferably of course in the *recent* past) using the older 'standard treatment'. This means that all the patients actively involved in such trials are on the experimental treatment and the investigators know this. Of course there is always an attempt, and, in the more sophisticated historically controlled trials, a very great attempt, to match the historical controls with those currently being treated with respect to known prognostic factors for the condition at issue. However, it is unavoidable that all the patients actively involved in such a trial are given the experimental treatment and the clinicians know this. This seems to make the possibility of at least treatment bias inevitable in such trials.

Randomization as usually performed eliminates selection bias in this sense. So long as the protocol is followed, the decision about which group a particular patient belongs to is taken out of the clinicians' hands and is made instead by some random process (usually whether the next number in a table of random numbers is even or odd). Moreover if the trial is performed double blind then, so long as 'blind is maintained', neither the particular patient nor, more importantly in this respect, the clinicians know which group that patient is in, and so the possibility of treatment bias seems to be ruled out. (There are issues – often overlooked – about how long blind is in fact maintained in most clinical trials. But let's leave these issues aside for present purposes.) Notice however that there is nothing special about the randomization in this regard – nothing in the toss of the coin or the random number table really plays a role, instead randomizing is simply one way of taking control out of the hands of clinicians; and blinding (if effective) makes it impossible for the clinician to identify securely those in the treatment group and so no question of preferential treatment (treatment bias) arises. Notice also that nothing in the argument shows that the *only* way that selection bias can be eliminated is through randomization.[7]

27.3 How Large an Effect Is Selection Bias Likely to Produce?

Suppose I am right that controlling for selection bias is randomization's only unambiguous epistemic virtue. (I believe that this was in fact the view of Austin Bradford Hill, who is credited as the first to import Fisher's randomizing methodology into medicine.[8]) The next question – especially given that it is conceded on all sides that randomization may involve some 'ethical cost' – is surely: *how large* an effect is selection bias likely to produce if not controlled for? (Questions of likely effect size are very often underemphasised in medicine, I would argue, in favour of the question of simple statistical 'significance'.)

There has been increasing recognition, even amongst arch-advocates of RCTs, that the answer to this question may well be 'very small'. This recognition goes back at least to a brief letter to the Editor of the *British Medical Journal* in 1980 by Doll and Peto (1980). They allow that selection bias is 'hardly likely to produce a tenfold artefactual effect' though, they insist, it 'may well produce a twofold artefactual error'. It is unclear exactly what metric they are presupposing here, but we can get by just in qualitative terms: selection bias is likely to be quite small. I can only think that they are referring here to what might be called 'practically ineliminable' selection bias; since if performed badly enough trials subject to selection bias can be as biased as you like (think about the comparison mentioned above between surgical and medical interventions for the same condition). So we are talking about trials in which sensible efforts have been made to match the two groups,

---

[7] So for example Bartlett and colleagues who introduced ECMO as a treatment for PHSS simply switched from treating all babies admitted to their hospital (U of Michigan) with the condition with the previously standard treatment to treating all babies admitted to their hospital with ECMO. No selection! (Though certainly the issue of treatment bias is a genuine one.) See Worrall (2008).
[8] See Bradford Hill op. cit.

without randomizing. Doll and Peto's concession that selection bias is 'likely' to be small, at once admits some intuitive Bayesianism (we are allowed to judge prior likelihoods) and implicitly admits that randomization is not needed, that historically controlled trials may be sufficient when the treatment effect (revealed by the historically controlled trial) is large. (But this concession went unnoticed in several important cases including I would argue the famous ECMO case.[9])

The concession was made entirely explicit in Peto et al. (1995) which allows that "... randomized trials may be unnecessary... For example, randomization is not needed to show that prolonged cigarette smoking causes cancer..." (p. 32) And more recently and more explicitly still, Paul Glaziou et al. (2007) write "'Some treatments have such dramatic effects that biases can be ruled out without randomised trials." (p. 351) Of course these concessions were not made before time – it is easy to produce a long list of treatments that are (i) established in medicine, which (ii) no sane person could deny are effective and yet (iii) have never been subjected to an RCT.[10] Nonetheless the concessions are important and welcome.

27.4 How Doll, Peto and Others Turn the Smallness of Selection Bias into an Argument for RCTs

But the central aim of Doll and Peto (1980), of Yusuf et al. (1984) and of Peto et al. (1995) was not at all to argue for the virtues of some historically controlled trials. On the contrary, their aim was to use the fact that historically controlled trials are bound to suffer from the possibility of selection bias, *even if* that bias is small, as a further argument for the necessity of *RCT*s! Their aim was to argue in effect that we need RCTs even more not less. Hence the title of the (1984) paper: 'Why do we need some large, simple, randomized trials?' – from which I in turn took the title of this paper.

One crucial premise of the Doll/Peto argument is that the romantic age of medicine – the days of the great breakthroughs producing new treatments with 'dramatic' effects – is over (or at least very largely so). Doll and Peto wrote (op. cit., p. 44) "most of the really important therapeutic advances of the past decade have involved the recognition that some particular treatment for some common condition yields a *small but important* improvement in the proportion of favourable outcomes." To which Yusuf et al. (1984, p.410) added: "if any widely practicable intervention had a very large effect, ... then... these huge gains in therapy are likely to be identified more or less reliably by simple clinical observation, by 'historically controlled' comparisons, or by a variety of other informal or semi-formal non-randomized methods". Hence (op. cit., p.411) "if there remains some controversy about the efficacy of any widely practicable treatment, its effects on major endpoints may well be either nil, or moderate...."

---

[9] See Worrall (2008).
[10] See Worrall (2007b) and the list in Rawlins (2008).

So a clinical trial, in the current situation in medicine, will need to be able to distinguish between a null and a 'moderately' (they really mean small) positive effect. And this is exactly what an historically controlled trial cannot do – even though the (practically ineliminable) bias to which it is subject is admittedly itself small. Hence we need an RCT to do this. Moreover, we need *large* RCTs: "It is chiefly because one [nowadays] usually needs to be able to distinguish reliably between moderate and null effects that trials need to be *strictly* randomized ... and much, much larger than is currently usual" (Yusuf et al. 1984, p. 410). The need for the trial to be large is based on the fact that randomization, despite what some of its advocates often seem to claim, cannot be guaranteed to equalize the experimental and control groups in terms of other potentially prognostic factors. Trials no matter how carefully randomized are, instead, subject to 'random error' – and the claim is that 'random error' is likely to be small if, but only if, the trial is large. As Doll and Peto (op. cit.) put it "the small randomized trials that are regrettably commonplace nowadays have random errors which are often far larger than the real differences to be detected."

So we need large RCTs to distinguish the small effects that are all we can reasonably expect. In order to be *very* large, practically speaking they need to be multi-centre; and this in turn means that the trials need to have a very *simple* protocol since complexity may produce differences between treatment centres that obscure the true effects. Finally – and importantly – small effect sizes are not to be scoffed at: treatments yielding small effects on common conditions may well finish up saving more lives overall than dramatic treatments for much rarer conditions. For example, Peto et al. (1995) claim that the ISIS-2 study which found an absolute risk reduction of heart attacks of under 2% and whose results were published in 1988 had probably by 1995 "avoid[ed] about 100,000 vascular deaths in developed countries alone." (p. 26).

This, then, is 'Why we need some large, simple, randomized trials'. And this view has proved very influential – especially in cardiology. Large trials on reducing heart attacks and stroke, for example, have included (with numbers of patients in parentheses):

ASSET (5,200)
GISSI-2 (12,700)
GISSI-3 (19,500)
CURE (12,200)
ISIS-2 (17,000)
ISIS-4 (58,000)[11]

This is certainly an interesting and seemingly powerful argument for the special epistemic power of randomized trials – not one that is usually cited and not one that I analysed in my earlier papers.

---

[11] These numbers are taken from (and my treatment influenced by) Penston (2003).

27.5 Analysis of the Argument

Clearly a crucial premise of this argument is that it is at least unlikely that the treatments being tested in the current situation in medicine will have large effects. Doll and Peto and collaborators give no argument for this beyond the rather strange claim that any big effects out there would probably already have been discovered. This sounds rather like physicists in the nineteenth century holding the view that Newton had made the big breakthrough and that all that was left for physicists to do was to fill in details. There have been some recent quite major breakthroughs – for example in the treatment of leukaemia and of HIV Aids. And it is difficult to see the general grounds for the pessimism involved in their assumption. (And it should be carefully noted, I believe, that, despite being arch-advocates of the epistemic superiority of RCTs, they are admitting that (sophisticated) historically controlled trials are sufficient to reveal anything other than 'moderate' (really: small) effects.)

However their argument can be re-gigged so as to avoid reliance on this pessimistic premise by making it more local. There may of course *in particular cases* be good reasons to think that some proposed treatment aimed at, say, reducing the risk of myocardial infarction or strokes is unlikely to have a really 'dramatic' effect. Hence the argument would now suggest that, in those cases where we have good prior reason to think the effect of the treatment, if any, is 'moderate', we need to perform RCTs, since the selection bias inherent in non-randomized studies may produce effects of the same order of magnitude as (or higher than) the likely effect.

Can there be any reason to question even this more measured claim? There seem to me to be two such reasons.

27.5.1 *The Issue of 'External Validity'*

As is frequently conceded, the issue of 'external validity' is one that can always be raised for any trial – though the conceder usually then goes on to categorise external validity as a difficult problem and practically to ignore it! Suppose it is agreed that the RCT is the most reliable means of arriving at the 'right' result so far as the set of patients in the study in the study is concerned. (This is usually called 'internal validity'.) It is still reasonable to question whether that result is likely to generalise to the 'target population' (that is, the set of people who will be treated if the treatment is declared 'effective' in the trial). It should be noted that, contrary to a fairly widespread myth, there is no guarantee (even of a classical statistical sort) that a randomized study's result will generalise in this way, since there is no sense in which the initial study group is a random sample from any specified population.

Standardly, research reports in the medical journals will have titles like (taken from a randomly chosen recent edition of the *Lancet*) "Efficacy and safety of ustekinumab... in patients with psoriasis..." or "Active symptom control with or without chemotherapy in the treatment of patients with malignant pleural mesothelioma...".[12] They will then report (usually randomized) trials on some *selected group* of patients – where the selection involves a number of exclusion criteria (often over 65s will be excluded, so will those exhibiting risk factors for various conditions, those exhibiting certain co-morbidities and so on). The trials will generally involve some *very precise treatment regimen* which the trialists are not allowed to alter or adjust and will generally run for some *relatively brief period* (as Michael Rawlins, the head of the UK National Institute for Clinical Excellence, reports "Most RCTs, even for interventions that are likely to be used by patients for many years, are of only six to 24 months duration."[13]) And the study will report that administration of substance S is (or is not) effective – meaning more (or no more) effective than the treatment given to the control group (often placebo, sometimes the currently accepted treatment for the condition at hand).

So, assume that the trial outcome is positive, and that the trial is a pharmaceutical one testing substance S for efficacy in treating condition C. Which exact theory has actually been tested? Not the (dangerously vague) claim that, say, substance S is effective for condition C, but rather the more specific claim that substance S when administered in a very particular way to a very particular set of patients for a particular length of time is more effective[14] than some comparator treatment (often, as I say, placebo). This is the claim for which the RCT provides evidence – let's assume for present purposes impeccable evidence.

But this is not, of course, the claim that the practising physician would like to have evidence for. She would like to know whether the treatment is effective (in a wide sense that certainly involves factoring in any side-effects, whether short or long term) when prescribed to the sorts of patients she would like to prescribe it to. This 'target population' is not very precisely characterised but will certainly include many types of patient excluded from the trial (the elderly perhaps, or those with significant co-mordibity). Moreover there will be the possibility of adjusting the dose in the light of individual patient's reactions. In the trial, care may be taken that the patient receives the allotted treatment; in 'the wild' patients forget. Finally, if the condition is a chronic one then the physician may want to prescribe S for a long time – certainly much longer than the trial itself is likely to have lasted.

Note that the issue of external validity is not what is sometimes dismissively called a 'purely philosophical' one. We are not here asking something on a par with 'does the fact that the sun has always risen in the past give us good grounds for thinking it will tomorrow?' Unlike David Hume's case, we often know on good specific grounds that the trial population and the target population are different. For example, a study by Bartlett et al. (2005) looked at 25 recent RCTs on NSAIDs and 27 recent RCTs on Statins and found that older people, women and ethnic minorities were (quite significantly) under-represented compared to the general (and therefore also presumably the 'target' population). Moreover not only do we know

---

[12] Lancet **371**, 2008, pp. 1665 and 1685.
[13] Rawlins (op. cit., p. 16).
[14] Of course 'effectiveness' is a tricky notion too – positive effect on the 'target disorder' is only part of the story, side effects need to be taken into account too.

that there are such differences, background knowledge, largely in the form of previous experience, lends good grounds for thinking that those differences may result in differences in outcome (and it lends no reason to think that such differences will be small).

Nor do we need appeal here merely to logical possibility: there are a number of real cases in which a treatment endorsed by an RCT had to be withdrawn later because of significantly deleterious overall outcome. One such case involved Benoxaprofen (Opren). This was an NSAID developed in the early 1980s for arthritis/musculo-skeletal pain. Its big attraction over other NSAIDs was that it was to be taken only once a day and hence was likely greatly to increase patient compliance. A large RCT was performed in a trial restricted to 18–65 year olds. The trial had an impressively positive result; Opren was very aggressively promoted and duly cornered the market. Now, it is a fact that the population of people who suffer from arthritis and musculo-skeletal pain has an average age much higher than that of the general population. It turned that in the elderly (who had not been represented in the trial population), Benaxaprofen has a significantly deleterious effect – causing a significant number of deaths from hepato-renal failure for example – and the drug was duly withdrawn. Michael Rawlins cites a total of 22 drugs that have been approved by RCTs in recent years only to be later withdrawn for safety reasons (2008, p. 22).

The issue of external validity arises especially sharply, I believe, in the case of the very large randomized trials recommended by Doll and Peto. If you are performing a large trial, you are (as Doll and Peto suggest) expecting no more than a small effect (and a trial would in practice never get to be large if the effect were itself at all large). While intuitively it's quite unlikely that a therapy that produces, say, a 50% reduction in absolute risk even in a small RCT will not prove of positive benefit in the target population as a whole, this seems altogether more plausible in the case of tiny "effects" "revealed" by mega-trials. Like all trials, these trials involve specific 'selection criteria' (partly, though far from exclusively, with ethical considerations in mind). Those meeting these criteria *may well* suffer from fewer side-effects or have a different response than is typical within the overall target population.

For example, in the GISSI-3 study assessing a proposed treatment for ischaemic heart disease, only 45% of the 43,047 people admitted to the coronary care units in the hospitals involved in the trial were randomized. A back-up study showed the excluded group to have roughly twice the mortality of the included group.[15] Notice that the absolute risk reduction allegedly found by the GISSI-3 study was 1.4%. It is not as if the reasons for exclusion are always clear-cut (so that the 'target population' could be more precisely defined on the basis of the study). For example, one exclusion 'criterion' employed in the ASSENT-2 trial was "any other disorder that the investigator judged would place the patient at increased risk"! The ISIS-2 trial listed any further reason for exclusion "not specified by the protocol but by the responsible physician".[16] These trials also generally involve a 'run in' period meant to test for compliance, side effects, and, in statin cases, increased creatinine and

---
[15] For details and references see Penston (2003).
[16] Taken from Penston op. cit.

hyperkalaemia. It may *well* be that a therapy that has a tiny positive effect even in a large trial population that is unusually compliant, shows fewer immediate side-effects, have normal levels of creatinine and hyperkalaemia, etc., has a negative effect in a population where partial compliance, existence of side-effects and so on is the norm.

Of course on any account – Bayesian, as well as classical frequentist, and even commonsense – the larger the trial, other things being equal, the stronger the evidence. But other things never are equal, and here in particular the two factors (i) large population, but (ii) small effect pull in opposite directions. When we bring in the inclusion and exclusion criteria which lead to the study population satisfying special conditions not shared by the target population, it seems difficult to form a reasonable view about what the study result is telling us about the effect in the target population.

27.5.2 Are Such Small Effects Worth Having?

So, one problem with the argument of Doll and Peto (and collaborators) is the issue of external validity when such small effects (if any) are likely to be involved. The second issue is whether such small effects as may or may not be revealed in the mega-trials that they advocate are worth having if they exist at all.

For example, several such trials have investigated the effect of various statins on subsequent mortality from stroke and heart attack (LIPID, CARE, etc.). These have uniformly found absolute risk reductions of less than 2%. Here are some representative results.

Study	Outcome	Abs RR	NTI
LIPID	mortality	1.9%	98.1
CARE	stroke	1.2%	98.8
GISSI-3	composite	1.4%	98.6

Here the third column gives the absolute risk reduction. Put plainly, these results are telling us that if we go ahead and use these drugs for treatment, then even if the trial result happens to generalise (that is, the treated population turns out to reflect the study population – and this is certainly questionable, as we just saw), then more than 98% of those treated will get *no* benefit (see the fourth column representing 'Number Treated Ineffectively').

It is crucial when trying to make a serious assessment of the (likely) impact of some treatment on a condition to ignore all talk of *relative* risk reduction: one hears figures of 30% or even 50% risk reductions bandied about, which sound striking, but are in fact systematically misleading since they suppress the base rate. Suppose, unrealistically but for sake of a particularly telling example, only 1 in a million of those whom medics propose to treat with some prophylactic medicine will on average develop some outcome (say a stroke within the next 5 years) if left untreated.

Then, if that medicine reduces the average rate to zero then this will of course represent a 100% relative risk reduction. It by no means follows, however, once we factor in side effects, that this is a treatment that can rationally be recommended. It is always *absolute* risk reduction that we need to know in order to make a rational decision about the use of some treatment. An equivalent statistic sometimes (laudably) used is the NNT, standing for 'number needed to treat'. This is an expectation value: the number of patients you would need to treat on average in order to produce one positive event (recovery or amelioration of symptoms or whatever). So in the unrealistic example just cited, the NNT is 1 million. However surely the statistic (entirely analytically equivalent to either absolute risk reduction or NNT) that is likely to have most (rational) rhetorical impact is NTI – 'number treated ineffectively'. This is just NNT *minus* 1, and measures the average number of people who will be treated ineffectively in producing just one positive event. Hence the NTI column in the above table.

But what, will go up the cry, if you are that 1 in a 100 (or whatever) who will benefit? Surely if the benefit is no myocardial infarction or no stroke in the next 5 years you want to reap that benefit. And of course if there were no 'downside' then treatment would be the rational course even with such high NTIs. But there always is a downside and it is this that Doll and Peto entirely ignore when producing their plausible argument for the importance of even small positive effects from treatments.

Returning to the trials on statins, these trials were regarded as the justification for introducing mass prescription of statins as prophylaxis for stroke and heart attack. In 2003, well over 5% of the entire US population were taking statins as prophylactic medicine. According to our "best" evidence, 98% of those will get no benefit (even assuming the results generalise).[17] This is a lot of people and means pretty good business for the pharmaceutical companies!

Once you factor in side-effects (and surely just being on long-term medication should count as a side-effect), it is surely at least questionable whether this treatment policy is sustainable. Of course some side-effects will (generally) be revealed in the trial and can be taken into account in deciding whether to treat or not, but the worrying thing is surely longer term side effects that do not (cannot) show up in the trial. Remember that, as Michael Rawlins points out, almost all trials last for between 6 and 24 months (and relatively few seem to be even close to the upper end). But statins, like puppies, are for life!

Again we are not dealing here with mere "philosophers' logical possibilities". One particular statin, Cerivastatin, was 'sanctioned' in an RCT but then quickly withdrawn because of an unexpectedly high number of deaths amongst those treated. It seems to me sobering to think that of those who died probably more than 98%, even on the most favourable interpretation of the trials on the basis of which the drug was introduced, were receiving no benefit from the drug.

---

[17] Figures again taken from Penston op. cit.

Doll and Peto, as we saw, make the apparently very cogent point that many lives may be saved by discovering treatments that have small effects provided the condition is common. But they are – clearly – only doing half of what ought to be the *expected utility calculation*! They entirely ignore the 'downside'. Suppose that some drug is in reality 1% effective for some condition, then the expected utility of using it as a treatment for that condition is:

P(helps) × utility(helps) + P(doesn't help) × disutility(taking it ineffectively)

Given that the first probability is only 0.01 and the second 0.99 and given that there *is* a downside in terms of side effects, it cannot simply be assumed that this expected utility is positive. Medicine should surely beware the drive to treat at all costs.

27.6 Conclusion

Neither this, nor my earlier arguments about the evidential weight of various types of clinical trial, is at all aimed at denigrating RCTs in general, let alone questioning the application of scientific method in medicine. On the contrary they are aimed at encouraging the correct application of science in medicine. Randomization can sometimes be of epistemic value, so long as it is not regarded as an evidential *sine qua non*. The main thing is to keep one's critical, philosophical-commonsense faculties at full power: this new argument by Doll and Peto at least carries less weight than might first meet the eye.

References

Bartlett C, Doyal L, Ebrahim S, Davey P, Bachmann M, Egger M, Dieppe P (2005) The causes and effects of socio-demographic exclusions from clinical trials. Health Technol Assess 9:1–152
Doll R, Peto R (1980). Randomised controlled trials and retrospective controls. Br Med J 280:44
Glaziou P, Chalmers I, Rawlins M, McCulloch P (2007) When are randomised trials unnecessary? Picking signal from noise. Br Med J 334:349–351
Hill AB (1937) Principles of medical statistics, 1st edn. in 1937, 9th edn in 1971. Livingstone, London
Herbert V (1977) Acquiring new information while retaining old ethics. Science 198:690–693
Howson C (2000) Hume's problem: Induction and the justification of belief. Oxford University Press, Oxford
Penston J (2003) Fiction and fantasy in medical research. the large scale randomised trial. The London Press, London
Peto R, Collins R, Gray R (1995) Large scale randomized evidence: Large simple trials and overviews of trials. J Clin Epidemiol 48:23–40
Rawlins M (2008) De Testimonio: On the evidence for decisions about the use of therapeutic interventions. Royal College of Physicians. http://www.rcplondon.ac.uk/pubs/brochure.aspx? e = 262. Accessed 18 December 2008
Tukey JW (1977) Some thoughts on clinical trials, especially problems of multiplicity. Science 198:679–684
Worrall J (2002) What evidence in evidence-based medicine? Philos Sci 69:S316–S330

Worrall J (2007a) Why there's no cause to randomize. Br J Philos Sci 58:451–488
Worrall J (2007b) Evidence in medicine and evidence-based medicine. Philos Compass 2(6):981–1022
Worrall J (2008) Evidence and ethics in medicine. Perspect Biol Med 51:418–431
Yusuf S, Collins R, Peto R (1984) Why do we need some large, simple randomized trials? Statist Med 3:409–420

Chapter 28
Incontinence, Honouring Sunk Costs and Rationality

António Zilhão

28.1 Honouring Sunk Costs

According to a basic principle of rationality, the decision to engage in a course of action should be determined solely by the analysis of its consequences. Thus, considerations associated with previous use of resources should have no bearing on an agent's decision-making process. However, some times agents persist carrying on an activity they themselves judge to be nonoptimal under the circumstances because they have already allocated resources to that activity. When this is the case, they are said to be honouring sunk costs or displaying the sunk cost effect. This sort of behaviour has been observed to occur frequently and in a variety of different situations. Moreover, agents who exhibit it tend to do so consistently.

In general, the psychological literature hypothesizes that this effect is a cognitive bias that manifests itself as a "robust judgment error" (Arkes and Blumer 1985). This diagnosis considers all cases of the sunk cost effect to be of an intrinsically irrational nature in that they all consist of a systematic departure from the normative model. The question that is left to be addressed is then the question of finding out the deviant psychological mechanism underlying such maladaptive departures from rationality.

Arkes, a leading researcher in the psychology of sunk costs, suggested that this mechanism consists of a overwhelming desire not to *appear* wasteful subjects exhibit (Arkes 1996). It is important to underline the verb 'to appear' here. As a matter of fact, the desire not to waste resources is in itself sensible and presumably evolutionarily justified. However, according to Arkes, such a desire on its own could not possibly account for the effect of honouring sunk costs. Let me explain this.

Rationally speaking, the presence of a desire simply not to waste resources should manifest itself in the carefulness with which subjects make decisions concerning allocation of resources; but once such a decision has been carefully made,

---

A. Zilhão (✉)
Philosophy Department, University of Lisbon, Alameda da Universidade,
1600-214 Lisboa, Portugal
e-mail: AntonioZilhao@fl.ul.pt

and the resources have been spent, if the investment proves *ex post facto* to have been unsuccessful for reasons that could not have been anticipated, then there should be no reason for regret. *A fortiori*, there should be no reason why subjects should want to stick to the original decision when better options become perceptible. Therefore, there should be no reason either to honour sunk costs in order to avoid being wasteful.

It is only when the original decision was careless that feelings of regret for having been wasteful are justified. But, once the resources have been irretrievably spent, it is already too late to do anything to avoid them. Thus, given the above mentioned principle of rationality, even when the original decision was careless and wasteful, it should still be less wasteful to abandon the initial failed investment and change course than to stick to it. That is, and unsurprisingly, the desire not to waste resources is, in both cases, best satisfied by the display of a fully normative behaviour.

According to Arkes, when the public analyses investment decisions made in everyday situations, they tend to use an informal folk-notion of 'waste' that has an absolute rather than a relative character. In this usage, 'to be wasteful' is perceived to mean something like'not to fully utilize the thing or things one has acquired' rather than something like 'not making the best use of our resources taken as a whole' (Arkes 1996). Still according to him, it is in order to avoid being judged by others to have been wasteful, in the former sense of the term, that we develop a desire not to *appear* wasteful (also in the former sense of the term). But behaving in agreement with this desire is frequently not congruent with behaving rationally. And it is in such incongruous cases that the sunk cost effect emerges (Arkes 1996). Typically, the effect is observed when a previous decision made us spend an important amount of resources and, in order to avoid risking being judged by others to have been wasteful, we behave as if that decision were an optimal one, even when it was not. Paradoxically, our fear of wastefulness makes us display an objectively wasteful behaviour, as is vividly expressed in the formula frequently used to refer to this effect, namely, 'throwing good money after bad' (Arkes and Blumer 1985; Bornstein and Chapman 1995).

28.2 Honouring Sunk Costs, and the Two Systems of Reasoning View

Arkes substantiates his analysis with a rich collection of data. No doubt, its consideration leads us to establish a connection between the sunk cost effect and some sort of aversion people display towards wastefulness. In some cases, this aversion can presumably be accounted for in terms of the subjects having an understanding of the concept of "being wasteful" akin to the one Arkes identifies. However, I beg to disagree with the idea that the desire not *to appear* wasteful may be taken to be the default explanation for the sunk cost effect.

The opposition Arkes assumes there to obtain between an individual default mechanism that tends to make each of us behave as a rational man, on the one hand,

and a desire for social appraisal that makes us deviate from our default course of action and leads us to make irrational judgments of wastefulness, on the other hand, seems to me to be inherently implausible. After all, "the others" means nothing else but the sum of each of us. On what grounds could such a mistaken social concept of "waste" have originated in a population, if behaving in agreement with it would go so conspicuously against the natural tendency all members of the population were supposed to be endowed with? There's a mystery here and I do not see how it could be solved.

Moreover, given our assumed default rational nature and given the fact that in general we know whether our original decision was carefully or carelessly made, we should in general be able to know whether or not a charge of having wasted resources was indeed appropriate; but this knowledge of ours would then turn at least some of the displays of the sunk cost effect into cases of deceit or even "mauvaise foi" rather than into cases of a true "robust judgment error".

Fortunately, an interesting alternative explanation for at least some of the cases of this effect is mentioned in the relevant literature – the 'learn a lesson' justification (Bornstein and Chapman 1995). As a matter of fact, some subjects report that they pursued a course of action they themselves thought was of lower expected utility than some other because they felt they had been careless with the use of their resources and therefore thought they should put up with the consequences of their bad decision in order to teach themselves a lesson and not to repeat a similar mistake in the future.

The empirical validity of *ex post facto* justifications is, in general, questionable. Nevertheless, let us consider this particular one for a moment. It has two very interesting features. First, it shows that subjects are concerned with wastefulness itself, as they should, given the evolutionary value of avoiding it, rather than with simply preventing their *appearing* to be wasteful. Second, given the fact the self-teaching they mention would use the experience of the unpleasant consequences of past careless decisions in order to make agents improve their own future decision-making, the mechanism generating it would actually be driven by a rational care not to waste precious resources in the future rather than by an irrational concern with resources irretrievably lost in the past. Therefore, this pattern of behaviour would make good evolutionary sense.

Now, the consistency of a proposed explanation with a wider explanatory framework is by itself no proof of its empirical validity. However, Bornstein and Chapman, the proponents of this explanation, report that in a number of cases it tends to square better with the available evidence than the alternatives. Interestingly enough, despite having suggested it themselves, they underscored their own explanation by stressing what they consider to be one major objection against it – the objection that, "in order to teach oneself a lesson, one must already know it" and this makes the teaching in question "paradoxical" (Bornstein and Chapman 1995). But the paradox might be easily accommodated and dissolved through the adoption of a multiple self view, as they themselves acknowledge (Bornstein and Chapman 1995). The question then is whether or not there is independent evidence for this view.

The idea of multiple selves admits being interpreted in multiple ways. One of these interpretations is the two systems of reasoning view. Steven Sloman, a supporter of this view, did in fact present some independent evidence supporting it (Sloman 1996a). According to him, there is a peculiar set of reasoning problems that is characterized by the fact that they all satisfy what he calls Criterion S. Following his own definition, a reasoning problem satisfies Criterion S if it causes subjects to believe *simultaneously* two contradictory responses. Thus, reasoning problems satisfying Criterion S have a perceptual analog in the Müller-Lyer illusion in which two lines *appear* to us to be of different lengths despite the fact that, at the very same time, we already *know* that they are of the same length. Given the fact that, in principle, a system of reasoning can output only one response at each time, Sloman argues that satisfaction of Criterion S does show that two independent systems of reasoning exist, that they were both mobilized to solve the reasoning problem in question, and that they provided the subject with two different responses to the same task (Sloman 1996a).

Appealing to independent evidence concerning the existence of two systems of reasoning allows me to alleviate Bornstein and Chapman's qualms regarding the paradoxical consequences of the explanation they suggested. But I think the introduction of Sloman's Criterion S in this debate allows me to do better than that. Indeed, I think it can help me adjudicate in a number of cases between the two competing explanations for the sunk cost effect mentioned above. I will show this below.

After introducing Criterion S, Sloman draws an important contrast between reasoning problems satisfying it and reasoning problems also revealing factors affecting cognitive performance but not satisfying it. The latter are thus not aptly characterized as highlighting the existence of two systems of reasoning. Conspicuous among these reasoning problems are those which originate conflicting responses that are perceived as correct *sequentially* but *not simultaneously*. These problems admit being seen as having a perceptual analog in the Necker cube or the duck/rabbit type of figures which subjects are typically able to see now this way now that way but not the two ways simultaneously. Typical examples are the cases in which the conflicting responses are due to conflicting linguistic interpretations of a term or expression. According to Sloman, psychological evidence suggests that when a subject has one semantic interpretation in mind then the other interpretation that conflicts with it is not held simultaneously (Sloman 1996b). This is consistent with the hypothesis that the two semantic interpretations are generated within the same system of reasoning.

Now, if the emergence of the sunk cost effect were, under all circumstances, just a consequence of one's desire to comply with a faulty, socially induced, folk-concept of wastefulness, then the effect should tend to fade away in normal individuals subsequently to some exposure to explicit teaching in economics. However, this is not what happens in a significative proportion of cases. But, besides showing this, the empirical reports show also two other things. First, they show that an important proportion of those subjects that respond to reasoning problems according to the sunk cost effect, when debriefed and explained why their reasoning is faulty, tend to accept the normative solution whilst at the same time contending that, despite having understood and accepted the explanation, their previous response remains associated to a feeling of being right attached to it. Secondly, they show also that an important number of the subjects that do respond normatively confess subsequently that, somehow, the sunk cost response also seemed right to them, despite their knowing the reasons why it should not be so (Stanovich 1999). That is, apparently, once subjects become aware of the existence of a normative response different from the more spontaneous non normative response, the two responses tend to pop up in their minds simultaneously, not subsequently.

Therefore, and assuming Sloman's interpretation of the relevant evidence to be correct, it seems to make sense to include at least some of the reasoning problems associated with the sunk cost effect within the class identified by him as the class of problems that satisfy Criterion S, that is, the class of problems that indicates the underlying presence of two competing systems of reasoning. At the same time, it is also clear that if Arkes's explanation were the default explanation for this effect, then the conflicting responses given by the subjects should have originated in a semantic ambiguity associated with the terms 'waste' and 'wasteful', and what is typical of these cases is that the conflicting responses should pop up in the subject's minds sequentially, and not simultaneously.

In sum, there is a subset of the reasoning problems designed to test the sunk cost effect, namely the subset that satisfies Criterion S, for which the 'learn a lesson' explanation is actually more consistent with the wider explanatory framework suggested by the evidence than the 'desire not to appear wasteful' explanation.

28.3 Incontinence

Let me now introduce into the discussion the topic of incontinent action. The modern *locus classicus* of the explication of the concepts of *continence* and *incontinence* is Donald Davidson's paper "How is Weakness of Will Possible?". There, Davidson tells us that an agent is continent if and only if he searches exhaustively his belief set in order to make sure that *all* relevant reasons were appropriately weighed up and taken into account in forming his own best judgment and if he acts in agreement with it (Davidson 1970). Davidson's account of continent action is thus basically an account of rational action *qua* instantiation of a model of pure unbounded rationality. But continent action is also taken by Davidson to be rational action in a psychologically relevant sense.

On the other hand, an agent is deemed by Davidson to be incontinent if and only if he *neglects* or *does not attend to* a relevant part of his own beliefs. And this is what turns his action into an irrational action. However, he does attend to some of them. And this is what makes it an intentional action. The cognitive arguments against the descriptive validity of a model of pure unbounded rationality are well known. Thus, there is no need to rehearse them here. Assuming they are basically sound, Davidson's explication of the concepts of continence and incontinence entails the consequence that all real human action is incontinent. If this is the case, the concept pair continent/incontinent ceases to have any explanatory relevance for psychological purposes.

In opposition to Davidson, Gary Watson claims that, according to the common concept of incontinence, allegedly incontinent agents behaving against their own best judgment are, in reality, indistinguishable from agents acting under some sort of compulsion. Thus, no case of so-called incontinent action is in reality free or intentional or both (Watson 1977). Therefore, from Watson's standpoint, if an action was free and intentional, then it was continent, in Davidson's sense of the term; if it was not continent, in the same sense, then it was either not free or not intentional or neither of these, and therefore it was no incontinent action either. Thus, the common concept of incontinent action cuts no ice.

Whoever has the intuition that there really are both continent and incontinent actions, in the sense of free and intentional actions performed, respectively, in agreement with the agent's best judgment and in disagreement with it, must feel dissatisfied with the outcome of this discussion. Davidson provides us with an interesting and ingenious development of the idea of how incontinent actions are possible; but from his account, together with some well established assumptions concerning our cognitive architecture, it follows that the concept of continent action cannot but refer a normative ideal never to be attained in practice. Watson, in turn, tells us that this intuition is basically wrong and that the common concept of incontinent action is a non starter. According to the former, we are all incontinent; according to the latter, we are all continent. Despite standing in opposition to each other, the consequences of Davidson's and Watson's perspectives lead us inexorably in the same direction: the pair of concepts continence and incontinence is useless for psychological purposes.

I believe this concept pair to be empirically meaningful; thus, I think it deserves to be preserved in psychological theorizing. In order to support this belief, I'll put forth below an alternative description of the phenomena allegedly captured by these concepts. I identify the sources of intentional action thus. On the one hand, there are judgments resulting from explicit processes of deliberative reasoning, regardless of the proportion of the agent's reasons effectively taken into account in the reasoning process and regardless of the inferential strategy underlying their production. On the other hand, there are judgments resulting from the deployment of a different system of reasoning containing a wide set of fast and frugal heuristics (Gigerenzer et al. 1999). I call the former kind of judgments *slow judgments*; I call the latter kind of judgments *fast judgments*.

I conceive of fast judgments as being geared to action in a more straightforward way than slow judgments. Furthermore, I assume that the mind has a modular arrangement. It thus makes sense to suppose that slow judgments and fast judgments originate in different structures of the mind. In agreement with Sloman's view, I thus conceive of cases in which both these structures get mobilized for responding to the same problem.

For instance, suppose that, in a situation in which he is faced with a particular problem, an agent forms by means of explicit deliberation a best slow judgment on how to act and that he intends to act on such a judgment; however, given both the *domain* the problem belongs to and *the structure of the environment*, a particular heuristic harboured in the agent's cognitive apparatus is also triggered when the moment of action approaches. As a result, a fast judgment is quickly formed and, *without having given up his slow judgment*, the agent acts in a way that is not the one he contemplated as a result of his explicit deliberative reasoning. Thus, he acted in agreement with a fast judgment and in disagreement with his best slow judgment. In other words, the agent acted against what he himself identifies sincerely as his own best judgment. Typically, when this happens, the agent feels surprise towards his own action. Together, these two features are typical of incontinent action. Therefore, I propose to analyse the concept of incontinent action thus: an incontinent action is a free and intentional action that is determined by a fast judgment the content of which disagrees with the content of the agent's own best slow judgment but that nevertheless overrides it.

On the other hand, I appeal to the concept of an action triggered by a slow best judgment in order to mark out the lines defining continent action. I contend that the formulation of these judgments typically involves the mobilization of the resources of the agent's language faculty; in general, this also means their being accessible to consciousness. As a consequence, when he acts continently, the agent feels no surprise towards his own behaviour. These, I take it, are far more realistic concepts of continent and incontinent action than Davidson's or Watson's.

28.4 Honouring Sunk Costs, Incontinence, and Rationality

It seems to me that the performance of the informed subjects that did not respond normatively to some of the reasoning problems designed to test the sunk cost effect inspite of the fact that they "should have known better", and that reported feeling baffled when confronted with their own non normative response, fits rather well within the definition of incontinent action I presented. That is, it seems appropriate to say about those subjects that, to their own surprise, they responded intentionally against their own better reason to respond normatively. And I am able to identify a pattern here that can be found across different sets of data associated with the testing of quite disparate cognitive phenomena (Zilhão 2006).

In the case at hand, I hypothesize that teaching oneself a lesson is indeed responsible for an important subset of the cases of the sunk cost effect (but not for all) and that this behaviour gets triggered as the result of a heuristics that reacts at a deeper level against our indulging in wasteful behaviours and issues in a fast judgment that makes us honour sunk costs. If this hypothesis is correct, then, the following three consequences should follow from it: acting contrary to this heuristics should be difficult and should require some effort and cognitive expertise; it should be expected that even subjects that respond normatively could not help feeling a tendency to honour sunk costs, in a Müller-Lyer illusion sort of way; and it should be expected that there should be subjects that do not respond normatively inspite of the fact that they "should have known better", and that report feeling baffled when confronted with their own non normative response. Each of these three consequences is indeed reported in the relevant empirical literature.

I thus contend that in a number of situations (although by no means in all) the peculiar character of baffling actions of honouring sunk costs performed by informed agents is best captured by viewing them as incontinent actions, that is, actions triggered by fast judgments, in which one of our deep seated heuristics takes the upper hand and overrides locally more effective slow judgments. And I contend further that these actions, rather than mere indicators of the underlying existence of irrational cognitive biases, frequently express a kind of adaptive rationality (Gigerenzer 2000) that needs to be taken into account in psychological theorizing. In the particular case of the sunk cost effect, it should be bourne in mind that, as long as the long term benefits of learning the lesson not to be wasteful in a particular situation may outweigh the costs of not changing now the initial decision, an incontinent agent (according to my definition) may be considered to be acting in a more reasonable way in that situation than a corresponding continent agent (according to my definition). That is, what, at an atomic level of observation, may appear to be a case of irrational behaviour might in fact be, when looked at from a more strategic viewpoint, a perfectly well designed pattern of action dictated by a meaningful heuristics planted in us by an evolutionarily justified concern with not wasting precious resources in the future.

References

Arkes HR (1996) The psychology of waste. J Behav Decis Making 9:213–224
Arkes HR, Blumer C (1985) The psychology of sunk cost. Organ Behav Hum Decis Process 35:124–140
Bornstein BH, Chapman GB (1995) Learning lessons from sunk costs. J Exp Psychol Appl 1(4):251–269
Davidson D (1970) How is weakness of the will possible? In: Feinberg J (ed) Moral concepts, Oxford University Press, Oxford. Reprinted in Davidson, D. 1980 Essays on actions and events. Clarendon, Oxford, pp 21–42
Gigerenzer G (2000) Adaptive thinking – rationality in the real world. Oxford University Press, Oxford
Gigerenzer G, Todd PM, the ABC Research Group (1999) Simple heuristics that make us smart. Oxford University Press, Oxford
Sloman SA (1996a) The empirical case for two systems of reasoning. Psychol Bull 119(1) 3–22
Sloman SA (1996b) The probative value of simultaneous contradictory belief: reply to Gigerenzer and Regier. Psychol Bull 119(1):27–30
Stanovich KE (1999) Who is rational? Studies of individual differences in reasoning. Lawrence Erlbaum, Mahwah, NJ
Watson G (1977) Skepticism about Weakness of Will. Philos Rev 86:316–339
Zilhão A (2006) Incontinence, fast and frugal heuristics and probability matching. In: Manrique FM, Peris-Viñé LM (eds) Actas del V Congreso de la Sociedad de Lógica, Metodología y Filosofía de la Ciencia en España – Granada, 29 Noviembre – 1 Diciembre 2006. Ediciones Sider, Granada, pp 242–246

Chapter 29
Causal Fundamentalism in Physics

Henrik Zinkernagel

29.1 Introduction

Causality in physics has had bad press in philosophy at least since Russell's famous 1913 remark: "The law of causality, I believe, like much that passes muster among philosophers, is a relic of a bygone age, surviving, like the monarchy, only because it is erroneously supposed to do no harm" (Russell 1913, p. 1). Recently Norton (2003, 2006) has launched what would seem to be the definite burial of causality in physics. Norton argues that causation is merely a useful folk concept, and that it fails to hold for some simple systems even in the supposed paradigm case of a causal physical theory – namely Newtonian mechanics.

The purpose of this article is to argue against this devaluation of causality in physics. I shall try to defend that Norton's charges against causality in Newtonian mechanics are flawed, and I will also suggest how the central causal message of Newtonian mechanics may proliferate into its supposed successor theories, namely special (and to some extent general) relativity and quantum mechanics. My main argument is that Norton's (2003) alleged counterexample to causality (all events have causes) within standard Newtonian physics fails to obey what I shall call the causal core of Newtonian mechanics (essential parts of the first and second law). In particular, I argue, Norton's example is not in conformity with Newton's first law – and his attempt to reformulate this first law (in order to make it conform to his example) results in an impoverished theory which lacks important features of Newtonian mechanics (in particular, in this 'Nortonian mechanics', the notion of inertial frames lacks a physical justification, and the close connection between the first law and the notion of time is lost).

H. Zinkernagel (✉)
Department of Philosophy I, Campus de Cartuja, University of Granada, Granada 18071, Spain
e-mail: zink@ugr.es

29.2 The Dome and the Alleged Failure of Determinism (and Causality)

Norton (2003) provides an interesting example of a system – a mass on a dome – which seems to conform to Newtonian mechanics, yet fail to be deterministic. This indeterminacy comes about due to an uncaused change in the state of motion of the mass on the dome, and can therefore also be seen as a failure of causality (all events or, more precisely in Newtonian mechanics, all changes of states of motion have a cause). As Norton points out, this example is striking as no reference to exotic features, such as space invaders appearing with unbounded speed from spatial infinity or an infinite number of interacting particles, are needed to produce the indeterminism.[1] If Norton is right that his example is an example of a Newtonian system then not all Newtonian systems are causal and, as Norton emphasizes, Newtonian mechanics cannot therefore (in general) license a principle or law of causality.

In the example, we are invited to consider a unit mass point which, under the influence of gravity, can slide frictionlessly over the surface of a dome in which a radial coordinate r is inscribed (Fig. 29.1). The shape of the dome is given by the height function $h = (2/3g)r^{3/2}$ which specifies the vertical distance of each point below the apex at $r = 0$. At any point on the dome, the gravitational force tangent to the surface (equal to the net force acting on the mass point) is $F = r^{1/2}$ which implies that Newton's second law for the mass point takes the form:

$$F = ma \Rightarrow d^2r/dt^2 = \sqrt{r} \qquad (29.1)$$

The remarkable feature of the example is that if one starts with the mass at rest at the apex (that is, with $r(t = 0) = r'(0) = 0$), Eq. 29.1 does not only have the expected solution in which the mass remains at the apex forever ($r(t) = 0$ for all t), but also solutions corresponding to the mass taking off spontaneously at an arbitrary time T in an arbitrary radial direction:

$$\begin{aligned} r(t) &= (1/144)(t - T)^4 \text{ for } t \geq T \\ r(t) &= 0 \text{ for } t \leq T \end{aligned} \qquad (29.2)$$

There is no doubt that Norton has found an interesting example of a differential equation without a unique solution. It is interesting because it apparently corresponds to a physical (even if idealized) situation within Newtonian mechanics.

---

[1] Both space-invader examples and the so-called supertasks (involving an infinite number of particles) may be argued to be less troublesome for determinism in Newtonian mechanics e.g. since they involve non-conservation of energy (see, e.g., Alper and Bridger cited in Norton 2006, p. 13) and, in the space-invader case, non-conservation of particle number (and hence an ambiguity in the very specification of the physical system), see, e.g., Malament (2007).

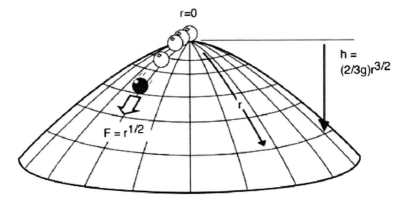

Fig. 29.1 The mass on the dome (From Norton 2006)

But the question is: is the mass on the dome a genuine Newtonian system – that is, a system which satisfies Newton's laws of motion?[2]

29.3 Is Newton's First Law Satisfied for the Mass on the Dome?

As Norton recognizes, the spontaneous motion of the mass described by Eq. 29.2 may make one wonder whether Newton's first law is satisfied at the crucial moment $t = T$ (the last moment at which the mass is at rest). Nevertheless, Norton (2003, 2006) argues that the mass on the dome example does indeed satisfy this law both in an 'instantaneous' form (if $F = 0$, then $a = 0$) and in the 'non-instantaneous' form which Newton originally proposed (see below). This is so, Norton says, since for any moment up to and including $t = T$, there is no acceleration of the mass point, the net force acting on the mass vanishes, and the mass can be said to be in 'uniform motion' (in a state of rest). As far as I can see, however, this conclusion is much too fast.

Newton's original formulation (in the 1729 translation of the *Principia*) of his first law is as follows:

(N1): Every body continues in its state of rest, or of uniform motion in a right line, unless it is compelled to change that state by forces impressed upon it.

There are at least two problems in maintaining that Norton's mass on the dome satisfies (N1). First, take a look at the word "continues" in (N1): If a body is not compelled to change (at t) then it *continues* in its state of motion. Can that mean anything but: if it is not compelled to change at some moment t, then at least in

---

[2] For a detailed mathematical analysis of the dome example, see Malament (2007). Norton (2006) discusses a list of possible objections to the dome being a Newtonian system but rejects each of these. As will be clear in the following, I disagree with some of his arguments.

the very next moment (even if that moment is infinitesimally close) the state is unchanged? If this is right then the motion in Norton's system does not conform to Newton's first law at $t = T$, since – for any $\varepsilon > 0$ – it is not true that the mass *continues* in its original state at $t + \varepsilon$. This issue is closely related to Norton's own, and in my view correct, suspicion that the phrasing "compelled to change" in (N1) "...suggests that changes of motion must be brought about by forces acting at the same time as the change, if not even earlier" and so it suggests that "...forces must be first causes" (Norton 2006, p. 6).[3] There is no first cause which forces the particle to move and so, according to (N1), it remains at the apex.[4]

The second (though related) problem concerns Norton's claim that the mass at $t = T$ is in a state of rest (a particular case of a state of uniform motion). Clearly, Norton needs to say that the mass is moving uniformly (being at rest) at $t = T$ since, by Newton's first law, it must be so since it has been at rest for all $t < T$ and since there are no forces acting on the mass at this moment. But the question here is what exactly is meant by uniform motion. Uniform motion means travelling equal (for instance, zero) distances in equal times – so that moving uniformly at some instant t seems to require considering, at least small, intervals of time at both sides of the instant t. Another way to state this point is that whereas one may speak of instantaneous velocity, or velocity at a specific moment (by the usual limiting procedure, $v = \Delta x/\Delta t$, $\Delta t \to 0$, the limit taken either from above or below), it makes little sense to speak of *constant* velocity at some specific moment unless reference is made to moments or time intervals on both sides (before and after) the moment in question.

As an illustration of this point, compare Norton's mass on the dome with a standard example in Newtonian physics in which a white ball at rest on a billiard table is hit at $t = T$ by a red ball. At any moment $t < T$ the white ball is 'moving uniformly' (being in a state of rest). Moreover, for any moment $t < T$ we can find a successive moment $t + \varepsilon$ ($<T$) in which the ball is also moving uniformly (so that intervals on both sides of any moment $t < T$ can be found in which the velocity is zero). But even though the velocity of the white ball is still zero at the time of collision ($t = T$), we are not saying that it is moving uniformly at this moment. Of course, in this case we can attribute the non-uniform motion to the fact that a force acts on the ball – but it is equally true that the velocity at any moment $t > T$ is non-zero and therefore that the velocity is not *constant* at $t = T$.

---

[3] See also the *Note added in proof* at the end of the paper.
[4] As far as I can see, this point also accounts for why Norton's "time reversal trick" (Norton 2003, p. 16) does not support that the acausal mass on the dome is a Newtonian system. The 'reversed motion' in which a mass point with a precisely adjusted initial velocity slides up the dome and halts exactly at the apex *is* consistent with Newtonian mechanics. In this reversed case, at $t = T$ the net force on the mass point vanishes, and for all $t > T$ the mass point is at rest and in uniform motion, and so no conflict arises with (N1). But insofar as forces should be first causes we cannot generate Norton's motion from this allowed reversed case (when time is "run backwards" from some $t > T$, (N1) – understood as including the constraint that forces must be first causes – demands that the mass remains at rest for all t).

Norton notes that if one were to insist (as I do) that $r'(T) = r''(T) = 0$ is not a sufficient condition for identifying a 'state of rest' or 'uniform motion', one would be "creating difficulties with other canonical examples" (2006, p. 6). For instance, Norton alludes to a harmonic oscillator, like a mass attached to a spring, passing through the origin: "We normally think of the mass at just that one moment as moving inertially – there is no net impressed force, so the velocity is constant, in the sense that the acceleration vanishes." (2006, p. 7). But Norton does not explain (nor cite evidence for) why this is supposed to be an instance of 'normal thinking'. Indeed, I believe that a more natural reaction to such examples is to say that the mass is *not* moving inertially when it passes through the origin (nor at any other moment), precisely because it is only at that moment that velocity, acceleration and net-impressed force vanish.[5]

29.4 Is Newton's First Law Really Needed? Inertial Frames and the Notion of Time

In Norton (2003), it is argued that we do not really need Newton's first law in order to do Newtonian physics or, at least, that we only need it as a special case of the second law. Thus, even if my critical remarks in the last section – concerning whether Newton's first law in its original form is satisfied – are accepted, Norton could respond that the mass on the dome is nevertheless a Newtonian system satisfying Newton's laws. More specifically, Norton claims that all a system needs to comply with in order to be Newtonian is an *instantaneous* version of Newton's first law:

(NI*): In the absence of a net external force, a body is unaccelerated.

This law is indeed satisfied for the mass on the dome since for all $t \leq T$ (including $t = T$), $a(t) = r''(t) = 0$. Now, (NI*) amounts to reducing Newton's first law to a special case of the second, $F = ma = 0$, so the question is whether this move – making Newton's first law redundant – is legitimate within Newtonian physics (even if it is often done in physics textbooks). I see at least two (related) reasons to resist the idea that (NI*) is enough for Newtonian physics, and therefore two reasons to resist that the first law is redundant in the sense of being merely a special case of the second law.[6] The first reason has to do with inertial frames and the second with the notion of time in Newtonian mechanics.

---

[5] Note that the third derivative of the position is non-vanishing at the moment when the harmonic oscillator passes through the origin. In the mass on the dome case, the fourth derivative is ill-defined at $t = T$. Uniform motion would seem to require, then, the vanishing also of higher (>2) order time derivatives of position.

[6] Newton himself, of course, stated (N1) as an independent law and some modern authors, e.g., Anderson (1990, 1192), likewise consider it erroneous to regard the first law as a special case of the second. See also discussion and references in Brown (2006, p. 15 ff.).

29.4.1 Inertial Frames

According to a tradition going back to Neumann and Lange (see, e.g., DiSalle 2008) Newton's first law is closely related to the notion of an inertial frame of reference. It is only in such inertial frames that Newton's second law can be applied in its standard form, $F = ma$, in which F is an impressed (and not a fictitious) force. And it is precisely this standard form of Newton's second law that Norton invokes on the dome, so his example implicitly assumes that the dome is neither rotating nor accelerating in some direction. More generally, the use of the standard form of Newton's second law in the dome example presupposes that an inertial frame of reference exists. Now, one way to understand Newton's first law in a non-redundant way is precisely to construct it as an existence claim (see, e.g., Friedman 1983, p. 117). On this account, Newton's first law implies that inertial frames (in which any putatively free particle will move in a straight line or stay at rest) exist.[7]

Newton himself did not need the first law to secure the existence of an inertial frame as he independently (of the laws) assumed the existence of absolute space. But if the notion of absolute space is rejected (either due to its inobservability, on some relational account of space – holding, e.g., that the notion of space is necessarily bound up with that of laws – or with a view to relativity), Newton's first law in its original (non-instantaneous) form seems to be the only physical justification for assuming the existence of an inertial frame – and this justification is absent in 'Nortonian mechanics' without the full version of the law. In 'Nortonian mechanics', the existence of inertial frames will have to be postulated on an ad hoc basis and it may therefore be argued that the transition from Newtonian to Nortonian mechanics implies a loss of explanatory power.

However, it has been claimed, e.g., by Earman and Friedman (1973, p. 337), that a four-dimensional approach to Newtonian mechanics does indeed make (N1) redundant insofar as inertial frames – and, correspondingly, the trajectories of freely moving particles – are determined directly by the (affine) structure of spacetime. In this approach, then, it is thus not necessary to postulate the existence of inertial frames independently. Nevertheless, this objection might be countered by questioning whether the postulation of absolute spacetime structure explains the existence, e.g., of inertial frames and the trajectories of force-free bodies. As Brown (2006, p. 24) notes:

> In what sense is the postulation of the absolute space-time structure doing more explanatory work than Moliére's famous dormative virtue in opium? ... It is simply more natural and economical – better philosophy in short – to consider absolute space-time structure as a codification of certain key aspects of the behaviour of particles (and/or fields).

---

[7] One could perhaps argue that Newton's second law could likewise be understood as an existence claim concerning inertial frames or that both laws of motion (or all three) jointly assert the existence of inertial frames (see, e.g., DiSalle 2008, p. 6). Still, the connection between the first law and inertial frames might be more fundamental, e.g., because free particles can be used, at least in principle, to construct such frames ("We must define an inertial system as one in which at least three non-collinear free particles move in noncoplanar straight lines; then we can state the law of inertia as the claim that, relative to an inertial system so defined, the motion of any fourth particle, or arbitrarily many particles, will be rectilinear" [DiSalle 2008, p. 5]).

29.4.2 The Notion of Time

The second reason to question whether Newton's first law is redundant in mechanics has to do with the role and notion of time in the theory. Recall first Newton's famous distinction:

> Absolute, true, and mathematical time, of itself, and from its own nature, flows equably without relation to anything external, and by another name is called duration: relative, apparent, and common time, is some sensible and external (whether accurate or unequable) measure of duration by the means of motion, which is commonly used instead of true time; such as an hour, a day, a month, a year. (Newton 1729, p. 6).

As argued in detail in Zinkernagel (2008), it is reasonable to question whether this notion of absolute time is at all intelligible. In particular, it is very hard, if not impossible, to specify the meaning of 'flow' (or 'equable flow') without relation to anything external. Indeed, 'equable flow' involves the idea of steady or uniform motion, which more than suggests a reference to the motion of a physical system. In turn, the notion of uniform motion is instantiated by the free particles described by Newton's first law of motion. Now, as Newton himself notes in the Scholium, there might not be any real systems in uniform motion (due to friction and the universality of gravitational attraction) – hence uniform motion is an idealization. But just as Newton's 'common time' refers to real physical systems which can be used as clocks, it can be argued that for his absolute time to make sense, it must be seen to refer to idealized physical systems which can be used as perfect clocks, such as a free particle in uniform motion. In this way, (N1) and freely moving particles provide an implicit definition – and thus a sensible notion – of absolute time in Newtonian mechanics.[8] Insofar as a sensible notion of absolute time is part of the theory, it is not clear that we can do Newtonian mechanics without (N1).

Of course, the motions of ideal clocks in Newtonian mechanics may be either uniform or non-uniform (accelerated) so one might well ask whether Newton's second law – and associated idealized physical systems such as the mathematical pendulum – could not, after all, be sufficient for an implicit definition of time in the theory. However, I think there are reasons to believe that this suggestion for making redundant (N1) will not work. First, as argued above, Newton's second law can only be applied in its standard form (and so provide a measure of absolute time via, e.g., a mathematical pendulum) if the system is at rest in an inertial frame, and such a frame may need (N1) for its justification. Second, and in accordance with the above discussion, the notions of an "equable flow of absolute time" and of "equality of time intervals" can be seen to be interdependently defined by (N1) since equal intervals of time are those in which a free body moves equal distances.[9] In other words,

---

[8] This is consistent with a version of relationism about time in which time is dependent upon, but not reducible to, physical systems which can serve as clocks (see Zinkernagel 2008 and Rugh and Zinkernagel 2009a for detailed discussions of this version of relationism).

[9] The close relation between (N1) and absolute time in Newtonian mechanics has been emphasized also e.g. by Barbour: "... the law of inertia [N1] itself has two quite distinct parts: the rectilinearity

absolute time and uniformly moving systems (including uniformly rotating systems; see below) may be taken to be mutually dependent notions.

However, it could still be argued that absolute time might just as well be implicitly defined through Newton's second law and systems involving forces – e.g., systems involving uniform circular motion, such as a satellite orbiting a planet, in which equal time intervals can be defined as those in which the satellite moves equal distances (or through equal angles). Nevertheless, uniform circular motion is not uniform motion and the latter notion may be argued to have conceptual priority regarding the definition of time. In part because force (and forced motion) is defined by Newton (1729, p. 4) in terms of deviations from uniform motion; and in part because application of the second (instantaneous) law of motion seems to require a prior specification or notion of time (e.g., in order to use the orbiting satellite as an ideal clock one presupposes that the magnitude of the centripetal force is constant in time).[10] This suggests that the relation between (N1) and time may be more fundamental than that between time and Newton's second law.

29.5 Causality in Newtonian Mechanics and Beyond

The notion of time (and that of ideal clocks) in Newtonian mechanics is also central to understand the 'causal message' of the theory. Norton's overall idea is that causality plays no fundamental role in modern physics and, in particular, no such role in Newtonian mechanics. By contrast, my view is that there is a clear 'causality content' in Newton's first two laws which can be captured as follows:

'Causal core' of Newtonian mechanics.

A body in uniform motion continues its motion unless the body is caused (by a force) to change its motion (accelerate). The same causes (forces) acting in the same circumstances will have the same effects.

The first part of this causal core is slightly more general than Newton's first law since uniform motion is not restricted to be straight line motion. This is in conformity with Newton's own remarks in the *Principia* (after stating his first law) which allude to uniformly rotating systems such as spinning planets.[11]

---

of the motion and the uniformity of the motion. These correspond, respectively, to absolute space and absolute time" (Barbour 1989, p. 28).

[10] This point is closely related to one made earlier, namely that Norton's instantaneous form of Newton's (first or) second law in which $F = ma = 0$ is not a sufficient condition for uniform motion. In consequence, one cannot define absolute time from this special case of Newton's second law (since a time interval, in which the acceleration is constantly equal to zero, must be presupposed when integrating up $F = ma = 0$ to get uniform motion).

[11] Right after stating the first law, Newton writes: "A top, whose parts by their cohesion are perpetually drawn aside from rectilinear motions, does not cease its rotation, otherwise than as it is retarded by the air. The greater bodies of the planets and comets, meeting with less resistance in more free spaces, preserve their motions both progressive and circular for a much longer time" (Newton 1729, p. 14).

The second part of the causal core is an abstract way of formulating the causality content in Newton's second law, which does not presuppose the precise form of this law. This second part is particularly important when discussing non-uniform periodic systems as it implies that a body in non-uniform periodic motion repeats its earlier states of motion unless caused to change by additional forces. Thus, for instance, whenever the 'mathematical pendulum' reaches its top position, it will start falling. One direct consequence of the causal core is thus that it guarantees the continuity of a system in motion – and in particular the continuing motion of an ideal clock – and therefore also the continuing 'flowing' of absolute time.

Norton (2003) claims that a causal fundamentalist is confronted with the dilemma that either a causal principle restricts the content of our science or it does not. This is a dilemma according to Norton since no restricting causal principle which holds for all of science is forthcoming whereas a causal principle which does not imply such restrictions would be an "empty honorific". But, at least as far as Newtonian mechanics is concerned, the causal core does impose restriction on science: It rules out non-continuous motion in the sense described above and, in particular, it rules out examples like Norton's mass on a dome to be admitted into Newtonian mechanics.

If the relationism about time mentioned above – which tie the notion of time to that of physical processes which can be used as clocks – is on the right track, Newtonian mechanics embodies a close link between time and the causal core. This close link can be argued to proliferate into Newtonian mechanics' supposed successor theories and in this sense pave the way for a more general (i.e., beyond Newton) causal fundamentalism. There is no room here for arguing this point in detail so what follows is merely a few notes and references suggesting how such an argument might be constructed.

In special relativity, the causal core of Newtonian mechanics is valid as it stands, and, just as in Newtonian mechanics, the physical basis for the notion of time in this theory may be argued to rest on physical processes which can be used as clocks. Indeed, Einstein himself argued that the physical meaning of the (space and) time coordinate(s) is (are) given in terms of (measuring rods and) standard clocks. A standard clock in an inertial system in special relativity is just like the (idealized) clocks in Newtonian mechanics and such standard clocks obey the causal core.

In general relativity things are less clear: On the standard interpretation, gravity is not seen as a force (but rather as curvature of space-time) and (N1) does not hold in general so one might well ask whether there is room in the theory for the causal core. Still, the physical interpretation – and the correspondence with empirical tests – of the theory is established in terms of (rods and) standard clocks. In particular, a relationist account of (space-)time in GR – in which a sensible notion of time is (non-reductively) dependent on physical clock systems may be argued to be more satisfactory than substantivalist alternatives (see Rugh and Zinkernagel 2009a).

As regards quantum mechanics, the principle of 'same causes, same effects…' in the causal core does not hold in general due to the probabilistic nature of the theory. Nevertheless, the time dependence of wave functions, and hence the reference to the evolution of a quantum system, refers to the ordinary (classical) conception of time. So if the suggested relationism for time is accepted then a case might be made for the necessity of classical described clocks in (the interpretation of) quantum theory (see Rugh and Zinkernagel 2009b). A similar point can be made for quantum field theory in which both time and space are treated as classical background variables. Indeed, from a relationist premise, Teller (1999, p. 321) argues in the context of quantum field theory that "...space-time facts are facts about actual or potential space-time relations between physical bodies. The presupposed space-time relations are classical, exact valued; so the presupposed physical bodies between which the relations do or would have to be taken, in this respect, are classical too".

29.6 Summary and Conclusions

I have argued that Norton's mass on the dome system fails to obey Newton's first law in its standard formulation. Moreover, Norton's instantaneous formulation of Newton's first law seems insufficient as a replacement for the original version since the notion of inertial frames in such a modified theory lacks a physical justification and since an intelligible notion of time in Newtonian mechanics appears to be closely tied to Newton's first law in its standard form. I therefore claimed that the mass on the dome is not an acausal Newtonian system because it is not a Newtonian system. I suggested (but in no way proved) how the causal content (or causal core) of Newtonian mechanics – given its close connection to the notion of time – may play a central role also in relativistic and quantum theories. In this sense, the possibility remains that a principle of causality (captured by the causal core) – *pace* Norton – plays a fundamental role in physics.

Note added in proof (footnote 3)
Carl Hoefer notes (in private communication) that my understanding of (N1) would seem to rule out also standard examples in Newtonian mechanics in which a force is "turned on smoothly". In such cases, $F = 0$ up to some moment $t = T$, and $F > 0$ for $t > T$, so (N1) is apparently violated at $t = T$ (since the system does not continue in its original state for any $t > T$ even though $F = 0$ for $t \leq T$). Moreover, as Norton (2006, p. 8) hints, the standard use of continuously varying forces (and continuously varying trajectories) in Newtonian mechanics seems to weaken the demand for forces being first causes. However, in standard examples involving continuously varying forces the physical situation may just as well be described via a sequence of discrete (first cause) forces, as can be seen, e.g., in Newton's analysis (mentioned by Norton) of planetary orbits using polygonal trajectories in which a series of discrete forces act momentarily at the beginning of each segment (and the limit of vanishing segment size is taken at the end). This equivalence between a continuously varying force and a sequence of discrete forces is *absent* in the dome case: Physically, we cannot attribute any first cause to the dome motion whereas a smoothly turned-on force can be described in first cause terms (the "turning on" can

be seen as a first cause of the change of motion, even if it can be modelled also by a continuously varying force).[12] Mathematically, the lack of equivalence in the dome case is reflected by the fact that the Newtonian *difference* equation (for a polygonal path of motion) corresponding to Norton's *differential* equation (29.1), in contradistinction to this latter equation, does have a unique solution, namely $r(t) = 0$ for all t.[13] Thus, as no first cause can be associated with Norton's particle at the apex it stays put.

Acknowledgements It is a pleasure to thank Carl Hoefer and Svend Rugh for comments and discussion. I also thank the audiences at the "Singular causality, counterfactuals and mental causation" workshop in Granada and at the I EPSA Madrid conference for comments during presentations of this work. Financial support from the Spanish Ministry of Education and Science (project HUM2005–07187-C03–03) is gratefully acknowledged.

References

Anderson JL (1990) Newton's first two laws of motion are not definitions. Am J Phys 58:1192–1195
Barbour J (1989) Absolute or relative motion?, vol. 1: The discovery of dynamics. Cambridge University Press, Cambridge
Brown H (2006) Physical relativity. Oxford University Press, New York
DiSalle R (2008) Space and time: Inertial frames. In: Zalta EN (ed) The Stanford encyclopedia of philosophy (Fall 2008 Edition), URL = <http://plato.stanford.edu/archives/fall2008/entries/spacetime-iframes/>
Earman J, Friedman M (1973) The meaning and status of Newton's law of inertia and the nature of gravitational forces, Philos Sci 40:329–59
Friedman M (1983) Foundations of space–time theories. Princeton University Press, Princeton, NJ
Malament D (2007) Norton's slippery slope, Manuscript, http://philsci-archive.pitt.edu/archive/00003195/
Norton JD (2003) Causation as folk science. Philosophers' imprint vol. 3, no. 4 http://www.philosophersimprint.org/003004/; reprinted in Price H, Corry R (eds) Causation and the constitution of reality. Oxford University Press, Oxford, 2007

---

[12] When the "turning on" is modelled by a series of discrete forces, the first (or indeed any) non-vanishing discrete force in the series can be identified with a first cause, and my point is that in standard cases – but not in Norton's – we can use both the continuously varying force approach and that of a series of discrete (first cause) forces.

[13] If one takes $r(t + \Delta t) = r(t) + v(t)\Delta t$; $v(t + \Delta t) = v(t) + r(t)^{1/2}\Delta t$; and $r(0) = v(0) = 0$, the difference equation for the mass on the dome is found to be $(r(t_{n+2}) - 2r(t_{n+1}) + r(t_n))/h^2 = (r(t_n))^{1/2}$ in which $h = \Delta t$ (step size) and $t_n = nh$. Since $r(t_0) = r(t_1) = 0$, this equation has the unique solution $r(t_n) = 0$ for all n (and thus the same solution in the limit $h \to 0$). I first saw this equation on an anonymous internet forum post by a user named "jason1990". Hans Henrik Rugh points out (in private discussion) that there are many ways to "discretize" a differential equation, and in this sense the differential equation quoted here is not unique for the dome. However, the difference equation approach is arguably the more fundamental one for Newton and, as far as I can see, the demand of forces as first causes acting at the beginning of each segment does in any case single out the quoted difference equation.

Norton JD (2006) The Dome: An unexpectedly simple failure of determinism, Manuscript. philsci-archive.pitt.edu/archive/00002943

Newton I (1729) Mathematical principles of natural philosophy. Trans. by A. Motte and F. Cajori. University of California Press, Berkeley, 1962

Rugh SE, Zinkernagel H (2009a) On the physical basis of cosmic time. Stud Hist Philos Modern Phys 40:1–19

Rugh SE, Zinkernagel H (2009b) Time and the cosmic measurement problem (in preparation)

Russell B (1913) On the notion of cause. Proc Aristotel Soc 13:1–26

Teller P (1999) The ineliminable classical face of quantum field theory. In: Cao TY (ed) Conceptual foundations of quantum field theory. Cambridge University Press, Cambridge, pp 314–323

Zinkernagel H (2008) Did time have a beginning?, Int Stud Philos Sci 22(3):237–258

Index

A

Absolute objects and general relativity
    Anderson–Friedman program, 198, 199, 240–244, 246–248
        covariance, 197–207, 241, 244–245
        spacetime theory classification, 242
        symmetry, physical notion, 241
    invariance group
        definitions, 244–245
        diffeomorphism invariance, 199, 202,243
        Minkowski metric, 215, 243
        Rosen's theory, 243
    scalar density, 246–247
Absolute time, 264, 265
Adams, J.C., 276
Aharonov, Y., 87
Albert, R., 161
Alexandrov–Zeeman theorem, 264
Anderson–Friedman program
    covariance, 241
    spacetime theory classification, 242
    symmetry, physical notion, 241
Anderson, J.L., 199, 240–244, 246, 248
Anderson's definition, 247
Anti-Machian principle, 173
Anti-psychological approach, 66
Arago, François, 276
Arbitrage theory of option pricing, 138
Aristotle, 212
Arkes, H.R., 303, 304
Austin, J., 214

B

Bartlett, C., 296
Basu, P.K., 268, 269, 271
Beatty, J., 2, 3, 6
Bechtel, W., 146
Bell, 182
Bell-type correlations, *See* Upper probability
Bell-type inequalities, common causes
    common screener-off, 89–90
    correlated events, 88–89
    "genuine" separate screener-offs, 91
    relative minimality of derivations, 90–91
    spin particles separation, 87
Belousek, D.W., 184
Belze, L., 132
Berger, R., 43, 44, 47
Biological networks
    chemical reactions, 155
    discrete models, 158
    functional modularization, 156, 157, 159, 160
    graph theory, 160
    interaction motifs, 160
    kinetic models, 158
    molecular mechanism, 159
    network decomposition, 155
    structural modularization, 157
Black, F., 129–133, 135–138
Black, M., 21, 22, 25, 27
Black powder, *See* Iron oxide
Bogen, J., 267, 268, 272–276
Bohm, D., 32, 33, 87
Bohm's theory, 32
Bohr, N., 33, 35, 37
Boolean network models, 155, 156, 158
Born, M., 32–34
Bornstein, B.H., 305, 306
Bose-Einstein statistics, 176
Bosonic systems, 177
Bottom-up experiments, 142–143
Bruggeman, F.J., 154, 155, 158, 161

C

Callebaut, W., 3
Call option, 131

Canonical bijection, 261
Cantor, G., 61–63
Cartesian coordinates, 260
Cartwright, N., 75–78
Cassirer, 224
Cattaneo, L., 235
Causal closure principle, 144
Causal explanation
  mechanistic analysis, 141–142, 150
  scientific explanation, 141
  structural explanation, 51–52
Causal fundamentalism
  determinism, 312–313
  Newtonian Mechanics, 318–320
  Newton's first law
    inertial frames, 316
    mass, dome, 313–315
    notion of time, 317–318
Causal models
  boundary conditions, 82
  Cartwright's view, 76
  causal constraints, 80
  causal discourse, 78, 79
  dynamical laws, 79, 82
  Humean accounts, 77, 78
  interventionist accounts of causation, 82
  intrinsic asymmetry, 83
  manipulation and control, 83
  Russell's view, 77
  state-space models, 76–77, 79, 80
  temporal asymmetry, 81
  van Fraassen's view, 76–80, 83
Cavaillès, J., 61
Cayley, A., 120, 121
Chapman, G.B., 305, 306
Chemical bond
  chemical structure theory
    Berzelian formulae, 117
    Daltonian idea, 118
    Dewar's molecules, 119
    elemental composition, 117–118
    epistemic caution, 120
    Frankland's work, organometallic compounds, 118
    'graphic notation' system, 119
    Hofmann's glyptic formulae, 119–120
    isomers, 117–118
    Kekulé's presentations, 118–119
    Propylic alcohol and Friedel's alcohol isomerism, 119
    'sausage formulae,' 118–119
    stability and reactivity, 121
    stereochemical theories, 121
    structural formulae, 121
  electron
    covalent and ionic bonds, 123
    electrochemical theory, 123
    Ingold, organic reactions, 124
    Lewis view, 121–124
    non-polar and inorganic compounds, 122
    polar bond, potassium chloride, 122
    tautomerism, 122
    Valence, 123
  quantum mechanics
    molecular-orbital approach, 124
    quantum-statistical objection, 125
    'resonance hybrids,' canonical structures, 124
    Schrödinger equation, 124
    semi-empirical wavefunctions, 124–125
    structural view, Lewis and Pauling work, 125
    valence formulae, 126
Chern-Simons numbers, 111–112
Citric acid cycle, 157
Clauser–Horne inequality, 90, 91
Clauser-Horne-Shimony-Holt inequalities, 95, 100
Clauser, J.F., 90, 91, 95
Clifton, R., 44, 47, 49
Common screener-off (C-SCR), 89–91
Complex vector field, 11, 12
Compton, A.H., 31
Compton effect, 31, 33
Conservation laws, 41
Contessa, G., 48
Continuous state models, 155
Continuous-time/mathematical finance, 129
Conventionalist philosophy of geometry, 70–72
Conway Morris, S., 4
Copenhagen interpretation (of quantum mechanics), 33
Cosmological constant, 246
Costa de Beauregard, O., 211, 212
Costantini, M., 233
Coulson, C., 125
Craver, C.F., 146
Crum Brown, A., 119, 120
Csibra, G., 227–236

D

Daltonian idea, 118, 119
Darwin, 188

Index 325

Dauben, J., 61, 63
Davidson, D., 307–309
Davies, P., 211
Davisson, C., 31, 34
Dawid, R., 9
Dawkins, 193
de Broglie, L., 31, 34
de Broglie's hypothesis, 31
Decoherence, 99
Dedekind, R., 61
Degrees of generality, 232, 233, 236
de la Vega, J.P., 132
Devaney Chaos
  condition-justification, 283–284
  natural-world-justification, 282–283
  redundancy-justification, 284
Devaney, R., 282
Dewar, J., 119, 120
Dirac, P.A.M., 10, 13–17, 113
Diffeomorphism invariance, 199, 202, 243
Direct-matching hypothesis (DMH), 228, 229, 231
Dobzhansky, T., 1
Dodd, D.L., 133
Dolev, Y., 212–216
Doll, R., 292–295, 297–300
Downward causation
  causal closure principle, 144
  conceivability, 144
  downward constraints
    causal relations, 148
    degrees of freedom, 147
    deterministic chaos, 148
    partial constraint, 147
    reasoning and decision making., 149
    regulatory mechanism, 147
    system laws, 149
  supervenience basis, 144
Downward/top-down experiments, 142–143
Duality transformation, 11
Dual resonance model, 9, 18
Duhemian dictum, 267
Duhem, P., 272, 275
Dürr, S., 37
Dynamics of Reason, 66

E
Earman, J., 199, 201–204, 208, 244, 316
Einstein, A., 16, 31–35, 65–73, 87, 164, 165, 171–173, 197, 211, 212, 251
Einstein field equations, 163, 168
Einstein, H., 171, 172

Einstein–Podolsky–Rosen–Bohm (EPRB)
  experiment, 87, 88
Einstein's light quantum hypothesis, 31
Einstein's recoiling slit, 35
Electromagnetic duality
  classical electrodynamics
    duality transformation and rotation, 11, 12
    energy and momentum densities, electromagnetic field, 11
    magnetic monopoles, 12–13
    magnetic source terms, 12
    Maxwell's equation, 10–12
    symmetry, 10, 11, 13
  meaning
    description, 15–16
    Faraday's discovery, electromagnetic induction, 16
    quantization condition, 16
  quantum electrodynamics
    consistency problem, 13
    Dirac quantization condition, 14
    fine structure constant, 15
    global topology, 14
    magnetic pole, 13
    quantization, electric charge, 14
    quantum mechanics, 13, 14
Elevation process, 71
Empirical symmetry, 105–107
Entanglement of quantum state, 99
Epple, M., 61
EPR state
  decay of, 100–102
  decoherence, 99
  joint probability distribution, 99–101
  time evolution of, 99–100
Equivalence principle, 71
Ethane acid, 119
Euclid, 56
Euclidean constructions, 67
Euclidean geometry, 65, 68, 71
Euclidean geometry, Tarski's axioms, 258
Euclidean plane, 22
Euler, L.P., 56
Euler–Lagrange equation, 203, 208
Evidence Based Medicine (EBM), 289
Evolution and directionality
  adaptive landscape metaphor, 194
  additive genetic variance, 188, 190
  average fitness, population, 189
  contingencies, nature, 195
  environmental change, 193
  environmental deterioration, 191
  environmental fixity, 193

The Fundamental Theorem of Natural
  Selection (FTNS), 188–195
The Genetical Theory of Natural Selection,
  188
human-induced habitat destruction, 195
macroevolutionary, 188
microevolutionary dynamics, 188
progressiveness, 188
thermodynamical process, 187
time-asymmetry, 187, 188
Evolutionary developmental biology
  existence problem, 2
  historical contingency
    Cambrian explosion, 3
    causal dependency, 4
    generalizations, 2, 3
    laws of nature, 3
    memoryless stochastic processes, 4
    minefield, 3
    special constraint, 2
    time lags, 4
  inherency *vs.* contingency
    blueprint/program notions, 5
    cell interactions, 5–6
    center of gravity, 4
    evolvability, 5
    Müller and Newman's emphasis, 4
    organismal systems approach (OSA), 5
    phenotypic evolution, 5
  modern synthesis, 1
Exclusion Principle, fermions, 176
Expectation of measurement, 94

F
Falkenburg, B., 31, 37
Fama, E.F., 130, 134
Faraday, M., 10, 107
Faraday's cube, 107–108
Faraday's discovery, electromagnetic
  induction, 16
Feferman, S., 285
Fermi-Dirac statistics, 176
Ferrari, P.F., 232, 233
Ferreirós, J., 58, 61
Feynman, R.P., 35
Fichte, J.G., 224
Finkelstein, D., 246
Fisher, I., 133, 134
Fisher, R.A., 1, 188–195, 292
Fock space formalism, 179
Fogassi, L., 229, 234
Formic acid, 119
Fort, H., 115–116

Frankland, E., 118–121
Free mobility principle, 68, 71
Frege, G., 56
Fresnel, A., 275, 276
Fresnel's wave theory, 276
Friedman's definition, 247
Friedman, M., 242, 243, 246, 248, 316
Fundamental Postulate of Statistical
  Mechanics (FPSM), 178

G
Galilei, G., 105–107
Galileo's cube, 107–108
Galle, J.G., 276
Gambini, R., 115–116
Garman, M.B., 137
Gauss constraint functions, 113
General covariance
  action–reaction principle, 207
  active *vs.* passive diffeomorphisms, 206
  Anderson–Friedman programme, 198, 199,
    240–244, 246–248
  Anderson's definition, 244
  coordinate transformations, 197
  diffeomorphisms, 199, 202, 243
  spacetime structure, 244
  symmetry principles, 244
Geometrie und Erfahrung, 72
Geometrodynamics, 219–226
Germer, L.H., 31, 34
Geroch, R., 240, 246
Gibbs paradox, 177, 178
Giere, R.N., 3
Giulini, D., 113–114, 240, 244, 246
Glaziou, P., 293
Glennan, S., 142
Global gauge transformations, 113–114
Gould, S.J., 3–5, 188
Grafen, A., 192
Graham, B., 133
Graßhoff, G., 89–91

H
Hacking, I., 40
Hacking's reality criterion, 40
Hamilton's principle, 203, 204
Harrison, J.M., 138
Hauser, M.D., 227
Hawley, K., 22
Healey, R., 83, 84
Heisenberg's matrix mechanics, 18
Heisenberg's uncertainty principle, 33, 37

Index

Heisenberg, W., 32, 33, 37, 45, 46
Heitler, W., 124
Herbert, V., 289
Hierarchical conception, mathematical sciences, 68
Hilbert spaces, 26
Hitchcock, C., 77, 80
Hofer-Szabó, G., 89–92
Hofmann, A., 119–120
Honouring sunk costs
　folk-notion of waste, 304
　incontinence, 309–310
　rationality, 304, 309–310
　robust judment error, 303
　two systems, reasoning view
　　Criterion S, 306, 307
　　ex post facto justifications, 305
　　mistaken social concept of waste, 305
　　Müller-Lyer illusion, 306
　　wastefulness, 304
Horne, M.A., 90, 91
Huggett, N., 178
Hughes, R.I.G., 45, 47, 48, 51
Hume, D., 296
Husserl, E., 224
Huxley, J., 1

I
Iacoboni, M., 229
'Immobil' compounds, 122
Incontinence
　honouring sunk costs and rationality, 309–310
　incontinent action, 308
　slow and fast judgments, 308
Inertial–kinematical structure, 71
Inertial motion
　Alexandrov–Zeeman theorem, 264
　distance and time, 262
　force concept, 263
　Lorentz transformation, 264
　operational/logical circularity, 263
　reference frames, rigid bodies, 262
　space-time and inertial trajectory, 263
Ingold, C., 124
Interference fringes, 37
Interlevel causation analysis
　bottom-up relation, 146
　constitution, three features, 145
　rigid designators, 145

Intrinsic value/fundamental value, 133
Iron oxide, 269, 272, 273
Itô, K., 129, 136, 137

J
Jacob, P., 227
Jammer, M., 32
Joint probability distribution, 99–101
Jourdain, P., 61

K
Kaluza-Klein theory, 173
Kandel, E., 143
Kant, I., 25, 27, 65–73
Kekulé, A., 118, 120
Ketterle, W., 34
Kim, J., 144, 145, 148, 149
Kimura, M., 190
Kitcher, P., 49, 55–57, 193
Kitcher's quintuple, 55
Klein–Gordon theory, 200, 203, 206
Knockout technique, 143
Kohler, R., 121
Kohlhagen, S.W., 137
Kreps, D.M., 138
Kretschmann, E., 197, 239, 241
Kripke, S., 145
Kuhn, T.S., 1, 55, 57

L
Lakatos, I., 279, 280, 285–286
Lambert, K., 49
Lange, L., 316
Larvor, B., 285, 286
Laudan, L., 271
Lavoisier, A., 268
Lavoisier–Priestley controversy
　iron oxide, 269, 272, 273
　Lavoisier's theory, 269
　partition concept, 271
　phlogiston theory, 269
　Priestley's theory, 269, 272
　universal generalisation, 271
Lead melting point, 273–274
Lehmkuhl, D., 164
Leibniz, G.W., 67, 221
Leng, M., 286
Leplin, J., 271
le Verrier, U.J.J., 276
Lewis, D., 170
Lewis, G.N., 117, 121–126

Lewontin, R.C., 3
Lintner, J., 133
Local gauge symmetry
   empirical symmetry, 108
   Faraday's cube, 107–108
   Galileo's cube, 107–108
   small *vs.* large, 108–109
London, F., 124
Lorentz contraction, 72
Lorentz transformation, 264
Lorenz, K., 3

M
Mach, E., 66, 69, 72, 73
Mach's principle, 164
MacKenzie, D., 139
Maddy, P., 62
Malament, D., 168
Mancosu, P., 57
Margrabe, W., 137
Market portfolio, 133
Markowitz, H.M., 129, 133
Martingale theory, 138
Mass–energy–momentum density tensor, 163–166, 169, 170
Mathematical definitions
   Devaney Chaos
      condition-justification, 283–284
      natural-world-justification, 282–283
      redundancy-justification, 284
   justification, role, 284–285
   proof-generated definitions, 285–286
   topological definitions, 280–281
Mathematical knowledge and practices
   "applicability," 58
   cognitive abilities, 56
   couple *Framework–Agent*, 56–57
   elementary and advanced mathematics
      Euclidean geometry, 59
      Hypothetical Conception, 59–60
      hypothetico-deductive view, 60
      "notional champs," 61
      technical practices, 59
   Eudoxos' theory of proportions, 58
   fraction arithmetic, 58
   key epistemological issues, 57
   Kitcher's pioneering views, 55–56
   Kuhn's ideas, "normal science" and "revolutions," 55
   levels, 57
   links, frameworks, 58
   objectivity
      axioms of Infinity and Power Sets, 61–62
      Bolzano's principle, 63
      Cantor's proof, non-denumerability theorem, 62, 63
      Euclid's proof, 62
      natural numbers, 62
   quadruple elements, 56
   "rational transitions," 55
   reckoning arithmetic, 57, 58
Mathematics role, financial theory
   American option, 131
   Arthur D. Little (ADL), consulting firm, 135
   Brownian motion, 130, 137
   call option, 131
   Chicago Board Options Exchange (CBOE), 129–130
   consolidation, first phase (1973–1983), 137–138
   derivative instruments, 131
   dynamic hedging, 136
   European option, 131
   financial and economic context
      Beta, multiplying factor, 134
      capital, 133
      Capital Asset Pricing Model (CAPM), 133, 134
      Efficient Markets Hypothesis (EMH), 134
      financial derivatives, 132
      intertemporal approach, consumer choice process, 133
      option trading, 132
      risk-free asset, 133
      share price, 134
      two-fund separation theorem, 133
      Woods agreements, 132
   financial instrument, 131
   Itô's calculus, 137
   models, financial markets, 138–139
   option pricing problem, 135–137
   premium option, 131
   randomness, price process, 130
   stock price, 135
   supply and demand, 139
Matrix mechanics, 32
Maudlin, T., 201
Maxwell-Boltzmann statistics, 176, 177
Maxwell field, 39
Maxwell, J.C., 10–12, 15
Maxwell's equation, 10–12
Maynard Smith, J., 187, 188, 190–192

Mayo, D.G., 275
Mayr, E., 1
Mazur, B., 60
Mechanist causation, 142
Mehrling, P., 135
Merton, R.C., 129, 135–137
Metabolic pathway analysis, 155
Metric density, 246
Mie, G., 220
Mielke, E.W., 222
Milky Way, 69
Miller, M., 134
Mills, E., 144
Minkowski, H., 263
Minkowski spacetime, 201, 215, 243
Mirror neurons
  direct-matching hypothesis (DMH), 228, 229, 231
  emotional expression/context, 227
  insula and cingulate cortex, 227
  lateral convexity, brain, 227
  low-level motor duplication, 230
  low-level resonance mechanism, 229
  mirror mechanism, primary function, 227
  motor goals
    action mirroring, 230–232
    motor intentions, 232–235
  overt motor act, 228
  single cell recordings, monkey, 227
  superior temporal sulcus (STS), 236
Modigliani, F., 134
Molecular structure, 273–274
Monod, J., 5
Montonen, C., 9
Moran, P.A.P., 190
Mossin, J., 133
Motor goals
  action mirroring
    degree of generality, 232
    experiment, macaque monkeys, 230
    infero-parietal lobule (IPL), 230
    low-level motor mirroring, 232
    mirror neuron congruency, 232
    sound-producing motor acts, 231
    top-down mechanism, 231
    ventral premotor cortex, 230
  motor intentions
    autistic spectrum disorders (ASD), 235
    degrees of generality, 233
    hand-and mouth-grasping action, 233
    low-level mirroring phenomenon, 234
    mouth-opening mylohyoid muscle, 235
    observer's motor repertoire, 233
    pre-wired motor chain, 235
    single IPL neurons, 234
    traditionally developed (TD) children, 235
    visual system, 233
Müller, G.B., 4–6

N

Neumann, C., 316
Newman, S.A., 4–6
Newton, I., 172, 316–318, 320
Newtonian gravitation theory, 71
Newtonian laws of motion, 65, 70
Nixon, R., 132
Noether, E., 203, 204
non-Abelian Yang-Mills guage theory, 108–109
  gauge transformations, 112–113
Norton, J.D., 197–202, 311–315, 319–320
Novelty of predictions, 275

O

Olive, D., 9, 14
Ontic version of structural realism (OSR), 50–51
Organismal systems approach (OSA), 5
Oyama, S., 4, 5

P

Particle detector, 34
Pauli, W., 221
Pauli matrices, 99–100
Pauling, L., 124, 125
Perfect anticorrelations (PCORR), 91
Performativity of economics, 139
Perrett, D.I., 236
Peto, R., 292–295, 297–300
Philosophy of geometry, 70
Photon field, 39
Physical dualities, intertheoretic relations
  dual resonance model, 9, 18
  electromagnetic duality
    classical electrodynamics, 10–13
    meaning, 15–17
    quantum electrodynamics, 13–15
    invariance principles of physics, 17–18
    'theory of magnetic poles,' 10
Pilot waves, 32
Pirani, F., 165
Pitts, J.B., 240, 243, 246
Planck–Einstein relation, 31

Pliska, S.R., 138
Podolsky, B., 87
Poincaré group, 204
Poincaré, H., 66, 68–73
Poisson, S.D., 275, 276
Poisson spot, 275
Polchinski, J., 18
Portfolio selection theory, 133
Portmann, S., 90, 91
Price, G., 190–192, 194, 195
Price, H., 84
Priestley, J., 268
Principle of relative motion, 69–70
Principle of the identity of indiscernibles (PII), 21, 22
Putnam, H., 211

Q
Quantised Singularities in the Electromagnetic Field, 13
Quantization of electricity, 14
Quantum field theory, 32, 34, 35, 39, 105–116
Quantum mechanical predictions, 90
Quantum optics
    atom interferometer, 37
    double slit experiment, 36, 38
    eraser, path information, 36
    generalised complementarity, 37
    interference patterns, 38
    polarization, 38
Quantum particles
    anti-symmetry, 27
    arbitrary system, 29
    Black's spheres, 25
    classical physics, 25
    correlation, 28
    distance relations, 25
    Euros, 24, 26
    fermions, 27
    gauge system, 29
    hermitean operators, 26
    Hilbert spaces, 26
    identical particles, 26
    indices, 29
    individuality, 24, 25, 27
    irreflexive relations, 24, 27
    Kant's universe, 25
    Leibnizean objecthood, 24
    Leibniz's principle, 21
    physically meaningful mapping, 25
    quantum mechanics, 26, 28
    quantum states, 28
    spins directions, 27
    structuralist analyses, 24
    tensor product formalism, 26
    weak discernibility, 22–23
Quantum probability, 41, 93–104
Quantum statistics
    Bohmian mechanics, 183, 184
    canonical monadic properties, 180
    classical statistics, 176–177
    Fock space formalism, 179
    FPSM, 178
    Gibbs paradox, 177, 178
    haecceitism, 178
    irreducible properties, 180
    N! factor, 177
    non-classical ontological setting, 178
    non-individuality, 177, 178
    particle's identities, 179
    possible reactions
        Eigenstate-Eigenvalue Link, 183
        irreducible R-type properties, 181
        quantum entanglement, 181
        relational holism, 182
    Principle of Indifference, 178
    relational holism, 182
Quantum θ-vacuum, 108–109
    in loop representation, 115–116
Quine, W.V., 23

R
Randomized clinical trial (RCT)
    absolute risk reductions, 298, 299
    cerivastatin, 299
    classical statistical significance test, 290
    double blind trial, 289
    Evidence Based Medicine (EBM), 289
    evidential *sine qua non*, 300
    external validity issue
        active symptom control, 295
        ASSENT-2 and ISIS-2 trial, 297
        benoxaprofen (opren), 297
        efficacy and safety, ustekinumab, 295
        GISSI-3 study, 297
        target population, 295–298
    real causal connection *vs.* mere correlation, 290
    relative risk reduction, 298, 299
    selection bias
        cardiology, 294
        dramatic effects, 293
        effect, 292–293
        random error, 294
    special epistemic weight, 290
    treatment bias, 291–292
    weight of evidence, 290

Rawlins, M., 297, 299
Redhead, M., 178, 179
Reductio ad absurdum argument, 87, 89
Reichenbach, H., 65, 66, 88, 89
Reichenbach's common-cause principle, 88
Relativism, 264
Relativity of motion, 68–69
Relativity theory, 65–74, 163–174, 197–218, 239–266
   Alexandroff present, 216
   constraint dependence, 168–169
   general theory of relativity, 197–202, 208
   interpretational dependence, 169–170
   inertial frames in, 213
   mass density, 171
   mass-energy momentum, 171, 172
   matter fields, 166–168, 172
   metric dependence, 165–166
   Minkowski spacetime, 201, 215, 243
   special theory of relativity, 198–205, 207, 208
Rempe, G., 37
Reiss, J., x
Rest time sequence, 256
Ricci curvature tensor, 163, 164
Rice, S., 190, 192
Robertson, H.P., 46, 49
Robertson's general derivation, 46
Rochat, M., 233
Rocke, A.J., 118, 120
Rosen, N., 87
Rovelli, C., 206, 207
Rubakov, V., 111
Rucker, R., 211
Ruse, M., 188
Russell, B., 56, 75–78, 311

S
Samuelson, P.A., 135, 136
San Pedro, I., x
Saunders, S., 23, 27, 29, 178
Scale invariant, theory, 247
Scholes, M., 129–133, 135–138
Schrödinger, E., 32
Schrödinger equation, 35
Schurz, G., 49
Science and Hypothesis, 70
Scientific explanation
   analogical models, 47, 48
   causal accounts, 44
   causal explanation, 47
   computational explanations, 43
   definition, theoretical explanation, 47, 48
   mathematical modelling, 43, 47
   phenomenonal explanation, 48, 49
   quantum mechanical phenomena, 44
   structural explanation (SE)
     causality, 51–52
     Kitcher's theory, 49
     Robertson's derivation, 49
     structural realism, 50–51
     uncertainty relations, 45–47
     understanding process, physical phenomenon, 49, 50
     unificationist theory of understanding, 49
   surrogative reasoning, 48
   two-slit interference experiment, 48
Scientific knowledge, 66–67
Scully, M.O., 35–37
Self-fulfilling prophecy, 131
Separate screener-offs (S-SCR), 91
Shapiro, S., 50
Sharpe, W., 133, 134
Simpson, G.G., 1
Sloman, S.A., 305–308
S-matrix, 32
Smith, P., 283
Sober, E., 4
Spieser, P., 132
State-space models, 76–77, 79, 80
Stemwedel, J., 125
Sterelny, K., 193
Stochastic integration theory, 138
Strong CP problem, 115–116
Structural explanation (SE)
   causality, 51–52
   Kitcher's theory, 49
   Robertson's derivation, 49
   structural realism, 50–51
   uncertainty relations
     common formal properties, 47
     explanandum phenomenon, 46
     fundamental structure, quantum theory, 46
     Heisenberg's uncertainty relations, 45, 46
     Hilbert space, 45
     mathematical property, Fourier transform, 46
     quantum mechanics, 45
   understanding process, physical phenomenon, 49, 50
   unificationist theory of understanding, 49

Suárez, M., 48
Superselection operator, 112
Suppes, P., 77, 80, 93
Swoyer, C., 48
Symmetry
  empirical, 105–107
  of structure, 105
  theoretical, 105–106

T
Tarski, A., 258
Teller, P., 178, 179, 182, 320
The Fundamental Theorem of Natural Selection (FTNS), 188–195
Theoretical symmetry, 105–106
The Theory of Magnetic Poles, 13
Time-reversal invariant theory, 80
Tobin, J., 133
Touhey, P., 283
Transcendental philosophy, 65–67, 70, 73
Transition probabilities, 40
Translation operator, 109
Treynor, J., 133, 135
Trivial semantical convention, 264
Tukey, J.W., 289

U
Umiltà, M.A., 231, 233
Uncertainty principle, 33
Unger, P., 145
Unimodular Relativity (UR), 246, 248
Upper probability
  definition, 94
  in quantum mechanics, 94–99

V
van Fraassen, B.C., 3, 75–80, 83
van't Hoff, J., 121
Volume element, 246, 247
von Baeyer, A., 121
von Helmholtz, H., 66–69, 71–73
von Neumann, J., 32, 33

W
Walborn, St.P., 38
Walden, S., 144
Walley, P., 93
Watson, G., 307–309
Wave function, 32
Wave-like quantum phenomenon, 34

Wave particle duality, 31–42
Weatherson, B., 163
Weisberg, M., 125
Werndl, C., 280, 283, 286
Westerhoff, H.V., 154, 155, 158, 161
Weyl, H., 219–224
Weylian approach
  agency theory, 219
  freedom and constraint, wavering
    Philosophy of Mathematics and Natural Science, 223
    silent flow, four-dimensional space-time, 223
    symbolic construction, third realm, 224
  matter, early modern times
    electromagnetism, 220
    extramundane, 221
    Field and Matter, 220
    general relativity, 220
    mass and extension, 220
    Philosophy of Mathematics and Natural Science, 219
    pure field theory, 220
    quantum monads, 221
  post-Weylian applications
    Gauss's theorem, 223
    general relativity, 222
    geometrodynamics, 221, 222
    holographic principle, 223
    Weyl's agens theory, 222
    wormhole, space-time, 222
Wheeler, J.A., 32, 221, 222
Wigner, E.P., 41
Williams, G.C., 2, 5
Williams, J.B., 133
Winding number, 109
Wislicenus, J., 121
Wood, J.N., 227
Woodward, J., 77, 143, 267, 268, 272–276
Wright, L., 189, 191
Wright, S., 1
Wüthrich, A., 90, 91

Y
Yang-Mills guage theory, 108–109
Yusuf, S., 293

Z
Zanotti, M., 93
Zeh, H.D., 99
Zinkernagel, H., 317
Zurek, W.H., 32

Breinigsville, PA USA
19 May 2010
238318BV00006B/20/P